The Ciba Foundation is an international scientific and educational charity. It was established in 1947 by the Swiss chemical and pharmaceutical company of CIBA Limited—now CIBA-GEIGY Limited. The Foundation operates independently in London under English trust law.

The Ciba Foundation exists to promote international cooperation in biological, medical and chemical research. It organizes about eight international multidisciplinary symposia each year on topics that seem ready for discussion by a small group of research workers. The papers and discussions are published in the Ciba Foundation symposium series. The Foundation also holds many shorter meetings (not published), organized by the Foundation itself or by outside scientific organizations. The staff always welcome suggestions for future meetings.

The Foundation's house at 41 Portland Place, London W1N 4BN, provides facilities for meetings of all kinds. Its Media Resource Service supplies information to journalists on all scientific and technological topics. The library, open five days a week to any graduate in science or medicine, also provides information on scientific meetings throughout the world and answers general enquiries on biomedical and chemical subjects. Scientists from any part of the world may stay in the house during working visits to London.

PROTO-ONCOGENES
IN CELL
DEVELOPMENT

PROTO-ONCOGENES IN CELL DEVELOPMENT

A Wiley-Interscience Publication

1990

JOHN WILEY & SONS

Chichester · New York · Brisbane · Toronto · Singapore

Published in 1990 by John Wiley & Sons Ltd.
Baffins Lane, Chichester
West Sussex PO19 1UD, England

RC 268.415 P76 1990

Other Wiley Editorial Offices

John Wiley & Sons, Inc., 605 Third Avenue,
New York, NY 10158-0012, USA

Jacaranda Wiley Ltd, G.P.O. Box 859, Brisbane,
Queensland 4001, Australia

John Wiley & Sons (Canada) Ltd, 22 Worcester Road,
Rexdale, Ontario M9W 1L1, Canada

John Wiley & Sons (SEA) Pte Ltd, 37 Jalan Pemimpin 05-04,
Block B, Union Industrial Building, Singapore 2057

Suggested series entry for library catalogues:
Ciba Foundation Symposia

Ciba Foundation Symposium
295 pages, 51 figures, 10 tables

Library of Congress Cataloging-in-Publication Data
Proto-oncogenes in cell development.
    p.   cm.—(Ciba Foundation symposium; 150)
  Editors: Greg Bock and Joan Marsh.
  'Symposium on Proto-oncogenes in Cell Development, held at the
Ciba Foundation, London, 19–21 September 1989'—Contents p.
  'A Wiley–Interscience publication.'
  Includes bibliographical references.
  ISBN 0 471 92686 8
  1. Proto-oncogenes—Congresses.   2. Oncogenes—Congresses.
3. Cell transformation—Congresses.   4. Developmental cytology—
Congresses.   I. Bock, Gregory.   II. Marsh, Joan.   III. Symposium
on Proto-oncogenes in Cell Development (1989: Ciba Foundation)
IV. Series.
  [DNLM: 1. Cells—congresses.   2. Gene Expression Regulation—
congresses.   3. Proto-Oncogenes—congresses.   W3 C161F v. 150/QZ
202 P9675 1989]
RC268.415.P76 1990
591.87'6 – dc20
DNLM/DLC
for Library of Congress   90-11986
                                                 CIP

British Library Cataloguing in Publication Data
  Proto-oncogenes in cell development.
  1. Animals. Cells. Development
  I. Bock, Greg   II. Marsh, Joan *1960–*
  591.8761

  ISBN 0 471 92686 8

Phototypeset by Dobbie Typesetting Limited, Devon.
Printed and bound in Great Britain by Biddles Ltd., Guildford.

Contents

Participants

S. Alemà   Istituto di Biologia Cellulare, CNR, viale Marx 43, I-00137 Rome, Italy

P. Bentley   PO Box CIBA-GEIGY AG, CH 4002, Basle, Switzerland

J. S. Brugge   Department of Microbiology, University of Pennsylvania School of Medicine, 209 Johnson Pavilion, Philadelphia, Pennsylvania 19104-6076, USA

J. R. Feramisco   Department of Medicine, University of California at San Diego, Theodore Gildred Cancer Facility, T-011, 225 Dickinson Street, San Diego, California 92093, USA

R. Greil (*Ciba Foundation Bursar*)   Department of Internal Medicine, University of Innsbruck, Anichstrasse 35, A-6020 Innsbruck, Austria

E. Hafen   Zoological Institute, University of Zürich, Winterthurerstrasse 190, CH-8057 Zürich, Switzerland

M. R. Hanley   MRC Molecular Biology Unit, University of Cambridge Medical School, Hills Road, Cambridge CB2 2QH, UK

E. Harlow   Cold Spring Harbor Laboratory, PO Box 100, Cold Spring Harbor, New York 11724, USA

J. K. Heath   Department of Biochemistry, University of Oxford, South Parks Road, Oxford OX1 3QU, UK

T. Hunter (*Chairman*)   Molecular Biology & Virology Laboratory, The Salk Institute, PO Box 85800, San Diego, California 92138-9216, USA

T. M. Jessell   Neurobiology & Behaviour Center, College of Physicians & Surgeons of Columbia University, 722 West 168th Street, New York 10032, USA

J. L. Knopf   Department of Molecular Biology, Genetics Institute Inc, 87 Cambridge Park Drive, Cambridge, Massachusetts 02140, USA

H. Land   Imperial Cancer Research Fund, PO Box 123, 44 Lincoln's Inn Fields, London WC2A 3PX, UK

P. F. Maness   Department of Biochemistry & Nutrition 505 FLOB, University of North Carolina School of Medicine, Building 231H, Chapel Hill, North Carolina 27599-7231, USA

A. P. McMahon   Roche Institute of Molecular Biology, Nutley, New Jersey 07110, USA

M. Méchali   Institut Jacques Monod, CNRS, Université Paris VII, Tour 43, 2 Place Jussieu, F-75251 Paris Cedex 05, France

W. H. Moolenaar   Division of Cellular Biochemistry, The Netherlands Cancer Institute, Plesmanlaan 121, 1066 CX Amsterdam, The Netherlands

M. Noble   Ludwig Institute for Cancer Research, (Middlesex Hospital/ University College Branch), Courtauld Building, 91 Riding House Street, London W1P 8BT, UK

C. Norbury   ICRF Cell Cycle Group, Microbiology Unit, Department of Biochemistry, University of Oxford, South Parks Road, Oxford OX1 3QU, UK

P. Nurse   ICRF Cell Cycle Group, Microbiology Unit, Department of Biochemistry, University of Oxford, South Parks Road, Oxford OX1 3QU, UK

R. Nusse   Division of Molecular Biology, Netherlands Cancer Institute, Plesmanlaan 121, 1066 CX Amsterdam, The Netherlands

C. J. Sherr   Howard Hughes Medical Institute Research Laboratories, Department of Tumour Cell Biology, St Jude Children's Research Hospital, 332 N Lauderdale, Memphis, Tennessee 38101-0318, USA

T. Sugimura   National Cancer Center, 1-1 Tsukiji 5-chome, Chuo-ku, Tokyo 104, Japan

G. F. Vande Woude   BRI Basic Research Program, Frederick Cancer Research Facility, PO Box B, Frederick, Maryland 21701, USA

I. M. Verma The Salk Institute, PO Box 85800, San Diego, California 92138-9216, USA

E. Wagner Research Institute of Molecular Pathology, Dr Bohr-Gasse 7, A-1030 Vienna, Austria

M. Waterfield Ludwig Institute for Cancer Research, (Middlesex Hospital/University College Branch), Courtauld Building, 91 Riding House Street, London W1P 8BT, UK

B. Westermark Department of Pathology, University of Uppsala, Akademiska Sjukhuset, S-751 85 Uppsala, Sweden

Introduction

T. Hunter

Molecular Biology & Virology Laboratory, The Salk Institute, PO Box 85800, San Diego, CA 92138-9216, USA

The title of this Ciba Foundation Symposium is 'Proto-oncogenes in cell development', which is indeed a timely topic. There are now over fifty proto-oncogenes known, and from the intensive analysis of this type of gene over the past decade it has become clear that proto-oncogene products play central roles in cellular and organismic physiology. It is largely through the study of proto-oncogene products and their activated oncogenic counterparts that we have begun to uncover some of the fundamental processes that regulate cell growth and differentiation, both at a single cell level and at the level of the organism. What we now realize is that proto-oncogene products are all elements of a complex cellular signalling network, in which these proteins perform a variety of functions, including acting as ligands and growth factors outside the cell, as receptors in the plasma membrane and signal transducers in the cytoplasm, and as transcription factors in the nucleus. In general, proto-oncogenes have been highly conserved through evolution; in some cases they play analogous and essential roles in the most primitive single cell eukaryotes and in mammals. A good example is *ras*, which serves a membrane-signalling function in both yeast and humans. The power of genetics in invertebrates such as yeast, *Caenorhabditis elegans* and *Drosophila*, has allowed us to learn a great deal about mammalian proto-oncogene function through a study of the invertebrate homologues. Many intriguing identities have emerged between genes that were originally identified as developmental genes in simpler organisms and proto-oncogenes in vertebrates; e.g. the *Drosophila* gene *decapentaplegic* and the TGF-β1 gene, and *lin-12* in *C. elegans* and the epidermal growth factor (EGF) gene.

In this meeting we shall hear mostly about the functions of the normal cellular proto-oncogene products, but at the same time we should not lose sight of the fact that an analysis of their oncogenic counterparts has greatly advanced our understanding of proto-oncogene function, through the instructive nature of the mutations that have converted these normal cellular genes into constitutively active oncogenes. There will be papers on representative proto-oncogene products that function at every level in the cell. Starting from the outside of the cell, with proto-oncogenes encoding growth factors that act in endocrine, paracrine or even autocrine fashions, we will hear from Bengt Westermark about

1

the platelet-derived growth factor (PDGF)/*sis* family of growth factors, and from Takashi Sugimura about *hst*, a member of the fibroblast growth factor (FGF) family, which includes the *int-2* oncogene product. This will be a good starting point for John Heath to discuss the role of the different FGFs in mouse embryogenesis. Roel Nusse will talk about the *int-1/wingless* protein, which appears to act as a paracrine factor in the establishment of segment polarity in *Drosophila* embryogenesis. Andy McMahon's work on the role of *int-1* in mouse and *Xenopus* development will be relevant here.

At the cell surface there are several proto-oncogene products that act as growth factor or ligand receptors of various sorts, which transduce signals across the plasma membrane. Mike Hanley will tell us about *mas*, which is an angiotensin receptor and belongs to the family of receptors with seven transmembrane domains that are coupled to G proteins, and Ernst Hafen will describe the functions of the *sevenless* and the EGF receptor protein tyrosine kinases in *Drosophila* development. In this context, Bengt Westermark will give an account of the two receptors for PDGF, which are closely related protein tyrosine kinases, and I hope that we will have discussion contributions from Chuck Sherr about the CSF-1 receptor/c-*fms* protein and from Mike Waterfield about the EGF receptor tyrosine kinases. Another relevant topic will be the c-*kit* protein, a receptor protein tyrosine kinase encoded by the dominant white-spotting *W* locus in mice, mutations in which result in defects in stem cell development in several cell lineages.

As we move inside the cell, there are many immediate events induced by activation of surface receptors. Wouter Moolenaar is going to talk about lipid mitogens and their effects on various signalling processes like phosphatidylinositol turnover (PI) turnover. John Knopf will describe the enzymes involved in PI turnover, with particular reference to phospholipase Cγ and how this enzyme may be regulated directly by tyrosine phosphorylation by growth factor receptor kinases. Patricia Maness will tell us about pp60$^{c\text{-}src}$, the prototype of a class of protein tyrosine kinases that are associated with the inner face of the plasma membrane, and its role in neuronal development and function. In this context, we may also hear from Joan Brugge about the function of pp60$^{c\text{-}src}$ in the nervous system, and the distribution of the neuronal-specific, alternately spliced form of pp60$^{c\text{-}src}$. Work with p56^{lck}, another *src* family tyrosine kinase, suggests that these enzymes may serve as catalytic subunits for surface receptors, and it would be worth exploring to what type of receptor pp60$^{c\text{-}src}$ might be coupled.

In the arena of transmembrane signalling there are other areas worthy of discussion. One is PI-3 kinase, an enzyme that is a substrate for a number of the plasma membrane-associated protein tyrosine kinases, which generates a series of PI-3 phosphates from PI that are interesting molecules in search of a function: another is the inositol glycans, which are induced by insulin. It would also be valuable to discuss the *ras* family of GTP-binding proteins, a class of submembranous signalling molecules that may act through the GTPase activator protein (GAP) protein as effectors. The possible connections between pp60^{c-src} and p21^{c-ras} could be considered.

As we move away from the plasma membrane into the cytoplasm, signal transmission to the nucleus becomes an important function. George Vande Woude will talk about *mos*, a soluble cytoplasmic protein serine kinase, which may have this function. In this context, it will be appropriate to discuss *raf*, which is proving to be one of the more interesting proto-oncogene products, since it apparently transduces signals from growth factor receptors to the nucleus by physical translocation. The *raf* protein is phosphorylated on tyrosine and activated in PDGF-treated cells. It is tightly associated with the PDGF receptor after activation of that receptor. Other signalling molecules to consider are *pim-1*, which is also a protein serine kinase, some of the soluble protein tyrosine kinases, such as the *fps/fes* gene product, and potential protein tyrosine kinase regulators, such as *crk*, which may act through the SH2 region, a homologous region present in a number of cytoplasmic signalling molecules, including *crk* itself, phospholipase Cγ, pp60^{c-src} and GAP.

Ultimately, signals from the surface must reach the nucleus where gene induction events occur. Inder Verma will describe the association between the *fos* and *jun* proto-oncogene products, an interaction which provides a nice example of how a transcription factor network can be established using protein–protein interaction as a key regulatory mechanism. A number of other proto-oncogenes encode nuclear transactivators, such as *myc* and *myb*, and we may hear from Marcel Méchali with respect to the *myc* gene. Another type of transcription factor we should consider is the steroid receptor family, the members of which are sequence-specific ligand-regulated transactivators. The c-*erbA* proto-oncogene, which encodes the thyroxine receptor, is a member of this family and gave rise to the v-*erbA* oncoprotein, which cannot bind hormone and acts in a dominant negative fashion. We will also learn from Jim Feramisco about how to investigate the cAMP- and TPA-regulated gene induction pathways with microinjection techniques using purified proto-oncogene products and antibodies against these proteins.

Up to this point we have considered only proto-oncogenes that give rise to dominant oncogenes which incite continuous cell growth. However, in the past few years it has become apparent that there are negative regulators of cell growth whose loss can give the same phenotype as the dominant oncogenes. The prototype for these so-called recessive oncogenes or growth suppressor genes is the retinoblastoma (*Rb*) gene, whose product is a nuclear protein. Ed Harlow will talk about the *Rb* protein and how it may be a negative regulator of gene expression that affects cell cycle progression. In this category, there is also p53, which has recently changed its coat and become a negative rather than a positive regulator. Another interesting candidate for a nuclear regulator is the c-*abl* product, found to be localized in the nucleus.

The cell cycle is an area in which there has been considerable excitement since it has become clear that a single fundamental mechanism probably regulates the cell cycle in all eukaryotes. Chris Norbury will tell us about the *cdc2* protein kinase and what modulates its activity, and how *cdc2* in turn regulates progression through the cell cycle, probably at both the G1/S and the G2/M boundaries. George Vande Woude is going to talk about the role of *mos* in the meiotic cell cycle and how it may regulate *cdc2* function. There are other connections to be made between *cdc2* and proto-oncogene products, particularly *src*, *abl* and perhaps *Rb*.

Proto-oncogenes also play an important role in differentiation. We will hear from Mark Noble about how the common progenitor cells for astrocytes and oligodendrocytes decide which pathway to follow for differentiation, and how growth factors can influence this decision. Stefano Alemà will describe how myoblast differentiation can be affected by different classes of oncogene and proto-oncogene products, and how this in turn may be influenced by normal cells. Hucky Land with his work on the growth and differentiation of Schwann cells and Tom Jessell with his studies of cell patterning and neuronal recognition in vertebrate development may contribute to this discussion.

To look to the future, at the single cell level we will undoubtedly continue to study proto-oncogene function largely by reverse genetic techniques, by introducing activated and mutant forms of proto-oncogenes into cells and studying their effects, by creating dominant negative mutations to ask whether particular proto-oncogene functions are essential, by ablation of proto-oncogene expression through antisense vectors and antisense oligonucleotides or by microinjection of neutralizing monoclonal antibodies, and finally by homologous recombination to eliminate or alter gene function in cell lines.

Proto-oncogenes are also important at the organismic level. This was first shown by studies on the expression of proto-oncogenes in developing animals and adult animal tissues. A number of interesting patterns of expression were uncovered; for instance, the unique expression of c-*fms* in placental cytotrophoblasts, and the high levels of expression of pp60[c-src] and p21[c-ras] in the brain and of p75[c-myb] in haemopoietic cells all gave clues to the function of these proto-oncogene products. These early studies, which were largely done by Northern analysis, have been extended by some elegant *in situ* hybridization. For example, it was by this means that Andy McMahon showed that *int-1* expression occurred only in discrete regions of the nervous system during mouse development. These specific patterns of expression already tell us a lot about the functions of these proteins, but to progress further we will have to rely heavily on genetically tractable organisms such as *Drosophila*, where homologues of vertebrate proto-oncogenes like *raf* have been identified through a study of genes involved in development. Analysis of the effects of mutations in these genes will be important, and the identification of suppressor genes may allow us to elucidate signal pathways. For instance, it has recently been shown by this means that *torso*, a receptor protein tyrosine kinase, and *lethal (1) polehole* the *Drosophila* homologue of *raf*, lie on the same pathway.

Transgenic mice will also continue to be an extremely useful tool in the study of proto-oncogene function. Transgenic mice that express activated versions of proto-oncogenes or normal proto-oncogene products ectopically have already yielded informative phenotypes, and Erwin Wagner will be able to give us the benefit of his experience in this area. In the near future it is going to be possible to make routine gene ablations and replacements in mice by homologous recombination. The first reported example is the introduction of a mutant *abl* gene into the germline, and presumably we will soon learn the effects of this mutation on mouse development and cellular differentiation. Another animal that will continue to be useful is *Xenopus*, particularly for the study of early development, since one can readily determine the effect of various reagents by microinjection into oocytes or embryos.

The study of proto-oncogenes and their products in cell development has already been enormously fruitful, and it is clear this field still has a great deal to offer. I am sure that the presentations and discussion at this meeting will prove that point.

Structural and functional aspects of platelet-derived growth factor and its receptors

Bengt Westermark*, Lena Claesson-Welsh† and Carl-Henrik Heldin†

*Department of Pathology, University Hospital, S-751 85 Uppsala and †Ludwig Institute for Cancer Research, Biomedical Center, S-751 23 Uppsala, Sweden

Abstract. Platelet-derived growth factor (PDGF) is a dimeric molecule that exists as homodimers or heterodimers of related polypeptide chains (A and B). Two types of PDGF receptor have been identified. The PDGF α-receptor binds all three isoforms with high affinity whereas the β-receptor binds only PDGF-BB with high affinity, PDGF-AB with low affinity and does not appear to bind PDGF-AA. The α- and β-receptors are structurally related, each having an intracellular protein tyrosine kinase domain. Ligand-induced functional activation of the receptors appears to involve receptor dimerization. Binding of PDGF to its receptor is followed by internalization and degradation of the ligand–receptor complex. Experiments with mutant receptors have shown that ligand-induced internalization is not absolutely dependent on the kinase activity of the β-receptor. The v-*sis* oncogene of simian sarcoma virus (SSV) is a retroviral version of the PDGF B chain gene and SSV-transformation is mediated by an autocrine PDGF-like growth factor. Formal evidence that the expression of the PDGF β-receptor is sufficient to confer susceptibility to SSV-transformation has been obtained using porcine endothelial cells expressing a recombinant human β-receptor. PDGF is a chemotactic agent for several cell types. Recent experiments have shown that the PDGF β-receptor mediates a chemotactic response and that this effect requires an intact protein tyrosine kinase activity.

1990 Proto-oncogenes in cell development. Wiley, Chichester (Ciba Foundation Symposium 150) p 6–22

Platelet-derived growth factor (PDGF) was originally discovered as a secretory product of thrombocytes. In recent years it has become apparent that PDGF is synthesized by several normal cell types, including endothelial cells (Dicorleto & Bowen-Pope 1983), vascular smooth muscle cells (Seifert et al 1984), macrophages (Shimokado et al 1985), placental cytotrophoblasts (Goustin et al 1985) and fibroblasts (Paulsson et al 1987). PDGF is a potent mitogen for mesenchymal cells and glia cells; although its physiological function has not been elucidated, it is generally believed to play an important role as a growth regulator in development, tissue repair and wound healing. The structural homology between PDGF and the oncogene product of simian sarcoma virus (SSV), the v-*sis* product (Devare et al 1983, Doolittle et al 1983, Waterfield et al 1983, Johnsson et al 1984), suggests that PDGF has a role in transformation and tumorigenesis. Indeed, SSV-induced transformation has been shown to be mediated by a PDGF-like growth factor via an autocrine mechanism (reviewed by Westermark et al 1987). The frequent expression of PDGF in tumour cell lines (Heldin et al 1980, Betsholtz et al 1984, Nistér et al 1984) has been taken as an indication of a similar role in the development of 'spontaneous' tumours. In this review we highlight the structural and functional properties of PDGF and its receptors with emphasis on recent developments. For more extensive reviews on PDGF, see Ross et al (1986), Heldin et al (1985), Heldin & Westermark (1989).

Structure of platelet-derived growth factor

PDGF is composed of two polypeptide chains, denoted A and B, that are linked by disulphide bonds (Johnsson et al 1982). The two chains are encoded by separate genes (Betsholtz et al 1986) and synthesized as prepropeptides; the mature products are formed after dimerization and proteolytic processing. cDNA and genomic cloning of the A chain has shown that it occurs as two splice variants. The three most C-terminal amino acid residues of the shorter variant are replaced by 18 different residues with a high content of basic amino acids (Betsholtz et al 1986, Bonthron et al 1988, Rorsman et al 1988).

All three dimeric forms of PDGF have been isolated from natural sources. PDGF-AA has been found in the conditioned medium of several tumour cell lines (Heldin et al 1986, Westermark et al 1986, Hammacher et al 1988a). Both pig (Stroobant & Waterfield 1984) and human (Hammacher et al 1988b) platelets contain PDGF-BB. PDGF-AB has been isolated from human platelets (Hammacher et al 1988b) and constitutes the majority of PDGF purified from this source. In addition, the clonal human glioma cell line U-343 MGa Cl2:6, that expresses mRNA for both chains, synthesizes all three isoforms of PDGF (Hammacher et al 1988a).

The biosynthesis of PDGF has been studied in CHO cells transfected with an expression vector containing the coding sequences for both subunits, driven by separate promoters (Östman et al 1988). These cells release all three isoforms of PDGF, of which PDGF-AB is the predominant species in material purified from conditioned media. Pulse-chase analyses showed that all isoforms were processed to species of M_r values of about 30 000; these were the forms of PDGF that accumulated in the conditioned medium. In addition, PDGF-BB was processed to a 24 kDa component which remained associated with the cells. This component is probably identical in structure to the 24 kDa v-*sis* product that has been found in SSV-transformed cells and that is thought to mediate

SSV transformation (Robbins et al 1983). The exact subcellular localization of this component has not been elucidated.

In vitro translation analyses have shown that the intact PDGF B chain messenger RNA is inefficiently translated. However, when the relatively large 5′ untranslated sequence (5′ UTS) is removed, translation efficiency is increased severalfold (Rao et al 1988). This result agrees with the finding that only minute quantities of B chains are produced by intact cells even though they express relatively large quantities of B chain mRNA (Hammacher et al 1988a). In this context, it is interesting that the first exon of the PDGF B chain gene, containing the 5′ UTS, is not represented in the v-*sis* gene; it is likely that the loss of the translation-repressing 5′ elements contributes significantly to the transforming activity of SSV. It is also notable that all studies on transformation induced by the cellular counterpart of v-*sis*, c-*sis* which encodes the PDGF B chain, have been performed using 5′ truncated versions.

Receptors for platelet-derived growth factor

Initial studies on human fibroblasts revealed the presence of a 180 kDa PDGF receptor endowed with a ligand-stimulatable protein tyrosine kinase activity (Ek et al 1982). Cross-competition binding experiments using all three isoforms of PDGF have shown that fibroblasts express two types of PDGF receptors that differ in ligand specificity (Heldin et al 1988, Hart et al 1988). The previously identified receptor (denoted PDGF B type or β-receptor) is restricted in its binding capacity such that it binds only PDGF-BB with high affinity. It binds PDGF-AB with a 10-fold lower affinity and does not appear to bind PDGF-AA with any appreciable affinity. The recently discovered PDGF A type or α-receptor is more promiscuous and binds all three isoforms with high affinity.

cDNA clones for both types of PDGF receptor have been isolated (Yarden et al 1986, Gronwald et al 1988, Claesson-Welsh et al 1988, 1989, Matsui et al 1989). The primary structures of the two receptors show distinct similarities. Each receptor has an extracellular part with a spacing of the cysteine residues that predicts five immunoglobulin-like domains. They have a single membrane-spanning segment. The intracellular protein tyrosine kinase domains are similarly organized, each with an insert of some 100 amino acids that has no homology to protein kinases. The overall amino acid sequence similarity is 44% with the highest score in the two segments of kinase domains (87% and 74%) and in the intracellular juxtamembrane segments (83%).

The receptor for colony-stimulating factor 1 (CSF-1) and the product of the c-*kit* gene have a similar structural organization to the two PDGF receptor types, which includes the five extracellular immunoglobulin-like domains and a split protein tyrosine kinase domain (Yarden et al 1988 and references therein). A transforming potential of the CSF-1 receptor gene and c-*kit* is proven by the fact that they both have been transduced as oncogenes (v-*fms* and v-*kit*, respectively) by acutely transforming retroviruses. Transforming variants of the PDGF receptors have not yet been described.

Ligand-dependent activation of the PDGF receptors

The mechanism by which binding of PDGF to the extracellular domain of the PDGF receptor activates the intracellular protein tyrosine kinase is not understood. A possible clue was provided by experiments on purified pig PDGF β-receptor; addition of PDGF-BB was found to induce receptor dimerization in parallel with the activation of the protein tyrosine kinase (Heldin et al 1989). The role of receptor dimerization in signal transduction has been extensively discussed with regard to the epidermal growth factor (EGF) receptor (Schlessinger 1988). The idea is that binding of EGF leads to a conformational change which stabilizes the receptor in a dimeric configuration. Given the fact that PDGF is a dimer, one might envisage a model in which PDGF dimerizes its receptor by a bivalent interaction such that one molecule of PDGF binds two receptor molecules. This notion is supported by the finding that PDGF-BB-induced receptor dimerization, and activation of the receptor kinase, decreased at higher concentrations of the ligand. Additional evidence for ligand-induced PDGF receptor dimerization has been presented (Seifert et al 1989, Bishayee et al 1989).

Given the known ligand specificities of the two types of PDGF receptor (see above) and provided the receptor molecules diffuse freely at the cell surface, the receptor dimerization model infers that PDGF-AA may dimerize only α-receptors (PDGFR-αα), PDGF-AB should favour dimerization of both types of receptor in two configurations (PDGFR-αα, PDGFR-αβ), whereas PDGF-BB may dimerize receptors in all three configurations (PDGFR-αα, PDGFR-αβ, PGFR-ββ). We have recently produced experimental data that fit this model (Hammacher et al 1989). We took advantage of the finding that only PDGF-AB and PDGF-BB, and not PDGF-AA, induce actin reorganization and circular membrane ruffling in human foreskin fibroblasts; these effects are thus likely to be mediated by the β-receptor only. Since, according to the model described above, PDGF-AB can dimerize the β-receptor only in a heterodimer configuration (PDGFR-αβ), we expected that a blockade or down-regulation of α-receptors would inhibit PDGF-AB-induced ruffling; this was indeed shown to be the case. We also found that an excess of PDGF-AB competed with PDGF-BB in the induction of ruffling activity in cells with down-regulated α-receptors; in this experimental situation, PDGF-AB acts as an antagonist, unlike the situation in naive cells in which PDGF-AB stimulates ruffling. Our interpretation is that the B chain of PDGF-AB blocks the β-receptor in its monomeric configuration when no α-receptors are present. We also found that PDGF-AB requires α-receptors in order to down-regulate β-receptors effectively and to induce autophosphorylation of the receptor. In conclusion, available evidence supports the notion that PDGF dimerizes its receptors and that this is a necessary event for the functional activation of the receptor kinase.

Role of the PDGF β-receptor protein tyrosine kinase in internalization, down-regulation and degradation of the receptor

We have transfected an established line of porcine endothelial cells with PDGF β-receptor expression vectors. The untransfected cells were devoid of PDGF receptors and completely unresponsive to PDGF. Two constructs were used for transfection, one with the full length human β-receptor cDNA, and one with a similar cDNA in which the lysine residue in the assumed ATP-binding site had been changed to alanine by site-directed mutagenesis (K634A mutant receptor) (Claesson-Welsh et al 1989). From metabolically labelled cells expressing the wild type receptor or the mutant receptor, both the 190 kDa cell surface form and the 160 kDa precursor form of the β-receptor could be precipitated using a specific antiserum (Claesson-Welsh et al 1989). In subsequent studies using PDGF-BB as ligand, we found that both wild-type and mutant receptors underwent ligand-induced internalization and degradation at 37 °C. However, this occurred with somewhat slower kinetics in cells transfected with the mutant receptor than in cells expressing the wild type receptor (A. Sorkin et al, personal communication).

PDGF and its receptors: putative role in cell transformation

As briefly outlined above, the v-*sis* oncogene of SSV is a retroviral version of the PDGF B chain gene. There is ample evidence that SSV-induced transformation is mediated by an autocrine PDGF-like growth factor (Robbins et al 1983, Johnsson et al 1985, Betsholtz et al 1986). Using the endothelial cell system described above, we have recently obtained evidence that the expression of the PDGF β-receptor is sufficient to confer susceptibility to transformation by SSV; no α-receptors are required (B. Westermark et al, unpublished observations). Untransfected endothelial cells were refractory to SSV transformation, as has also been shown previously (Leal et al 1985). When expressing the PDGF β-receptor, however, the cells were highly susceptible to transformation and transformed foci were visible 3–4 days after infection with SSV. Cells transfected with the K634A mutant receptor were not transformed, indicating an obligatory role of the protein tyrosine kinase in SSV transformation. We conclude from these experiments that the porcine endothelial cell system may be very useful for further studies on the role of PDGF and its receptor in autocrine transformation.

Role of the protein tyrosine kinase of the PDGF β-receptor in chemotaxis

PDGF is a chemoattractant for a variety of cell types (Grotendorst et al 1981, Siegbahn et al 1989). These findings suggest that PDGF has an important role in mediating cell trafficking in development, inflammation and tissue repair. Studies by our group have indicated that the chemotactic response in human foreskin fibroblasts is mediated by only the β-receptor; in fact, activation of the α-receptor by the addition of PDGF-AA inhibited the chemotactic response to PDGF-BB (Siegbahn et al 1989). These findings prompted us to study the role of the protein tyrosine kinase of the β-receptor in chemotaxis (Westermark et al 1990). PDGF-BB was found to induce chemotaxis in porcine endothelial cells expressing the wild-type β-receptor; thus, the endothelial cells have all the downstream elements required for PDGF-BB-induced chemotaxis, although they do not normally respond to PDGF. Cells expressing the K634A mutant receptor were completely unresponsive to the chemotactic effect of PDGF-BB. This study thus provides formal evidence that the PDGF β-receptor mediates a chemotactic response and that this effect requires an intact protein tyrosine kinase activity of the receptor. Very little is known about the mechanism of chemotaxis in mammalian cells; our finding that tyrosine phosphorylation is involved in the chemotactic response to PDGF may prove useful for further studies on signal transduction events in chemotaxis.

Perspectives

The recent discovery that PDGF consists of a family of dimeric isoforms that interact with two homologous receptors, probably in different dimeric configurations, is intriguing. It seems that PDGF and its receptors form a regulatory system that is designed for fine tuning; the cellular response is determined by the relative proportion of the three isoforms as well as by the relative proportion of α- and β-receptors. A detailed delineation of the signal pathways of the two receptors (or, rather, the three receptor dimers) in relation to their cellular consequences will hopefully lead to a better understanding of this interesting growth regulatory system.

Acknowledgement

Our own work cited in the text was in part supported by the Swedish Cancer Society.

References

Betsholtz C, Westermark B, Ek B, Heldin CH 1984 Coexpression of a PDGF-like growth factor and PDGF receptors in a human osteosarcoma cell line: implications for autocrine receptor activation. Cell 39:447–457

Betsholtz C, Johnsson A, Heldin CH, Westermark B 1986 Efficient reversion of SSV-transformation and inhibition of growth factor-induced mitogenesis by suramin. Proc Natl Acad Sci USA 83:6440–6444

Bishayee S, Majumdar S, Khire J, Das M 1989 Ligand-induced dimerization of the platelet-derived growth factor receptor. Monomer-dimer interconversion occurs independent of receptor phosphorylation. J Biol Chem 264:11699–11705

Bonthron DT, Morton CC, Orkin SH, Collins T 1988 Platelet-derived growth factor A chain: gene structure, chromosomal location, and basis for alternative mRNA splicing. Proc Natl Acad Sci USA 85:1492–1496

Claesson-Welsh L, Eriksson A, Morén A et al 1988 cDNA cloning and expression of a human platelet-derived growth factor (PDGF) receptor specific for B-chain-containing PDGF molecules. Mol Cell Biol 8:3476–3486

Claesson-Welsh L, Eriksson A, Westermark B, Heldin CH 1989 cDNA cloning and expression of the human A-type platelet-derived growth factor (PDGF) receptor establishes structural similarity to the B-type PDGF receptor. Proc Natl Acad Sci USA 86:4917–4921

Devare SG, Reddy EP, Law JD, Robbins KC, Aaronson SA 1983 Nucleotide sequence of the simian sarcoma virus genome; demonstration that its acquired cellular sequences encode the transforming gene product p28[sis]. Proc Natl Acad Sci USA 80:731–735

DiCorleto PE, Bowen-Pope DF 1983 Cultured endothelial cells produce a platelet-derived growth factor-like protein. Proc Natl Acad Sci USA 80:1919–1923

Doolittle RF, Hunkapiller MW, Hood LE et al 1983 Simian sarcoma virus oncogene v-sis, is derived from the gene (or genes) encoding a platelet-derived growth factor. Science (Wash DC) 221:275–277

Ek B, Westermark B, Wasteson Å, Heldin CH 1982 Stimulation of tyrosine-specific phosphorylation by platelet-derived growth factor. Nature (Lond) 295:419–420

Goustin AS, Betsholtz C, Pfeifer-Ohlsson S et al 1985 Coexpression of the sis and myc proto-oncogenes in human placenta suggest autocrine control of trophoblast growth. Cell 41:301–312

Gronwald RGK, Grant FJ, Haldeman BA et al 1988 Cloning and expression of a cDNA coding for the human platelet-derived growth factor receptor: evidence for more than one receptor class. Proc Natl Acad Sci USA 85:3435–3439

Grotendorst GR, Seppä HEJ, Kleinman H, Martin GR 1981 Attachment of smooth muscle cells to collagen and their migration toward platelet-derived growth factor. Proc Natl Acad Sci USA 78:3669–3672

Hammacher A, Nistér M, Westermark B, Heldin CH 1988a A human glioma cell line secretes three structurally and functionally different dimeric forms of platelet-derived growth factor. Eur J Biochem 176:179–186

Hammacher A, Hellman U, Johnsson A et al 1988b A major part of platelet-derived growth factor purified from human platelets is a heterodimer of one A and one B chain. J Biol Chem 263:16493–16498

Hammacher A, Mellström K, Heldin CH, Westermark B 1989 Isoform-specific induction of actin reorganization by platelet-derived growth factor suggests that the functionally active receptor is a dimer. EMBO (Eur Mol Biol Organ) J 8:2489–2495

Hart CE, Forstrom JW, Kelly JD et al 1988 Two classes of PDGF receptors recognize different isoforms of PDGF. Science (Wash DC) 240:1529–1531

Heldin CH, Westermark B 1989 Platelet-derived growth factor: three isoforms and two receptor types. Trends Genet 5:108–111

Heldin CH, Westermark B, Wasteson Å 1980 Chemical and biological properties of a growth factor from human cultured osteosarcoma cells: resemblance with platelet-derived growth factor. J Cell Physiol 105:235–246

Heldin CH, Wasteson Å, Westermark B 1985 Platelet-derived growth factor. Mol Cell Endocrinol 39:169–187

Heldin CH, Johnsson A, Wennergren S, Wernstedt C, Betsholtz C, Westermark B 1986 A human osteosarcoma cell line secretes a growth factor structurally related to a homodimer of PDGF A chains. Nature (Lond) 319:511–514

Heldin CH, Bäckström G, Östman A et al 1988 Binding of different dimeric forms of PDGF to human fibroblasts: evidence for two separate receptor types. EMBO (Eur Mol Biol Organ) J 7:1387–1394

Heldin CH, Ernlund A, Rorsman C, Rönnstrand L 1989 Dimerization of B-type platelet-derived growth factor receptors occurs after ligand binding and is closely associated with receptor kinase activation. J Biol Chem 264:8905–8912

Johnsson A, Heldin CH, Westermark B, Wasteson Å 1982 Platelet-derived growth factor: identification of constituent polypeptide chains. Biochem Biophys Res Commun 104:66–74

Johnsson A, Heldin CH, Wasteson Å et al 1984 The c-*sis* gene encodes a precursor of the B chain of platelet-derived growth factor. EMBO (Eur Mol Biol Organ) J 3:921–928

Johnsson A, Betsholtz C, Heldin CH, Westermark B 1985 Antibodies against platelet-derived growth factor inhibit acute transformation by simian sarcoma virus. Nature (Lond) 317:438–440

Leal F, Williams LT, Robbins KC, Aaronson SA 1985 Evidence that the v-sis gene product transforms by interaction with the receptor for platelet-derived growth factor. Science (Wash DC) 230:327–330

Matsui T, Heidaran M, Miki T et al 1989 Isolation of a novel receptor cDNA establishes the existence of two PDGF receptor genes. Science (Wash DC) 243:800–804

Nistér M, Heldin CH, Wasteson Å, Westermark B 1984 A glioma-derived analog to platelet-derived growth factor: demonstration of receptor-competing activity and immunological crossreactivity. Proc Natl Acad Sci USA 81:929–930

Östman A, Rall L, Hammacher A et al 1988 Synthesis and assembly of a functionally active recombinant platelet-derived growth factor AB heterodimer. J Biol Chem 263:16202–16208

Paulsson Y, Hammacher A, Heldin CH, Westermark B 1987 Possible positive autocrine feed back in the prereplicative phase of human fibroblasts. Nature (Lond) 328:715–717

Rao CD, Pech M, Robbins KC, Aaronson SA 1988 The 5′ untranslated sequence of the c-*sis*/platelet-derived growth factor 2 transcript is a potent translational inhibitor. Mol Cell Biol 8:284–292

Robbins KC, Antoniades HN, Devare SG, Hunkapiller MW, Aaronson SA 1983 Structural and immunological similarities between simian sarcoma virus gene product(s) and human platelet-derived growth factor. Nature (Lond) 305:605–608

Rorsman F, Bywater M, Knott TJ, Scott J, Betsholtz C 1988 Structural characterization of the human platelet-derived growth factor A-chain cDNA and gene: alternative exon usage predicts two different precursor proteins. Mol Cell Biol 8:571–577

Ross R, Raines EW, Bowen-Pope DF 1986 The biology of platelet-derived growth factor. Cell 46:155–169

Schlessinger J 1988 The epidermal growth factor receptor as a multifunctional allosteric protein. Biochemistry 27:3119–3123

Seifert RA, Schwartz SM, Bowen-Pope DF 1984 Developmentally regulated production of platelet-derived growth factor-like molecules. Nature (Lond) 311:669–671

Seifert RA, Hart CE, Phillips PE et al 1989 Two different subunits associate to create isoform-specific platelet-derived growth factor receptors. J Biol Chem 264:8771–8778

Shimokado K, Raines EW, Madtes DK, Barrett TB, Benditt EP, Ross R 1985 A significant part of macrophage-derived growth factor consists of at least two forms of PDGF. Cell 43:277–286

Siegbahn A, Hammacher A, Westermark B, Heldin CH 1989 Differential effects of the various isoforms of platelet-derived growth factor on chemotaxis of fibroblasts, monocytes and granulocytes. J Clin Invest, in press

Stroobant P, Waterfield MD 1984 Purification and properties of porcine platelet-derived growth factor. EMBO (Eur Mol Biol Organ) J 3:2963–2967

Waterfield MD, Scrace T, Whittle N et al 1983 Platelet-derived growth factor is structurally related to the putative transforming protein p28^{sis} of simian sarcoma virus. Nature (Lond) 304:35–39

Westermark B, Johnsson A, Paulsson Y et al 1986 Human melanoma cell lines of primary and metastic origin express the genes encoding the constituent chains of PDGF and produce a PDGF-like growth factor. Proc Natl Acad Sci USA 83:7197–7200

Westermark B, Betsholtz C, Johnsson A, Heldin CH 1987 Acute transformation by simian sarcoma virus is mediated by an externalized PDGF-like growth factor. In: Kjeldgaard NO, Forchhammer J (eds) Viral carcinogens. Munksgaard, Copenhagen, p 445–457

Westermark B, Siegbahn A, Heldin CH, Claesson-Welsh L 1990 The B type receptor for platelet-derived growth factor mediates a chemotactic response via ligand-induced activation of the receptor protein tyrosine kinase. Proc Natl Acad Sci USA 87:128–132

Yarden Y, Escobedo JA, Kuang-WJ et al 1986 Structure of the receptor for platelet-derived growth factor helps define a family of closely related growth factor receptors. Nature (Lond) 323:226–232

DISCUSSION

Sherr: Your conclusion therefore, based on this model of two PDGF receptors that have the ability to form heterodimers, is that PDGF-BB is a universal ligand and PDGF-AA can stimulate effects only in cells that express the α-receptor.

Westermark: That's right. PDGF-AA is very restricted, as you say; PDGF-AB would stimulate cells that have either only α-receptors or α- and β-receptors; PDGF-BB could stimulate cells that have only α-receptors, both types of receptor or only β-receptors.

Sherr: Concerning the down-modulation, does the degradation of the receptor—not internalization—require kinase activity or not?

Westermark: If you compare the kinase-negative mutant receptor with the wild-type receptor and use PDGF-BB as the ligand, internalization of the mutant form of the receptor is a little bit slower than of the wild-type. The time it takes for the internalized receptor to become degraded is also a little bit longer in the mutant cells. The time for degradation once transported to the lysosome is the same for both. Although the time is longer for the mutant receptor, it is 100% degraded under those conditions.

Sherr: Do you know the half-lives of pre-labelled receptor in the presence and absence of ligand for the mutant and wild-type molecules?

Westermark: Those time course experiments are being processed in the lab right now. All these experiments have been done by a visiting scientist from Leningrad, Alexander Sorkin.

Sherr: Joseph Schlessinger and Axel Ullrich reported that kinase activity was necessary for degradation of the EGF receptor (Honegger et al 1987). Although kinase-inactive forms of the receptor were internalized, the signal to direct the mutant receptor to the lysosome pathway required kinase activity. We have very similar results for the CSF-1 receptor, kinase-defective mutants are not degraded. Interestingly, when we co-expressed an epitopically marked kinase-defective receptor with another receptor mutant that had kinase activity, the kinase-defective molecules were redirected to the lysosome pathway. This again implies that kinase activity, even in *trans*, provides a signal for degradation. I am surprised that the PDGF receptor, which is more closely related to the CSF-1 receptor than to the EGF receptor, lacks this property.

Westermark: Yes, but it is definitely so. This can be shown in another way. If you add the ligand to the cells at 37 °C, the mutant receptor becomes as down-regulated as the wild-type receptor. Jossi Schlessinger has done experiments in which he had what he thought were heterodimers of EGF receptors, kinase-negative, kinase-positive (personal communication). These became sorted such that the kinase-negative receptors were recycled and the kinase-positive receptors were degraded. That is different from your results.

Sherr: Yes, we are seeing a dominant effect in *trans*; we are forcing the degradation of a kinase-defective receptor by co-expression of a receptor that has kinase activity.

Westermark: There is one more example in the literature. Rusty Williams has found that kinase-negative PDGF receptors are down regulated and degraded (Escobedo et al 1988); so it might be a difference between the types of receptor.

Hunter: What about the insulin receptor? That also requires kinase activity to be removed from the surface, isn't that right?

Westermark: That requires the kinase to be internalized, so it differs from the EGF receptor which becomes internalized but not degraded.

Hunter: That depends on who you listen to. Gordon Gill would argue that the kinase-negative EGF receptor does not become internalized (Glenney et al 1988), but it is clear that down-regulation of the insulin receptor needs the kinase activity (e.g. Russell et al 1987). So there's no commonality here.

Wagner: In terms of the endothelial biology, did you say that the β-receptors are present only in the capillary endothelial cells?

Westermark: Previous experiments done in several laboratories, including our own, have shown that endothelial cells lack PDGF receptors. Those studies used endothelial cells from the umbilical cord or the aorta, that is endothelial cells from large vessels; as far as I know, capillary endothelial cells have not been studied before. Our experiments show that the capillary cells have receptors for PDGF, but we don't know about other vessels, for example medium-sized vessels.

Wagner: Is PDGF produced by both types of endothelial cells?

Westermark: PDGF is produced by all types of endothelial cells that have been studied, including those from the porcine aorta.

Wagner: In your functional studies, the gene for the PDGF receptor was transfected into endothelial cells which lacked the receptor, then when you added PDGF it was sufficient to transform these endothelial cells. What happens when you add ligand to cells that have been transfected with the PDGF receptor? Would this be sufficient to transform the endothelial cells?

Westermark: That is a matter of definition. There might be a phenotypic transformation; you cannot tell by morphology whether you have added the virus or the ligand.

Wagner: But you can immortalize these primary endothelial cells?

Westermark: SSV or PDGF does not immortalize cells. This transformation is caused, in our opinion, by an autocrine loop; it has nothing to do with immortalization. Probably only one further step is required for tumorigenesis.

Hunter: Are the tumours caused by SSV polyclonal or monoclonal?

Westermark: Unfortunately that has never been studied. The experiments were done in Fritz Deinhardt's laboratory about ten years ago and it was only published as an abstract and there was no analysis of clonality.

Hunter: From your model, you would predict them to be polyclonal.

Westermark: I would predict that the first response is polyclonal, but the tumour that emerges is monoclonal.

Hunter: Could I come back to the issue of the difference in signalling function between the α- and β-type PDGF receptors? Do you think this is due to an intrinsic difference in substrate specificity or perhaps to some difference in surface distribution? Do these receptors co-distribute across the whole surface of the cell or could the β-receptors be localized near microvilli or something else on the surface which allows them to interact with certain substrates?

Westermark: We have been able to study only the distribution of the β-receptor because we do not have monoclonal antibodies directed against the α-receptor. We have so far looked at only a limited set of reactions, but it seems that in most respects the two types of receptor behave in a similar way. Both are kinases and both become autophosphorylated. Annet Hammacher in Carl Heldin's laboratory has recently demonstrated that the α-receptor becomes autophosphorylated (Hammacher et al 1989), but apparently these experiments were more difficult to perform than with the β-receptor. With regard to substrates, either identified substrates or bands on gels, we in our lab know nothing about this.

I want to clarify one point. Human foreskin fibroblasts have about four times fewer α-receptors than β-receptors. We do not rule out the possibility that the α-receptor could also mediate actin reorganization, provided that enough receptors were present.

Hunter: In terms of mitogenicity, are both receptors equally efficient?

Westermark: In cells that have equal numbers of α- and β-receptors, PDGF-AA and PDGF-BB are equally potent.

Verma: Where is the decision made in the cell to form homodimers or heterodimers of the growth factor? Is it made when both A and B chains are present? Is it the concentration of the subunits or just the level of transcription? Why are there three different types of PDGF molecule? Why is a PDGF-BB homodimer made in preference to an AB heterodimer?

Westermark: That has not been studied carefully enough to answer the question. In CHO cells, which express both types of chain, about 70% of the PDGF that is purified is in the heterodimeric form. So there might be preferential assembly of heterodimers over homodimers, but that has not been studied. If you purify PDGF from cultured cells, glioma, melanoma or sarcoma, even though the cells make both types of messenger RNA, the protein that is purified is 90% or more PDGF-AA. The main reason may be that although the B chain message is there, it is very inefficiently translated. Apparently, in the long untranslated 5' sequence there are sequences that repress translation of the downstream mRNA. SSV has deleted the entire first exon and removed these sequences. I think it is post-transcriptional control of the synthesis of the B chain that keeps the concentration of the protein down.

Brugge: The ruffling that takes place after the interaction of PDGF-BB with its receptor, do you know if this event requires the kinase insert domain?

Westermark: That has not been tested. The only mutant form of receptor that has been tested is the kinase-negative one.

Maness: Regarding the ruffling, have you done a time course correlating ruffling with the amount of phosphotyrosine-containing protein?

Westermark: Time course studies have been done; it is a transient phenomenon. We usually pre-incubate the cells with PDGF at 0 °C, then transfer the cells to 37 °C. The first ruffles appear within five minutes and disappear within 60 minutes. We have not correlated this with the amount of phosphotyrosine in proteins. We have used orthovanadate, which inhibits phosphotyrosine phosphatase activity, to see whether it affects ruffling. It appears that orthovanadate delays the disappearance of the ruffles. That is as far as we have got.

Maness: Have you ever seen any changes in the microtubules or cytoskeleton?

Westermark: We have never looked at the microtubules.

Hanley: Coming back to the insert domain, is the β-receptor able to form the same type of complexes as the α-receptor, with lipid kinases or the *raf* gene product, for example?

Hunter: The insert is not needed for association of the PDGF β-receptor with *raf* protein, only for phosphatidylinositol (PI) kinase (Morrison et al 1989, Coughlin et al 1989).

Westermark: We would like to do these experiments in systems where we could express either just the α- or just the β-receptor. We are trying to express the α-receptor cDNA in endothelial cells, but we have not succeeded yet.

Harlow: Is that distinction generally accepted—that all receptors that have kinase domains with inserts are associated with PI kinase?

Hunter: No, I meant that the kinase insert deletion mutant of the PDGF receptor still associates with *raf* protein but does not associate with the Type I PI-3 kinase.

Harlow: Is that true for other receptors?

Sherr: I don't think you can generalize from one to the other. For example, unlike the PDGF receptor, there is no demonstrable interaction between the CSF-1 receptor and phospholipase Cγ (PLCγ). One problem is that there are many isozymes of PLC, so that the CSF-1 receptor might associate with other PLC isoforms. I don't think that it has even been demonstrated where on the PDGF receptor PLCγ binds, or whether that's a function of the insert or not.

Hunter: Rusty Williams has shown that binding of PLCγ to PDGF receptors does not require the insert (Morrison et al 1990). However, we were talking about PI kinase. Does it associate with the CSF-1 receptor?

Sherr: That is not clear. I think it may associate with the CSF-1 receptor.

Hanley: Lewis Cantley believes the Type I PI-3 kinase is always associated with tyrosine kinase-type growth factor receptors, regardless of the presence or absence of insert domains. However, in the PDGF receptor, from Rusty Williams' work the insert seems to be essential for complex formation and enzymic activation of PI-3 kinase (Coughlin et al 1989).

To clarify the potential role of PI-3 kinase, I have to emphasize that we are discussing two different types of lipid events. One is an undisputed signal generating degradation of inositol lipid by PLC, the other is a modification reaction, of unknown significance, of PI and possibly the other phosphorylated inositol lipids by PI-3 kinase. The correlation that has attracted a lot of interest is the co-precipitation by immunoprecipitating antibodies of occupied PDGF receptor and the PI-3 kinase. When the insert domain of the PDGF receptor is deleted, co-precipitation of PI-3 kinase is prevented and the mutant PDGF receptor is no longer mitogenic, even though many early biochemical events, including activation of PLC, still occur (Coughlin et al 1989). The insert-containing wild-type PDGF receptor is claimed to associate with and activate by phosphorylation this lipid-modifying enzyme, PI-3 kinase, which makes a novel lipid, phosphatidylinositol-3-phosphate (PI-3-P) of unknown function or significance.

Harlow: And the insert negative receptors do not?

Hanley: They do not drive mitogenesis and they do not associate with the PI-3 kinase.

Hunter: But tyrosine kinases without inserts, such as pp60[c-src], can also associate with PI-3 kinase, so it is a little unclear what role the insert plays in association (Sugano & Hanafusa 1985, Fukui & Hanafusa 1989).

Hanley: That is exactly right!

Westermark: Do receptors other than PDGF receptors associate with c-*raf*?

Sherr: The CSF-1 receptor probably associates with c-*raf*—the data are not at all complete. There is a change in the mobility of the *raf* protein in response to stimulation with CSF-1, which suggests some interaction. However, most of the phosphorylation of *raf* is on serine residues; we have not done the kinds of detailed studies recently reported by Debbie Morrison (Morrison et al 1988, 1989), who has demonstrated tyrosine phosphorylation of *raf-1* in response to

PDGF. She has told me that she and David Kaplan have done experiments that have led them to believe there is a ligand-dependent physical association between *raf* protein and the CSF-1 receptor, but we have not done those experiments in our own laboratory.

Verma: It has been asked several times, what is the function of the insert in the kinase domain in the PDGF receptor. Bengt Westermark said it was not important for mitotic activity. Is the answer that the work being done in Rusty Williams' lab can't be repeated by other people? Or is it that they are using different mutants, different cell types or the assay systems are different?

Hunter: Several details of the experiments are different. Rusty Williams assays his kinase insert deletions in CHO cells, which are very hard to growth arrest, because they are partially transformed.

Verma: Do CHO cells have PDGF receptors?

Hunter: There's a disagreement about that. Some CHO clones apparently do have residual PDGF receptors, although whether this affects the reported results is not clear.

Westermark: There are some indications that CHO cells express endogenous PDGF receptors at a low level. Liv Severinsson in Carl-Henrik Heldin's laboratory made a similar construct to Rusty Williams', except that she deleted the whole insert. She found very low, maybe 10%, kinase activity. That is closer to what Axel Ullrich has reported.

Hunter: In that system, Jaime Escobedo and Rusty Williams saw about 90% loss of mitogenicity, but most of the immediate early events were normal as far as they could tell—calcium and pH changes, for example (Escobedo & Williams 1988). Axel Ullrich and Jossi Schlessinger have tested quite a different system. They made a chimaeric receptor with an EGF external binding domain and a PDGF receptor transmembrane and cytoplasmic domain. They deleted either the whole kinase insert or the piece that Rusty Williams had deleted, which is only 80 residues long, and tested those in NIH/3T3 cells. In that system, which is better in the sense that fibroblasts normally respond to PDGF, they see some reduction of mitogenicity, but not as great as 90%, and all of the early responses are apparently intact. That could be due to a difference in cell type.

In contrast, there is the CSF-1 receptor (c-*fms*), which was tested by Tony Pawson in fibroblasts, not its normal host cell, and apparently provides a perfectly normal mitogenic response without the kinase domain insert (Taylor et al 1989). v-*fms* also transforms fibroblasts without the insert. Has the activity of these mutants been tested in macrophages or myeloid precursors?

Sherr: It has not been tested in macrophages. One might argue that there could be special effects in a cell in which the receptor normally resides but there is no direct evidence about this.

Deborah Morrison made another point about this insert: insert-negative mutants of the PDGF receptor expressed in baculovirus bind the *raf* gene product but do not recognize it as a substrate (Morrison et al 1989). Those same mutants do not bind the PI-3 kinase. Those results are being taken to suggest that this region of the receptor is important in interacting with PI-3 kinase.

Harlow: Tony, were you implying that the results from Rusty's lab using CHO cells were not informative?

Hunter: I don't think CHO cells are an ideal system.

Harlow: But if you can find any cell type in which you can distinguish between the mitogenic activity and what you call the early events, doesn't that mean that they can be separated?

Hunter: My interpretation is that the kinase insert is probably necessary for the maximum mitogenic response in both CHO and NIH/3T3 cells. The results from Schlessinger and Ullrich definitely show some impairment of mitogenicity but it is certainly not completely abolished, in contrast to Rusty Williams' interpretation of his CHO data.

Verma: Have these data been quantified?

Hunter: There was quantitation of receptor protein by blotting or immunoprecipitation, and the wild-type and kinase insert-negative receptors were expressed at reasonably comparable levels in both cases.

Sherr: I think the critical issue is whether you can ascribe specific functions to particular regions of these receptors. Is there a modular design or not? What has been argued from Rusty Williams' paper is that there is a modular design and that the kinase insert represents a surface for substrate recognition so that specific substrates bind there and others do not. The substrates which bind to the insert are implied to play a major role in mitogenicity, whereas the ones that do not bind there are irrelevant to mitogenicity. That's the strongest interpretation and probably is an overinterpretation of the data based on the kinds of experiments that other people are doing. If you make a deletion in a receptor and you subtly impair, for example, the kinase activity, you would expect to impair mitogenicity. What is at stake is this concept of modularity.

Hunter: This is a very crude technique—removing the insert probably affects the folding of the whole receptor.

Waterfield: The best models for the kinase domain predict that the insert lies outside of the main body of the structure, but that is all you can say (Bajaj M, Blundell T, Waterfield MD, unpublished). Everybody who predicts structure blind gets it wrong. We simply don't have enough information. Even understanding the importance of dimerization is difficult.

Hunter: Because we don't know how dimerization works. For the PDGF receptor there are some hints that the transmembrane domain plays a critical role in its function. Rusty Williams' experiments indicate that you can't simply replace the transmembrane domain with any other transmembrane domain and reconstitute a functional PDGF receptor (Escobedo et al 1988). In addition, there is an extremely high degree of conservation between the transmembrane domains of the mouse and human β-type PDGF receptors.

Waterfield: But not between those of α and β.

Westermark: That's right, only 48%.

Hunter: Another feature of the kinase insert is that it has autophosphorylation sites. That's true for the PDGF receptor; Andrius Kazlauskas and John Cooper (1989) have shown that a mutation which places a non- phosphorylatable residue at that site affects the association of particular cellular proteins with the receptor. It is true also for the CSF-1 receptor from the work of Peter Tapley and Larry Rohrschneider in Seattle, and from Peter van der Geer in my own group. However, we don't know whether that is intermolecular or intramolecular phosphorylation nor do we know the function of those sites, especially in the CSF-1 receptor.

Sherr: Has anyone exchanged the insert domains between receptors to see whether that affects the substrate specificity or mitogenicity?

Hunter: No, but there isn't total agreement about whether deletion of the insert from the kinase domain of PDGF renders it non-mitogenic. Jossi Schlessinger and Axel Ullrich would say something different to Rusty Williams. Secondly, you can delete the kinase insert domain from both the CSF-1 receptor (c-*fms*) and v-*fms* and have a perfectly functional receptor. In that case the insert is dispensable for things that can be assayed readily *in vitro*.

Waterfield: Mike, do you have some data from experiments with temperature-sensitive mutants of *erbB*?

Hanley: We were curious about how many of the events we have studied historically in different contexts, such as activation of PLC and phosphorylation of inositol lipids, provide information in growth control. A lot of things go on, but which of the events are necessary for mitogenesis?

We wanted to investigate whether activation of PLC was linked to oncogenic tyrosine kinase action. We took advantage of the fact that chicken erythroblast lines had already been created by Thomas Graf by infection with avian erythroblastosis virus. These were temperature sensitive for growth and differentiation (Beug et al 1982), because of a ts mutant of the v-*erbB* gene. As v-*erbB* is a ligand-independent version of c-*erbB*, the EGF receptor, we could manipulate the activity of this oncogene by changing the temperature and see whether PLC activity was also temperature sensitive, which would suggest (as has been shown by several labs) that tyrosine kinases activate PLC by direct phosphorylation. There was a correlation between the activation of v-*erbB* at the permissive temperature and the sustained activation of inositol lipid hydrolysis by PLC. Elevation to the non-permissive temperature stopped inositol lipid breakdown. We did not then check how those events were related to growth or differentiation—it was just an interesting correlation. Similar results were reported later by Kato et al (1987).

References

Beug H, Palmieri S, Freudenstein C, Zentgraf H, Graf T 1982 Hormone-dependent

terminal differentiation in vitro of chicken erythroleukemia cells transformed by ts
mutants of avian erythroblastoma virus. Cell 28:907–919

Coughlin SR, Escobedo JA, Williams LT 1989 Role of phosphatidylinositol kinase in
PDGF receptor signal transduction. Science (Wash DC) 243:1191–1194

Escobedo JA, Williams LT 1988 A PDGF receptor domain essential for mitogenesis
but not for many other responses to PDGF. Nature (Lond) 335:85–87

Escobedo JA, Barr PJ, Williams LT 1988 Role of tyrosine kinase and membrane-spanning
domains in signal transduction by the platelet-derived growth factor receptor. Mol
Cell Biol 8:5126–5131

Fukui Y, Hanafusa H 1989 Phosphatidylinositol kinase activity associates with viral p60src
protein. Mol Cell Biol 9:1651–1658

Glenney JR, Chen WS, Lazar CS 1988 Ligand-induced endocytosis of the EGF receptor
is blocked by mutational inactivation and by microinjection of anti-phosphotyrosine
antibodies. Cell 52:675–684

Hammacher A, Nister M, Helding CH 1989 The A type receptor for platelet-derived
growth factor mediates protein tyrosine phosphorylation, receptor transmodulation
and a mitogenic response. Biochem J 264:15–20

Honegger AM, Dull TJ, Felder S et al 1987 Point mutation at the ATP binding site
of EGF receptor abolishes protein-tyrosine kinase activity and alters cellular routing.
Cell 51:199–209

Kato M, Kawai S, Takenawa T 1987 Altered signal transduction in erbB-transformed
cells. J Biol Chem 262:5696–5704

Kazlauskas A, Cooper JA 1989 Autophosphorylation of the PDGF receptor in the kinase
insert region regulates interactions with cell proteins. Cell 58:1121–1133

Morrison DK, Kaplan DR, Rapp U, Roberts TM 1988 Signal transduction from
membrane to cytoplasm: growth factors and membrane-bound oncogene products
increase raf-1 phosphorylation and associated protein kinase activity. Proc Natl Acad
Sci USA 85:8855–8859

Morrison DK, Kaplan DR, Escobedo JA, Rapp UR, Roberts TM, Williams LT 1989
Direct activation of the serine/threonine kinase activity of raf-1 through tyrosine
phosphorylation by the PDGF β-receptor. Cell 58:649–657

Morrison DK, Kaplan DR, Rhee SG, Williams LT 1990 PDGF-dependent association
of PLCγ with the PDGF receptor signalling complex. Mol Cell Biol, in press

Russell DS, Gharzi R, Johnson EL, Chou CK, Rosen OM 1987 The protein tyrosine
kinase activity of the insulin receptor is necessary for insulin mediated receptor down
regulation. J Biol Chem 262:1833–1840

Sugano S, Hanafusa H 1985 Phosphatidylinositol kinase activity in virus transformed
and non transformed cells. Mol Cell Biol 5:2399–2404

Taylor GR, Reeijk M, Rothwell V, Rohrschneider L, Pawson T 1989 The unique insert
of cellular and viral fms protein tyrosine kinase domains is dispensable for enzymatic
and transforming activities. EMBO (Eur Mol Biol Organ) J 8:2029–2037

The *mas* oncogene as a neural peptide receptor: expression, regulation and mechanism of action

M. R. Hanley, W. T. Cheung, P. Hawkins, D. Poyner, H. P. Benton, L. Blair*, T. R. Jackson† and M. Goedert†

*MRC Molecular Neurobiology Unit and †MRC Laboratory of Molecular Biology, MRC Centre, Hills Road, Cambridge CB2 2QH, UK, and *Department of Pharmacology, University of California at San Diego Medical School, La Jolla, California 92093, USA*

Abstract. The human *mas* oncogene, which renders transfected NIH/3T3 cells tumorigenic, was identified as a subtype of angiotensin receptor by transient expression in *Xenopus* oocytes and stable expression in the mammalian neuronal cell line, NG115-401L. The *mas* receptor preferentially recognizes angiotensin III, and is expressed at high levels in brain. The *mas*/angiotensin receptor functions through the breakdown of inositol lipids and can drive DNA synthesis, unlike another inositol-linked peptide receptor, that for bradykinin. Comparative analysis of several early biochemical events elicited by either angiotensin or bradykinin stimulation of *mas*-transfected cells has not indicated a specific difference correlated with mitogenic activity. In particular, the inositol lipid kinase, phosphatidyl-inositol-3-kinase, thought to be involved in the mitogenic mechanism of platelet-derived growth factor receptors, is unaffected by activation of *mas*. These results have shown that a proto-oncogene encodes a neural peptide receptor, indicating that peptide receptors may be involved in differentiation and proliferation processes, as are other identified proto-oncogenes.

1990 Proto-oncogenes in cell development. Wiley, Chichester (Ciba Foundation Symposium 150) p 23–46

Peptides in eukaryotic growth control

Peptides are recognized as having mitogenic activity on populations of vertebrate cells and cell lines (Hanley 1985, Zachary et al 1987) but in most instances the evidence has been confined to activities observed on cultured cells (Table 1). Only recently have the implications of this activity for novel physiological functions of peptides as regulators of growth and differentiation begun to be assessed. Peptides do, however, function physiologically as growth regulatory agents in the simple unicellular eukaryote, budding yeast, where peptide mediators, the mating factors, arrest growth as part of the yeast mating response

23

TABLE 1   Mitogenic actions of peptides on cultured cell populations

Peptides	Transduction system	Sensitive cells
Angiotensins	Phosphoinositide hydrolysis	BALB/c 3T3, adreno-cortical
Bombesin family	Phosphoinositide hydrolysis	Swiss 3T3, bronchial epithelial, small-cell lung carcinoma
Bradykinin	Phosphoinositide hydrolysis	Swiss 3T3, human fibroblasts
Cholecystokinin/gastrin	Phosphoinositide hydrolysis	Gastric carcinoma, pancreatic acinar
Tachykinin family (substance P, substance K)	Phosphoinositide hydrolysis	Human fibroblasts, rat smooth muscle, synoviocytes
Vasoactive intestinal polypeptide	Adenyl cyclase activation	Keratinocytes
Vasopressin	Phosphoinositide hydrolysis	Swiss 3T3, rat chondrocytes, BSC-1 renal epithelial

(Hershkowitz & Marsh 1987). Although peptides appear to stimulate, rather than inhibit, proliferation in higher eukaryotic cells, these two forms of peptide-mediated growth control have many formal similarities: both proceed through a common initiating event, activation of a GTP-binding protein (Hershkowitz & Marsh 1987), and many of the downstream interactions appear to be highly conserved from yeast to mammalian cells. These observations suggest that there may be structural and functional relationships between the products of oncogenes and peptide receptors, but there has been little direct evidence of this. In one specific situation, small-cell carcinoma of the lung, bombesin-related peptides were proposed to act as autocrine stimulants which exacerbated the severity of tumour growth (Cuttitta et al 1985). These observations suggest that mitogenic peptide receptors, such as the bombesin receptor, might themselves function as oncogenes. This raises the possibility that peptide receptors are involved extensively in processes of growth and development, as are other proto-oncogene products.

Identification of the *mas* oncogene product as an angiotensin receptor

The human *mas* oncogene was originally identified (Young et al 1986) by a version of the NIH/3T3 focus formation assay, in which DNA-transfected cells were injected into nude mice and tumour formation was assessed (Fasano et al 1984). Using human genomic DNA from an epidermoid carcinoma, a segment of human DNA with tumorigenic potential was rescued from tertiary transformant cells. This DNA could induce a low frequency of foci in transformation assays, which appeared to consist of a very high density of cells with normal morphology. The normal human genomic *mas* proto-oncogene was identified. The transformation assay indicated that it had no focus-forming activity, but was weakly tumorigenic, albeit the time course of tumour formation was much slower than for the activated oncogene. Restriction mapping and the construction of chimaeric genes indicated that the oncogenic activation of *mas* occurred approximately 3.5 kb upstream from the predicted start of translation in the coding region. This oncogenic activation was not found in the original tumour tissue and appears to have occurred spontaneously, in parallel with genetic amplification, during the serial transfection and tumour formation steps. More recently, exactly the same pattern of 5' sequence rearrangement, accompanied by gene amplification, was independently obtained by transfection of human myelocytic leukaemia cell DNA (Janssen et al 1988).

The predicted protein sequence from the open reading frame of the human *mas* cDNA had a unique structure for an oncogene product (Young et al 1986), having seven hydrophobic domains which were predicted to form transmembrane α helices (Fig. 1). This feature is found in an increasing number of cloned cell surface receptors that activate effector mechanisms through a class of homologous trimeric GTP-binding proteins (Ross 1989). Thus, on purely structural grounds, it is likely that *mas* encodes a member of this receptor family. Comparison of the primary sequence and predicted secondary structure of the *mas* oncogene product with other members of this family showed a specific resemblance to the substance K peptide receptor (Hanley & Jackson 1987), which is known to have mitogenic activity (Hultgardh-Nilsson et al 1988). The sequence similarities prompted two predictions; that the activating stimulus for *mas* would be a mitogenic peptide, and that *mas* would function, like the substance K receptor, through the hydrolysis of inositol phospholipids. These hypotheses focused assays on a limited range of candidate peptide stimulants and a precise mechanism of action.

When a synthetic transcript from a restriction fragment of the human *mas* oncogene containing the complete protein coding region was injected into *Xenopus* oocytes, a novel sensitivity to angiotensins appeared which was not found in uninjected defolliculated oocytes (Jackson et al 1988). When analysed under voltage-clamp conditions, *mas* transcripts induced the expression of an angiotensin-activated inward current (Fig. 2), which was identical to that seen with the cloned substance K receptor (Masu et al 1987). Expression of other extrinsic receptors in *Xenopus* oocytes, such as the serotonin 1c receptor (Julius et al 1988), has shown that this inward current is carried by a chloride channel which is activated by intracellular calcium discharge after hydrolysis of inositol phospholipids (Meyerhof et al 1988). Consequently, this electrophysiological response was early evidence that *mas* functioned via stimulation of phosphoinositide hydrolysis. The order of potency for mammalian angiotensins was angiotensin III $\geqslant$ angiotensin II $\gg$ angiotensin I. These peptides are derived

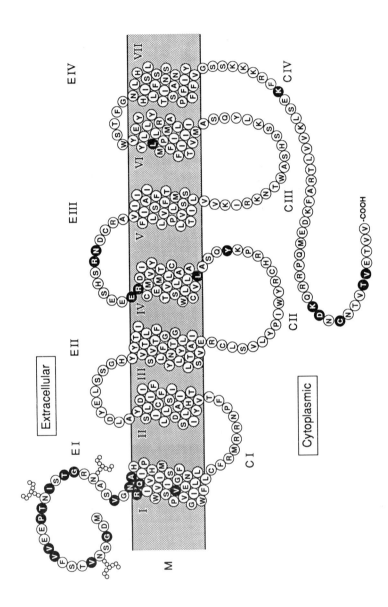

FIG. 1. The predicted protein sequence of the human *mas* oncogene (Young et al 1986) and the transmembrane organization suggested by its hydropathy plot. The residues in black are not conserved in the rat *mas* sequence (Young et al 1988). The N-terminus (EI) is placed extracellularly; it has three predicted sites of N-linked glycosylation, shown schematically. Extracellular domains are labelled EI to EIV, transmembrane domains are labelled MI to MVII, cytoplasmic domains are labelled CI to CIV.

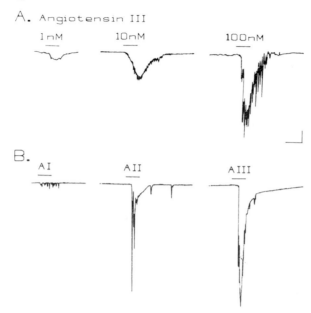

FIG. 2.   Voltage-clamp current records from *Xenopus laevis* oocytes microinjected with a synthetic transcript for the human *mas* coding sequence. (A) Dose dependency of bath application of angiotensin III. (B) Rank order of preference for the mammalian peptides angiotensin I (AI), angiotensin II (AII) and angiotensin III (AIII) at an application concentration of 100 nM. The vertical calibration bar corresponds to 100 nA in (A) and 50 nA in (B); the horizontal calibration bar represents one minute. The details are described in Jackson et al (1988).

by a sequence of processing steps from an initial product, which is released by the protease renin from the precursor protein, angiotensinogen (Fig. 3). Although angiotensin II has been regarded as the only physiologically relevant hormone in this family, endogenous angiotensin receptors in brain and adrenal cortex have the same preference for angiotensin III over angiotensin II (Peach 1977) as does that encoded by *mas*.

   Transfection of the neural cell line, NG115-401L, with the human *mas* oncogene gave a series of stable transfected lines, one of which, 401L-C3 (Jackson et al 1988), has been used for detailed mechanistic analysis of *mas*. Use of the intracellular calcium indicator, fura-2, confirmed that *mas* produced an angiotensin receptor which mobilized intracellular calcium with identical pharmacology to that seen when *mas* was expressed in *Xenopus* oocytes (Jackson et al 1988). This result shows that the specificity of the *mas* gene product is intrinsic and independent of the very different conditions of expression in the two systems (i.e. transient expression in a lower vertebrate cell and stable expression in a mammalian cell). Consequently, it is unlikely that *mas* expression is inducing endogenous angiotensin receptors. Moreover, the angiotensin

```
ASP-ARG-VAL-TYR-ILE-HIS-PRO-PHE-HIS-LEU- (ANGIOTENSINOGEN)
```

↓ RENIN

Angiotensin I.   ASP-ARG-VAL-TYR-ILE-HIS-PRO-PHE-HIS-LEU

↓ ANGIOTENSIN-CONVERTING ENZYME

Angiotensin II.   ASP-ARG-VAL-TYR-ILE-HIS-PRO-PHE

↓ AMINOPEPTIDASE

Angiotensin III.   ARG-VAL-TYR-ILE-HIS-PRO-PHE

↓ EXOPEPTIDASES

INACTIVE FRAGMENTS

FIG. 3.   Biosynthetic pathway and primary structures of mammalian angiotensins. The enzymes mediating each processing step are named to the right of the arrows.

receptor encoded by *mas* appears insensitive to large changes in membrane environment.

The stimulation of cytosolic calcium elevation in the 401L-C3 line is accompanied by production of inositol phosphates and activation of protein kinase C (Jackson & Hanley 1989), in accord with a mechanism involving inositol lipid hydrolysis.

To complete the formal identification of the angiotensin sensitivity with the *mas* gene product, we have used a monospecific antipeptide antiserum to a segment of the C-terminal tail (residues 301–316). This antiserum recognizes a single major protein of 45 kDa, which is slightly larger than the predicted molecular weight of 37.5 kDa, only in *mas*-transfectants and not in parent cells (Fig. 4). This immunoreactive protein can be labelled by cross-linking [³H]angiotensin II to intact cells, confirming that the predicted *mas* gene product is a cell surface angiotensin receptor.

Neural expression of the *mas* proto-oncogene

In the original description of the *mas* gene, one of the difficulties in establishing its possible functions was the failure to find the transcript in normal tissues

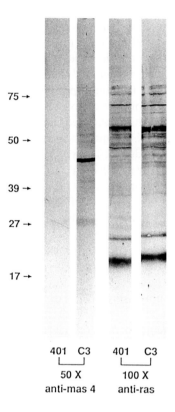

FIG. 4.   Immunoblot of membrane proteins (0.1 mg total each lane) from the NG115-401L cell line (401) and the 401L-C3 *mas*-transfected line (C3) using monospecific anti-human *mas* antisera or monospecific anti-*ras* antiserum that recognizes all forms of p21^{ras}. Antisera were raised in sheep to the *mas* C-terminal tail (residues 301-316) or the N-terminal region of human N-*ras*/c-Ha-*ras* (residues 1–18) by conjugation to a carrier protein. They were affinity purified on columns of the immobilized peptide and used at the dilutions shown. Immunoreactive bands were visualized by peroxidase-labelled anti-sheep antibodies. There are no *mas*-related antigens found in the parent line; a single major band is found in the C3 line. This band can be cross-linked to radioactive angiotensin, and corresponds to the *mas* gene product. For comparison, no changes in immunoreactive proteins are seen in another oncoprotein, *ras*.

or cell lines (Young et al 1986). It was suggested that *mas* might be expressed in neural tissues (Hanley & Jackson 1987); this was subsequently found to be the case for both the human (Jackson et al 1988) and the rat genes (Young et al 1988). Specifically, rat *mas* gene transcripts have been detected exclusively in brain, with highest levels in hippocampus, followed by cerebral cortex; in human brain the levels are highest in cerebral cortex (Jackson et al 1988). Although the neural cells expressing the gene have been established only recently in adult brain, in developing rat brain, *mas* transcripts have been localized by *in situ* hybridization to proliferating neuroblasts or specific populations of differentiating neurons (Martin et al 1989). Significantly, it has been difficult to establish unambiguously the presence or absence of the transcript in extraneural sites. In human peripheral tissues, such as liver and adrenal glands, very low levels of the transcript have been observed on Northern blots (Jackson et al 1988), but this result needs to be confirmed by more sensitive and specific assays, such as RNAse protection assays. In rat tissues, protection assays of *mas* mRNA have not detected *mas* expression in a variety of peripheral organs, including heart, kidney and liver (Young et al 1988). This suggests that *mas* may encode a specific neuronal subtype of angiotensin receptor, comparable to the preferential high level expression of the serotonin 1c receptor subtype in mammalian brain (Julius et al 1988). Given the sensitivity of a number of cardiovascular and non-vascular peripheral tissues to angiotensins, it is clear that *mas* is a member of a family of receptors. The precedent of the tachykinin receptors (Yokota et al 1989) implies that such a *mas*-related family of angiotensin receptors may have extensive sequence similarity and therefore be amenable to molecular cloning using probes or structural information derived from the *mas* gene.

Little is known of the acute or developmental tissue regulation of the *mas* gene, or its transcriptional or translational regulation in isolated cells. To pursue these questions, neuronal cells which constitutively express the *mas* gene product have been identified and are being compared in terms of gene regulation to transfected cell lines.

Regulation of angiotensin sensitivity

The availability of transfected cell populations expressing a single molecular type of angiotensin receptor provided an opportunity to investigate possible modes of control of acute sensitivity. The 401L-C3 clone has been examined for known forms of stimulation-dependent negative regulation (e.g. desensitization) using a combination of population and single cell approaches.

As activation of a constitutive bradykinin receptor had been shown previously to elicit intracellular calcium discharge (Jackson et al 1987), interactions between the angiotensin and bradykinin receptors in 401L-C3 cells were examined using fluorescence-detected calcium transients as an acute response. When angiotensin was added to 401L-C3 monolayers, a subsequent response to bradykinin could be elicited. However, when the order of addition was reversed, bradykinin abolished the angiotensin response (Jackson et al 1988). This transmodulation of one receptor by another has been observed with growth factor receptors and is termed 'heterologous desensitization'. As the activation of bradykinin receptors leads to stimulation of protein kinase C (Hanley et al 1988), an identified negative regulator of angiotensin receptors (Pfeilschifter et al 1989), the possible involvement of this pathway in the bradykinin effects was assessed. When 401L-C3 cells were briefly exposed to a phorbol diester, the angiotensin response was selectively abolished, with no effect on the bradykinin response (Jackson et al 1988). In the parent cells, protein kinase C can be down-regulated completely by overnight treatment with a stimulant (Mattingly et al 1987). Using such a pretreatment, the acute inhibitory effect of either a phorbol diester or bradykinin was eliminated, suggesting that this heterologous action of the bradykinin receptor on the *mas* response pathway might be mediated by direct phosphorylation of the *mas*-encoded angiotensin receptor. There is an important precedent for this kind of interaction in the structurally related cyclic AMP-stimulating receptors. Multiple kinase sites have been mapped on the β_2-receptor sequence, and protein kinase A sites are found on the third intracellular loop (Sibley et al 1987). Thus, the β_2-receptor can be transmodulated by any coexpressed receptor which activates protein kinase A, in a fashion directly analogous to the regulatory transmodulation of angiotensin sensitivity in the 401L-C3 cells. The analogy may be even more striking in that potential sites for protein kinase C phosphorylation are localized to the third intracellular loop in the *mas* protein, which is the same domain as the site of protein kinase A modification of the β_2-receptor.

These experiments were confined to cell populations, which may have a large degree of individual variability in calcium homeostasis. We have confirmed and extended the results on calcium transients and densensitization by examining single cell calcium levels by video imaging and single cell channel activity by whole cell patch-clamping. Fig. 5 shows fluorescent recording of calcium levels in single 401L-C3 cells. In Fig. 5A, sequential superfusion of bradykinin followed by angiotensin indicates that angiotensin calcium responses are desensitized by pre-exposure to bradykinin, as was found in population calcium measurements. In Fig. 5B, sequential addition of increasing doses of angiotensin abolishes its own calcium responses, an example of 'homologous desensitization'. As angiotensin activation, as well as bradykinin activation, of 401L-C3 cells stimulates protein kinase C (Jackson & Hanley 1989), it may be that this proceeds through the same mechanism as heterologous desensitization. However, a more interesting possibility is that there is a distinct form of agonist-induced desensitization mediated by a kinase distinct from protein kinase C, such as has been found for β-adrenergic receptors (Sibley et al 1987).

An unexpected heterogeneity has emerged in single cell measurements: only 50–60% of the total 401L-C3 population can respond to angiotensin, whereas all of the cells can respond to bradykinin. One possible explanation was that the 401L-C3 line was not clonal for *mas* expression. However, immunocytochemical staining with anti-*mas* antibodies confirmed that the cells were homogeneous for expression of the *mas* gene product. Therefore, another form of negative regulation of *mas* function may have been detected by single cell analyses. Since these cells have not been deliberately synchronized or

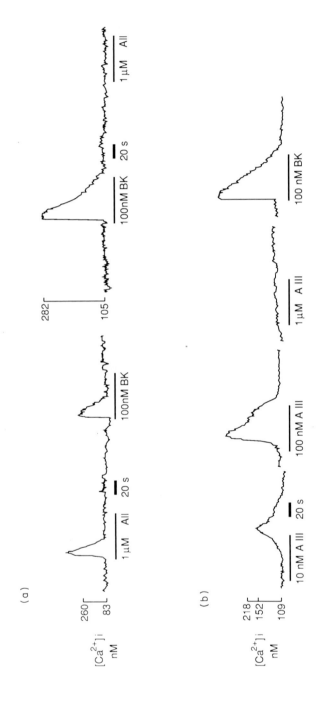

FIG. 5. Time courses of single 401L-C3 cell [Ca²⁺] responses to sequential superfusion of angiotensin II (AII), angiotensin III (AIII) or bradykinin (BK) at the concentrations shown. The heavy horizontal bars represent 20 seconds, the light horizontal bars indicate the periods of peptide exposure. Data are taken from photodiode recordings of video images of fluorescence recordings from monolayer fura-2-loaded cells. A more extensive description of the imaging system and technical details of the experimental design can be found in O'Sullivan et al (1989).

rendered quiescent, an appealing possibility is that there is some form of cell cycle control of *mas*. In view of the important precedent of cell cycle protein kinases, such as *cdc2* (see Norbury & Nurse, this volume), the specific hypothesis would be that a *cdc* homologue or a comparable regulatory kinase, would block *mas* activation, but not bradykinin receptor activation, by phosphorylation of the *mas* receptor or of an early component in its transduction cascade. This additional level of negative regulation of a mitogenic receptor, uncoupling its activity during specific phases of the cell cycle, may be a protective mechanism to ensure there is not a damaging signal for reinitiation of DNA synthesis when a cell has already experienced mitogenic stimulation.

Mechanism of action of *mas* in growth stimulation

The identification of *mas* as a specific peptide receptor acting via inositol lipid breakdown with the downstream consequences of calcium discharge and protein kinase C stimulation indicated that *mas* shared a common set of transduction events with the endogenous 401L-C3 bradykinin receptor. Yet, as Table 2 shows, angiotensin stimulates DNA synthesis but bradykinin does not. A number of early events (Rozengurt 1986) caused by occupation of the angiotensin and bradykinin receptors have been examined, such as activation of high affinity GTPase activity, spatial and temporal aspects of calcium changes, and regulation of cytosolic pH, but no significant differences have been seen.

However, a novel possibility for an obligate receptor-coupled event in mitogenesis has emerged from studies of mutant platelet-derived growth factor (PDGF) receptors in which the 'insert domain' in the tyrosine kinase region has been deleted (Escobedo & Williams 1988). When PDGF-occupied PDGF receptors are immunoprecipitated, other proteins are co-precipitated in a complex. One of these proteins has been identified as an inositol lipid-processing enzyme, phosphatidylinositol-3-kinase (PI-3 kinase; Type I PI kinase), which has been shown to make a novel inositol phospholipid, phosphatidyl-inositol-3-phosphate (PI-3-P; Whitman et al 1988). This enzyme appears

TABLE 2   Stimulation of DNA synthesis in 401L-C3 *mas* transfectants

Control (no additions)	100%
Bradykinin (1 μM)	$99 \pm 5\%$
Angiotensin III (1 μM)	$126 \pm 11\%$
Fetal calf serum (2.5%)	$224 \pm 21\%$
Insulin-like growth factor I (10 ng/ml)	$165 \pm 7\%$

Cells were placed in reduced serum (1%) for at least 24 hours before addition of the above stimulants for another 16 hours in the presence of [³H]thymidine. Labelled DNA was extracted and counted, giving control values of 15 000–20 000 cpm. Values quoted are the means of 3–8 independent determinations.

to be stimulated by PDGF acting through its receptor and is a substrate for the PDGF receptor tyrosine kinase (Kaplan et al 1987). The reason for intense interest in PI-3 kinase is that deletion of the insert domain of the PDGF receptor tyrosine kinase region gives receptors that function normally in many early events, such as activation of phospholipase C and tyrosine kinase activity, but are unable to drive DNA synthesis (Coughlin et al 1989). The only identified biochemical change is the inability of mutant PDGF receptors to co-precipitate PI-3 kinase activity. In addition, cellular protein complexes with the viral oncogene product, polyoma middle T antigen, which activates the tyrosine kinase c-*src*, also appear to co-precipitate activated PI-3 kinase (Courtneidge et al 1988). Thus, the activity of this enzyme, although mysterious in its possible mechanism, has been correlated in several ways with an essential role in cell growth.

The clonal cell expression of two peptide receptors sharing a common coupling to the phosphoinositide pathway, only one of which is able to drive DNA synthesis, provides a complementary system in which to ask whether activation of PI-3 kinase is a distinction between mitogenic and non-mitogenic pathways. In Fig. 6, the time courses of breakdown and conversions of inositol phospholipids in response to angiotensin are shown. Although there are detectable levels of PI-3-P in the 401L-C3 cells, the synthetic enzyme is not activated, nor is the PI-3-P broken down by phospholipase C. Thus, PI-3 kinase may be essential to mitogenic stimulation through the tyrosine kinase-type growth factor receptors, but may not be essential in the G protein-coupled class of receptors. This still leaves the question open of how the bradykinin and angiotensin response pathways differ. Using nystatin-permeabilized cell-attached patch-clamp, a potential early difference has been seen. When bradykinin receptors are activated, there is a small, reversible inhibition of calcium-dependent, voltage-activated potassium currents, but when the *mas*-type angiotensin receptors are activated, there is a time-dependent progression to complete abolition of voltage-activated potassium channels (L. Blair, unpublished observations). The angiotensin-stimulated silencing of these membrane channels is surprisingly long lived, lasting for hours. Thus, differential regulation of channel activity clearly reveals that there is some early difference at the plasma membrane which distinguishes these two receptors.

Implications for carcinogenesis and normal development

There is as yet no evidence that the *mas* oncogene is ever rearranged or activated in human tumours. Rather, it appears that the ectopic expression of this gene in an extraneural context may reveal a latent transforming potential, similar to that seen with the serotonin 1c receptor (Julius et al 1989). However, the location of human *mas* on the long arm of chromosome 6 (Rabin et al 1987) is in a region of several tumour-specific fragile sites associated with malignant melanoma and acute non-lymphocytic leukaemia; it is also a region where

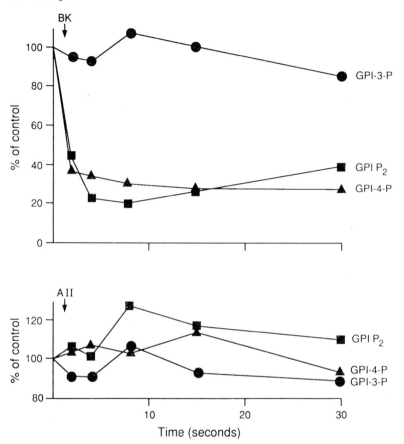

FIG. 6. Time courses of stimulated changes in [^{3}H]inositol-labelled phospholipids, expressed as a percentage of the control levels of labelled lipid. The individual lipids were resolved by high performance liquid chromatography as their deacylated inositol phosphates (described in Stephens et al 1989). Each species can be identified as its corresponding lipid as follows: GPI-3-P, phosphatidylinositol-3-phosphate; GPI-4-P, phosphatidylinositol-4-phosphate; GPI P2, phosphatidylinositol-4,5-bisphosphate. Monolayers of 401L-C3 cells were stimulated with supramaximal doses of bradykinin (BK; upper panel) or of angiotensin II (AII; lower panel) and perchloric acid-extracted samples were analysed in triplicate at the times shown.

characteristic karyotypic abnormalities of neuroblastomas are observed. Thus, there may be a targeted subset of tumours which will be profitable to examine for altered expression of *mas*.

In normal development, too little is known about the possible functions of neural peptides to assess whether the *mas* gene product acts as a developmental regulator. However, examination of the developmental appearance of angiotensin receptors has shown a transient expression of apparently functional receptors in mesenchymal and connective tissue sites, including skin fibroblasts and skeletal muscle (Millan et al 1989). *mas* or a related angiotensin receptor may function exclusively in developmental processes at these sites. In order to evaluate these possibilities, it will be necessary to develop preparations which can maintain a developmental programme *in vitro*. A model study that illustrates how such an approach can assess developmental actions of peptides is the use of isolated embryonic chick otic vesicles (Represa et al 1988). In this system, bombesin can trigger an ordered sequence of normal developmental events which approximate steps in the early development of the inner ear.

Peptides exhibit several tissue-specific trophic activities (Zachary et al 1987), suggesting that *mas* may also have extraproliferative trophic functions. *mas* may be one of many potential examples of multifunctional peptide receptors, acting in different ways in different cellular or developmental contexts. If, as has been suggested (Hanley 1989), all forms of cell growth employ a common panel of genes, then mitogenic receptors and transforming genes expressed in a non-dividing cell type, such as a cardiac myocyte or a neuron, may be co-opted to regulate other growth phenomena, such as somatic hypertrophy or nerve process extension. From this perspective, the *mas* gene product may have both developmental functions, in controlling neural cell proliferation and maturation, and differentiated functions, in regulating neuronal excitability and gene expression.

Acknowledgements

The authors thank Dr Ian Varndell (Cambridge Research Biochemicals) for assistance in preparing antisera, Drs M. Berridge, T. Cheek and R. Moreton (Department of Zoology, Cambridge) for generous help with single cell calcium imaging.

References

Coughlin SR, Escobedo JA, Williams LT 1989 Role of phosphatidylinositol kinase in PDGF receptor signal transduction. Science (Wash DC) 243:1191–1195

Courtneidge SA, Kypta RM, Ulug ET 1988 Interactions between the middle T antigen of polyomavirus and host cell proteins. Cold Spring Harbor Symp Quant Biol 53:153–160

Cuttitta F, Carney DN, Mulshine J et al 1985 Bombesin-like peptides can function as autocrine growth factors in human small-cell lung cancer. Nature (Lond) 316:823–826

Escobedo JA, Williams LT 1988 A PDGF receptor domain essential for mitogenesis but not for many other responses to PDGF. Nature (Lond) 335:85–87

Fasano O, Birnbaum D, Edlund L, Fogh J, Wigler M 1984 New human transforming genes detected by a tumorigenicity assay. Mol Cell Biol 4:1695–1705

Hanley MR 1985 Neuropeptides as mitogens. Nature (Lond) 315:14–15

Hanley MR 1989 Mitogenic neurotransmitters. Nature (Lond) 340:97–98

Hanley MR, Jackson TR 1987 Substance K receptor: return of the magnificent seven. Nature (Lond) 329:766–767

Hanley MR, Jackson TR, Cheung WT et al 1988 Molecular mechanisms of phospholipid signalling pathways in mammalian nerve cells. Cold Spring Harbor Symp Quant Biol 53:435–445.

Hershkowitz I, Marsh L 1987 Conservation of receptor/signal transduction system. Cell 50:995–996

Hultgardh-Nilsson A, Nilsson J, Jonzon B, Dalsgaard C-J 1988 Coupling between inositol phosphate formation and DNA synthesis in smooth muscle cells stimulated with neurokinin A. J Cell Physiol 137:141–145

Jackson TR, Hanley MR 1989 Tumor promoter 12-o-tetradecanoylphorbol 13-acetate inhibits mas/angiotensin receptor-stimulated inositol phosphate production and intracellular Ca^{2+} elevation in the 401L-C3 neuronal cell line. FEBS (Fed Eur Biochem Soc) Lett 251:27–30

Jackson TR, Hallam TJ, Downes CP, Hanley MR 1987 Receptor-coupled events in bradykinin action: rapid production of inositol phosphates and regulation of cytosolic free Ca^{2+} in a neural cell line. EMBO (Eur Mol Biol Organ) J 6:49–54

Jackson TR, Blair LAC, Marshall J, Goedert M, Hanley MR 1988 The mas oncogene encodes an angiotensin receptor. Nature (Lond) 335:437–440

Janssen JWG, Steenvoorden ACM, Schmidtberger M, Bartram CR 1988 Activation of the MAS oncogene during transfection of monoblastic cell line DNA. Leukemia 2:318–320

Julius D, MacDermott AB, Axel R, Jessell TM 1988 Molecular characterization of a functional cDNA encoding the serotonin 1c receptor. Science (Wash DC) 241:558–564

Julius D, Livelli TJ, Jessell TM, Axel R 1989 Ectopic expression of the serotonin 1c receptor and the triggering of malignant transformation. Science (Wash DC) 244:1057–1062

Kaplan DR, Whitman M, Schaffhausen B et al 1987 Common elements in growth factor stimulation and oncogenic transformation: 85kd phosphoprotein and phosphatidyl-inositol kinase activity. Cell 50:1021–1029

Martin KA, Grant SGN, Hockfield S 1989 The MAS oncogene is developmentally regulated in the rat central nervous system. Soc Neurosci Abs 15:501

Masu Y, Nakayama K, Tamaki H, Harada Y, Kuno M, Nakanishi S 1987 cDNA cloning of bovine substance K receptor through oocyte expression system. Nature (Lond) 329:836–838

Mattingly R, Dreher ML, Hanley MR 1987 Down regulation of phorbol diester binding to NG115-401L neuronal cells is dependent on structure, concentration and time. FEBS (Fed Eur Biochem Soc) Lett 223:11–14

Meyerhof W, Morley S, Schwarz J, Richter D 1988 Receptors for neuropeptides are induced by exogenous poly (A+) RNA in oocytes from *Xenopus laevis*. Proc Natl Acad Sci USA 85:714–717

Millan MA, Carvallo P, Izumi S-I, Zemel S, Catt KJ, Aguilera G 1989 Novel sites of expression of functional angiotensin II receptors in the late gestation fetus. Science (Wash DC) 244:1340–1342

Norbury C, Nurse P 1990 Controls of cell proliferation in yeast and animals. In: Proto-oncogenes in cell development. Wiley, Chichester (Ciba Found Symp 150) p 168–185

O'Sullivan AJ, Cheek TR, Moreton RB, Berridge MJ, Burgoyne RD 1989 Localization and heterogeneity of agonist-induced changes in cytosolic calcium concentration in single bovine adrenal chromaffin cells from video imaging of fura-2. EMBO (Eur Mol Biol Organ) J 8:401–412

Peach MJ 1977 Renin-angiotensin system: biochemistry and mechanisms of action. Physiol Rev 57:313–370

Pfeilschifter J, Ochsner M, Whitebread S, DeGasparo M 1989 Down regulation of protein kinase C potentiates angiotensin II-stimulated polyphosphoinositide hydrolysis in vascular smooth muscle cells. Biochem J 262:285–291

Rabin M, Birnbaum D, Young D, Birchmeier C, Wigler M, Ruddle FH 1987 Human ras1 and mas1 oncogenes located in regions of chromosome 6 associated with tumor-specific rearrangements. Oncogene Res 1:169–178

Represa JJ, Miner C, Barbosa E, Giraldez F 1988 Bombesin and other growth factors activate cell proliferation in chick embryo otic vesicles in culture. Development 102:87–96

Ross EM 1989 Signal sorting and amplification through G protein coupled-receptors. Neuron 3:141–152

Rozengurt E 1986 Early signals in the mitogenic response. Science (Wash DC) 234:161–166

Sibley DR, Benovic JL, Caron MG, Lefkowitz RJ 1987 Regulation of transmembrane signaling by receptor phosphorylation. Cell 48:913–922

Stephens L, Hawkins PT, Downes CP 1989 Metabolic and structural evidence of a third species of polyphosphoinositide in cells: D-phosphatidyl-myo-inositol 3-phosphate. Biochem J 249:283–292

Whitman M, Downes CP, Keeler M, Keller T, Cantley L 1988 Type I phosphatidylinositol kinase makes a novel inositol phospholipid, phosphatidylinositol-3-phosphate. Nature (Lond) 332:644–646

Yokota Y, Sasai Y, Tanaka K et al 1989 Molecular characterization of a functional cDNA for rat substance P receptor. J Biol Chem 264:17649–17652

Young D, Waitches G, Birchmeier C, Fasano O, Wigler M 1986 Isolation and characterization of a new cellular oncogene encoding a protein with multiple potential transmembrane domains. Cell 45:711–719

Young D, O'Neill K, Jessell TM, Wigler M 1988 Characterization of the rat mas oncogene and its high-level expression in the hippocampus and cerebral cortex of rat brain. Proc Natl Acad Sci USA 85:5339–5342

Zachary I, Woll PJ, Rozengurt E 1987 A role for neuropeptides in the control of cell proliferation. Dev Biol 124:295–308

DISCUSSION

Waterfield: Could you summarize which of the receptors that have this basic seven transmembrane domain structure have been shown to be involved in the control of proliferation and transformation?

Hanley: Amongst members of this family that have been tested, mitogenic potential has been found in all cases, as far as I know. First, the five members of the muscarinic receptor subfamily show varying degrees of carbachol-activated mitogenic activity (Ashkenazi et al 1989). Second, as I noted in my paper, the serotonin 1c and serotonin 2 receptors are mitogenic (Julius et al 1989). Third, the β2-adrenergic receptor activates growth in an appropriate cell population, FRTL cells, where the cAMP pathway is mitogenic (Hen et al 1989). Fourth, from our own studies, we would add the substance K receptor. Moreover, the list of mitogenic peptides (Table 1) includes a selection of additional predicted members of the 'seven transmembrane domain' family, for which structural information is not yet available but which have been shown to play a role in the control of proliferation.

As for transformation and tumorigenicity, only the serotonin receptors and *mas* have been linked to oncogene-like growth deregulation, such as ligand-dependent focus formation and ligand-dependent clone formation by cells in soft agar.

The thesis I am working on, which is purely intuition, is that the whole G protein-coupled receptor superfamily will turn out to be conditionally mitogenic in the right cell background and the right conditions. This notion is supported by the identification of an activating mutation of the G_s α chain in a subset of human pituitary tumours. The mutant G protein drives the mitogenic adenyl cyclase pathway in a receptor-independent manner (Landis et al 1989).

The most controversial issue at present follows from the assertion that the entire muscarinic family is mitogenic. Only those receptors that are coupled to phospholipase C show robust DNA synthesis responses. However, the other muscarinic receptors, which show only small stimulation of DNA synthesis, are all linked to the negative regulation of adenyl cyclase through the inhibitory G_i protein. Normally, this is not considered to be a mitogenic pathway, but Pouyssegur and his colleagues have claimed that activation of the G_i pathway via the serotonin 1b receptor pathway can drive DNA synthesis by an as-yet unidentified mechanism (Seuwen et al 1988). It seems unlikely that the biochemical action of inhibiting adenyl cyclase activity is involved in this proliferative effect.

Hunter: Is transformation by *mas* dependent on the ligand?

Hanley: Yes, but this is difficult to show because of the nature of the ligand. Angiotensin is one of the most aggressively degraded substances in the body. If you add a bolus of angiotensin to a plate of NIH/3T3 cells, within five minutes it has been destroyed by peptidases. Angiotensin and bradykinin are both destroyed locally as fast as they are made. To study the transforming ability of the *mas*/angiotensin receptor, you have to put a renin–angiotensinogen system in a population of cells transfected with *mas*. When you transfect NIH/3T3 cells with *mas*, the rather small, atypical foci that are obtained may be due to their limited stimulation by angiotensin. It remains to be seen whether or not those cells are producing angiotensinogen and renin, or whether anything is being provided in serum.

Hunter: If you use the non-peptide angiotensin antagonist, do you reverse transformation?

Hanley: In the absence of any positive information on the production of angiotensin by *mas*-transfected cells, I can only say that we have seen inhibition of growth by one non-peptide antagonist. We are a little uncomfortable with that because although we know what these 'antagonists' are claimed to do, we don't know all their biological activities. For example, the active antagonist contains an indole moiety, and it is possible that it has another activity related to its growth inhibitory properties.

Verma: You are saying that the *mas*-dependent transformation is ligand dependent, just like that induced by addition of serotonin to fibroblasts containing the serotonin receptor. You don't think it has anything to do with overexpression of *mas* in an inappropriate cell type, namely fibroblastic cells?

Hanley: I can't at this point say that the predisposition to transformation does not require overexpression of *mas*. Remember, this oncogene has a rearrangement that has not yet been sequenced and the significance of which is not really known. All the experiments I described have been done with the oncogene; since we don't know the difference between patterns of expression of the oncogene and the proto-oncogene, we may be studying a state of ectopic active expression of *mas*.

Hunter: As far as I can remember, there is no change in the coding sequence of *mas* between the proto-oncogene and the oncogene products.

Hanley: The coding sequence is not altered, but we don't know anything about changes in the levels or tissue specificity of expression. Overexpression could be very important because a ligand like angiotensin, that is produced according to need physiologically, very rarely operates at conditions in which the receptor is saturated. Thus, when you overexpress the receptor, you don't increase the magnitude of its effects, but you do change the cell's sensitivity. So *mas* overexpression may lower the threshold for stimulation of DNA synthesis by angiotensin.

Sherr: One issue is whether any receptor or any gene that plays a role in regulating cell growth can be classified as an oncogene. A more rigorous classification might rely on whether genetic alterations of the proto-oncogene convert it to a constitutively active growth-promoting molecule, and I am not sure that that's really been demonstrated in the case of *mas*.

Hanley: That certainly has not been demonstrated. We use the name *mas* from its original identification in the literature as an oncogene. If one accepts that potentially all receptors, even going further than the G protein-coupled family, may under some conditions regulate growth, then the definition of 'oncogene' as 'growth-promoting' becomes of little use.

Sherr: I am not sure actually. Is the PDGF receptor a proto-oncogene? Naively, we might like to say that it is, but so far, nobody has programmed any activating mutation into that receptor that renders it ligand independent, nor is there any natural example that I am aware of where it acts as such. However, it can transform cells by an autocrine mechanism.

Hunter: I don't think anyone has tried hard to mutate the PDGF receptor in that way. The insulin receptor gene can be converted into an oncogene by appropriate manipulation (Wang et al 1987).

Hanley: There is an interesting issue here. If deregulated growth control is a widespread, perhaps universal, potential property of members of the G protein class of receptors, is there any example where such a receptor has been transduced by a virus? Bill Sugden (1989) has said that the Epstein–Barr virus latent membrane protein looks like a truncated member of this family that has six rather than seven membrane spanning regions. We are now addressing whether this six transmembrane segment receptor-like gene product is a truncated homologue of a member of this family and whether it is constitutively driving biochemical events that are ligand dependent in its presumptive counterpart. That would qualify the latent membrane protein as an activated member of this family.

Hunter: Has anyone found a mutation in any member of this family that gives it constitutive activity?

Hanley: No, and people have looked very hard for that. To date no one has introduced point mutations or deletions into any seven transmembrane segment receptor that do anything other than alter receptor desensitization and exaggerate ligand sensitivity.

Heath: Is it certain that the *mas* product is the angiotensin receptor? From your dose–response curves, the affinity constant for *mas* was in the range 10^{-8}–10^{-7}. What is the affinity of purified receptor?

Hanley: The receptor has never been purified. Secondly, I often say that the 'angiotensin II receptor' was designed by a committee! I don't think a single receptor type actually exists, I think it is a composite of several receptor types, for which we are now getting genetic evidence. Therefore there are no data on a 'real' dose–reponse curve for a single, isolated angiotensin receptor species. However, when you express angiotensin receptors in *Xenopus* oocytes, using unfractionated total mRNA from any of several tissues, you observe induced angiotensin sensitivity very like that seen when the cloned *mas* gene is expressed (Meyerhof et al 1988). Moreover, the substance K receptor has exactly the same sort of affinity for its ligand as *mas* has for angiotensin III. These affinities are in the range most frequently observed for neural peptides and their receptors.

Heath: Is anything known about the level of angiotensin in the circulation?

Hanley: It is about 1–10 pg/ml extracted mammalian plasma, but it is very difficult to know what importance to attach to such measurements, because of the rapid turnover of the peptide. Angiotensin and bradykinin are not true circulating hormones in the sense that they persist in blood at steady levels; they are transient local hormones whose levels may be increased in the circulation by persistent activation of their biosynthesis.

Sugimura: Are there any human tumours where *mas* is known to be responsible for the malignant phenotype?

Hanley: A role for *mas* has been looked for in two types of pathology— tumours and high blood pressure. There are several human pedigrees in which genetic origins of high blood pressure have been traced. We have started looking at such pedigrees and we have also looked for an association of high blood pressure with predisposition to specific neoplasia. We have a few candidates for *mas* involvement, particularly adrenocortical tumours, but we have no evidence of genetic rearrangement. At this stage we don't have any evidence for *mas* involvement in human tumours.

Feramisco: Michael, is there any evidence using the synthetic antagonist in an animal model for *mas* tumorigenicity?

Hanley: I don't think anyone has done that.

Alemà: Have you ever tried to transfect primary cells or embryonic stem cells with *mas*?

Hanley: We have done no experiments of that type. We have used a transient expression vector in COS cells, which might be used for other cells, but we haven't done anything like that as yet.

Jessell: You showed that angiotensin III, via binding to the *mas*-encoded receptor, inhibits a potassium conductance in 401L cells. In the assays of calcium transients in neuroblastoma cells, what contribution of the calcium transient comes from the inositol signalling pathway and what comes from the voltage-dependent activation of Ca^{2+} channels? Can you distinguish those? If you put *mas* into a non-neuronal cell that is unlikely to have a voltage-sensitive potassium channel, would you expect to see a similar increase in intracellular calcium, which would imply that it was acting through the lipid signalling pathway?

Hanley: Activation of *mas* generated intracellular calcium transients exclusively through inositol lipid hydrolysis, as far as we can tell. We get identical results in the absence of extracellular calcium. There is no calcium influx component (as assessed by cytosolic elevation of Ca^{2+}) when we depolarize the cells by extracellular potassium or try to depolarize them pharmacologically. Thus, there seems to be no contribution to the intracellular calcium transient from channel activity. Nevertheless, as I describe, we see a low density of L-type channels in these cells, so under some conditions elevation of cytosolic Ca^{2+} may have a channel-mediated influx component.

To answer your second question, we are introducing *mas* into a variety of cell types, including a family of 3T3 fibroblast cell lines, so we can do better mitogenicity experiments. The NG115-401L and related neuroblastoma clones are very hard to growth arrest. We would like to do the growth assays in a Swiss 3T3 cell background. We would like to examine focus formation and tumorigenicity using the standard indicator cell, NIH/3T3. However, it seems that the mechanism of intracellular calcium discharge is independent of cell type and we see identical angiotensin-stimulated calcium increases when *mas* is expressed either stably or transiently in different cells.

Norbury: I am rising to the *cdc2* bait—have you any evidence for modifications, particularly changes in the phosphorylation state of the *mas* product, which might correlate with the ability of the cells to respond to angiotensin?

Hanley: We don't yet have that information. We tried to guess whether *mas* itself might be a substrate for a *cdc2*-type kinase, using reported phosphorylated peptides from RNA polymerase as models of predicted consensus sites. On that basis, *mas* would not be a substrate for a *cdc2* kinase. But I don't know how stringent the requirements for a potential *cdc2* kinase site are.

Norbury: I am not convinced that p34[cdc2] does phosphorylate RNA polymerase *in vivo*! Is *mas* a phosphoprotein?

Hanley: Yes, under stable and transient expression conditions there are phosphorylated serine and threonine residues. This has not yet been measured stoichiometrically, but there are increases in ^{32}P-labelling of *mas* protein after stimulation by angiotensin.

Hunter: Has it been tested as a substrate for p34[cdc2]?

Hanley: No, it could be done now.

Maness: Is the *mas* proto-oncogene product expressed in early neurogenesis, which would indicate a possible role in growth control of neural cell lineages?

Hanley: The transcript is present in embryonic rat brain at higher levels than in the adult, but we haven't done any more than that. Those studies should now be repeated using both antibodies to the *mas* product and RNAse protection assays for the *mas* transcripts. *mas* may have a developmental or proliferative role distinct from its function in the adult.

Méchali: When you say that *mas* is acting in proliferative control in early development, what do you mean by early development?

Hanley: In our work, we are looking at fairly late stages before neurons have completed differentiation, but after they have become post-mitotic.

Méchali: You do not expect to find the *mas* gene expressed before neurulation?

Hanley: We have not looked that early.

Noble: Why do you think it might be expressed before neurulation?

Méchali: If it has some other function in the control of proliferation, then it could be playing a role in the very early embryo.

Moolenaar: When you measured thymidine incorporation, the stimulation by angiotensin was only 1.5-fold, whereas with fibroblasts there was 50–70-fold stimulation.

Hanley: When smooth muscle cells are treated with angiotensin their DNA synthesis is increased by no more than 50% over basal activity (Campbell-Bowell & Robertson 1981); the major effect seems to be hypertrophy. There are two reasons why we get comparable DNA synthesis in our cells. When we established the C3 cell line, we cloned it from G418-resistant transfectants, so it had undergone quite a bit of growth before a stable population was cloned. There may be a selective advantage in amplifying the *mas* gene product, as the C3 line seems to have done. The basal level of cell growth is higher in the *mas-* transfected clones than in the parental cell line. If we take this elevated rate of DNA synthesis as 100%, there is not a very dramatic increment above that, even in the presence of serum. The percentage increase is extremely precise and statistically significant. People who work with Swiss 3T3 cells may have an exaggerated sense of how easy it is to make a cell quiescent. It is very difficult to growth arrest a nerve cell line.

The second reason for the apparently low mitogenic activity is that peptides are frequently assessed as co-mitogens with insulin; conditions that give much larger responses in, for example, Swiss 3T3 cells.

Moolenaar: Is insulin required for an optimal mitogenic effect?

Hanley: No. There is no synergism with insulin. We have identified in the NG115-401L cells and transfected subclones an insulin-like growth factor receptor and a TGF-β receptor. There may be synergism of angiotensin with TGF-β in eliciting increased DNA synthesis.

Heath: Are your data derived from liquid scintillation counting or autoradiography? What do they mean in terms of numbers of cells?

Hanley: The cell population is essentially homogeneous, so we measure incorporation of [3H]thymidine into DNA.

Heath: So what does 100% mean in terms of numbers of cells entering DNA synthesis?

Hanley: We have been doing some limited experiments with autoradiography but because the cells are dividing even when 'quiescent' you still see some nuclear labelling under those conditions. I cannot say what proportion is labelled.

Alemà: Do you know how the desensitization process is regulated?

Hanley: We believe that *mas* is directly phosphorylated by protein kinase C. To show that formally, we would like to have pure protein kinase C isoforms and purified *mas* protein and show that *mas* is a good substrate for one or more of these. We think the desensitization is likely to be mediated by more than one regulated kinase.

Sugimura: Do you have any explanation for why the *mas* product is concentrated in the hippocampus?

Hanley: That is a hard question to ask for any receptor, why it occurs in a particular area. These proto-oncogenes are clearly multifunctional and my guess is that growth control is only one window on how they work. I believe that their function in the post-mitotic neuron is simply that of activating an excitability state. In addition, they have the potential to change the pattern of gene expression in target cells over long periods.

Maness: Do you know from your expression studies whether the *mas* proto-oncogene is expressed in glia or neurons?

Hanley: That's a good question. We now have cellular resolution *in situ* hybridization and have found constitutive expression of *mas* mRNA only in neurons (Bunnemann et al 1990). Amongst cell lines, we have several which express a *mas* phenotype and these are all neuronal, not glial, in character. Similarly, primary neurons but not glia transcribe the *mas* gene.

Hunter: Do you know whether angiotensin induces tyrosine phosphorylation? I am thinking of the parallel with bombesin.

Hanley: The bombesin issue is still controversial. Ken Brown, working with you, came out with a resounding No (Isacke et al 1986), but there are also claims that bombesin does induce tyrosine phosphorylation (Cirillo et al 1986). The difficulty is whether downstream events are activated by bombesin that themselves cause tyrosine phosphorylation. We haven't evaluated that at all for angiotensin.

Hunter: We would agree with Paolo Comoglio that you can detect minor changes in tyrosine phosphorylation after treatment with bombesin in Swiss 3T3 cells, but nothing of the sort that he claimed originally. But that may depend very much on cell physiology. Has anyone identified the bombesin receptor or bradykinin receptor?

Waterfield: Nobody knows any details of receptor structure at present.

Hanley: At this stage, the structural domain pattern for G protein-coupled receptors is pretty well established. What we know about bombesin and bradykinin receptors, or other mitogenic peptide receptors, strongly implies that they belong to the G protein-activating class. Therefore, we would expect them to have the seven transmembrane domain motif, and even to show some explicit sequence similarities with known receptors. I think that if there is evidence of a G protein involvement, you are forced to predict this type of seven hydrophobic segment sequence pattern.

Sugimura: Pertussis toxin is a mono-ADP-ribosylation enzyme that acts on G protein. Could you say more about the results with pertussis toxin and TPA?

Hanley: When we treat *mas*-transfected C3 cells acutely with TPA, angiotensin sensitivity is blocked. We believe this effect occurs at the level of the receptor. When we treat C3 cells with pertussis toxin, there is no effect on angiotensin sensitivity, even though at least one protein is ADP-ribosylated (Jackson et al 1987). Therefore, the G protein linked to the *mas* protein cannot be one of the cloned species that are ADP-ribosylated by this toxin.

Hunter: The key question about angiotensin is that if it doesn't work simply through phospholipase C, what is it working on?

Hanley: I agree. Our most interesting working hypothesis involves STE 2, the yeast peptide receptor, which is growth inhibitory and induces the expression of a panel of genes as part of the mating response. The C-terminal tail can be taken off STE 2 and the mutated receptor is altered in certain properties, and becomes hypersensitive to mating pheromone. However, it looks like the tail might contribute some other functional aspect, besides regulating desensitization (Reneke et al 1988). Jeremy Thorner has shown that the tail can function in *trans*. If the tail coding region is put on a synthetic anchorage sequence from *src*, it functions completely unlinked from the truncated receptor and restores normal mating factor sensitivity when co-expressed with the tail-deleted mutant. He believes there is a protein interaction being specified by the tail, which may suggest that the C terminus is the place to look for novel functions.

Most of the people working on these seven segment transmembrane receptors are thinking in terms of modular proteins, and the C-terminal tail is a candidate for modular switching. We want to make tail deletions, to see whether the tail in isolation has any kind of protein recognition characteristics, and whether the function of *mas* changes after deletion of its tail. Can we restore normal function by adding back the homologous tail? Or the tail from another receptor?

Land: Did you inject *ras* antibodies into your cells in order to determine whether *mas* action is *ras* dependent?

Hanley: There was a time when *ras* was thought to be a coupling G protein for phospholipase C. That idea has been convincingly disproved, but we discovered that *mas*-induced cell growth is dependent on the presence of biologically active *ras* in the cell. Dennis Stacey has more or less concluded that all forms of mitogen-stimulated growth are *ras* dependent (Mulcahy et al 1985).

References

Ashkenazi A, Ramanchandran J, Capron DJ 1989 Acetylcholine analogue stimulates DNA synthesis in brain-derived cells via specific muscarinic receptor subtypes. Nature (Lond) 340:146–150

Bunnemann B, Fuxe K, Metzger R et al 1990 Autoradiographic localization of *mas* proto-oncogene mRNA in adult rat brain. Neurosci Lett, in press

Campbell-Boswell M, Robertson AL 1981 Effects of angiotensin II and vasopressin on human smooth muscle cells in vitro. Exp Mol Pathol 35:265–276

Cirillo DM, Gaudino G, Naldini L, Comoglio PM 1986 Receptor for bombesin with associated tyrosine kinase activity. Mol Cell Biol 6:4641–4649

Hen R, Axel R, Obici S 1989 Activation of the Beta2-adrenergic receptor promotes growth and differentiation in thyroid cells. Proc Natl Acad Sci USA 86:4785–4788

Isacke CM, Meisenhelder J, Brown KD, Gould KL, Gould SJ 1986 Early phosphorylation events following the treatment of Swiss 3T3 cells with bombesin and the mammalian bombesin-related peptide, gastrin related peptide. EMBO (Eur Mol Biol Organ) J 5:2889–2898

Jackson TR, Patterson SI, Wong YH, Hanley MR 1987 Bradykinin stimulation of inositol phosphate and calcium responses is insensitive to pertussis toxin in NG115-401L neuronal cells. Biochem Biophys Res Commun 148:412–416

Julius D, Livelli TJ, Jessell TM, Axel R 1989 Ectopic expression of the serotonin 1c receptor and the triggering of malignant transformation. Science (Wash DC) 244:1057–1062

Landis CA, Masters SB, Spada A, Pace AM, Bourne HR, Vallar V 1989 GTPase inhibiting mutations activate the alpha chain of Gs and stimulate adenyl cyclase in human pituitary tumours. Nature (Lond) 340:692–696

Meyerhof W, Morley S, Schwarz J, Richter D 1988 Receptors for neuropeptides are induced by exogenous poly(A)+ RNA in oocytes from Xenopus laevis. Proc Natl Acad Sci USA 85:714–717

Mulcahy LS, Smith MR, Stacey DW 1985 Requirement for ras proto-oncogene function during serum-stimulated growth of NIH 3T3 cells. Nature (Lond) 313:241–243

Reneke JE, Blumer KJ, Courchesne WE, Thorner J 1988 The carboxyl- terminal segment of the yeast alpha-factor receptor is a regulatory domain. Cell 55:221–234

Seuwem K, Magnaldo I, Pouyssegur J 1988 Serotonin stimulates DNA synthesis in fibroblasts acting through 5-HT1B receptors coupled to Gi protein. Nature (Lond) 335:254–256

Sugden B 1989 An intricate route to immortality. Cell 57:5–7

Wang LH, Lin B, Jong SMJ et al 1987 Activation of transforming potential of the human insulin receptor gene. Proc Natl Acad Sci USA 84:5725–5729

Construction of mammalian cell lines with indicator genes driven by regulated promoters

Judy Meinkoth, Arthur S. Alberts and James R. Feramisco

Cancer Center, University of California at San Diego, 220 Dickinson Street, San Diego, CA 92103, USA

Abstract. The regulation of gene expression is a central feature of cell growth and differentiation. An important question is how information received at the cell surface is ultimately transmitted to the nucleus and how it causes changes in the pattern of gene expression. To study the molecular basis of this process, we have established cell lines carrying marker genes under the control of known regulatory promoter elements. These cell lines are being used to investigate the effects of activating normal cellular second messenger systems or of microinjecting proteins hypothesized to function in signal transduction, such as oncogene products or subunits of enzymes.

1990 Proto-oncogenes in cell development. Wiley, Chichester (Ciba Foundation Symposium 150) p 47–56

The regulation of gene expression is a central feature of cell growth and differentiation. While most biological modifiers of cell growth and differentiation exert their primary effect at the cell membrane or in the cytoplasm, eventually the action of many of the modifiers impinges upon the nucleus, where changes in gene expression occur (for review, see Cold Spring Harbor Symposium on Quantitative Biology 1988). The cellular and biochemical events triggered by the biological modifiers comprise various signal transduction systems. In the case of mammalian cell growth control, the modifiers are usually growth factors that bind to cell surface receptors. Numerous biochemical changes are triggered by growth factor/receptor interaction, including alterations in second messenger levels in the cytoplasm and changes in gene transcription in the nucleus. In order to identify the biochemical messengers involved in signal transmission from the cell surface to the nucleus, we have established indicator cell lines responsive to a wide variety of biological modifiers. Using these indicator cell lines, we should be able to ascertain the role of cellular oncogenes in signalling pathways. In our initial studies, we analysed the ability of the *ras* oncogene protein, a plasma membrane component, to stimulate changes in gene expression in the nucleus. We have utilized needle microinjection as the primary method for introduction of the purified *ras* proteins into the cells (Bar-Sagi & Feramisco 1986). One gene whose expression is enhanced after injection of the *ras* oncogene protein is the c-*fos* proto-oncogene (Stacey et al 1987). Others have established that the regulation of *fos* expression is complex and includes both transcriptional and message stability mechanisms. The transcriptional control has been traced to various enhancer sequence motifs in the 5′ untranslated region of the promoter of the *fos* gene (Cold Spring Harbor Symposium on Quantitative Biology 1988). Elements defined include cAMP-dependent response elements (CREs), serum response elements (SREs), and tumour promoter response elements (TREs).

ras oncogenes are thought to activate the signal transduction pathways in cells that lead to enhanced expression of genes whose promoters contain SREs or TREs, but not CREs. Most of the results leading to these conclusions come from transient co-transfection experiments in which the expression of a marker gene under the control of a defined promoter enhancer sequence is measured following transfection of the cells with an oncogene expression construct (Schonthal et al 1988). As these experiments generally require 24 to 36 hours before analysis, there is a possibility that secondary effects in addition to primary effects of the oncogenes may result in altered marker gene expression. This may obscure conclusions to some degree.

In an attempt to minimize this problem, we have begun to examine the acute effect of microinjected oncogene proteins in mammalian cells. Through the construction of mammalian cell lines containing a marker gene, β-galactosidase, under the control of regulated promoter elements, we have analysed changes in gene expression in response to the microinjection of signalling proteins (Riabowol et al 1988). Using this system, changes in gene expression can be detected in as little as two hours, thereby minimizing the potential for secondary effects on the assay.

Characteristics of an indicator plasmid

The RSV/Z/NEO plasmid (Fig. 1) is the parental plasmid into which various enhancer elements were inserted. This plasmid contains four essential features. (1) It contains the *lacZ* marker gene under the control of a small fragment of the RSV promoter devoid of any enhancer activity. (2) It contains a unique *Xho*I restriction site adjacent to the minimal promoter for the insertion of enhancer elements capable of driving the regulated expression of β-galactosidase. (3) It contains the bacterial neomycin resistance gene, enabling its selection in mammalian cells using the drug, G418. (4) It contains the ampicillin resistance gene to allow for propagation in *Escherichia coli*.

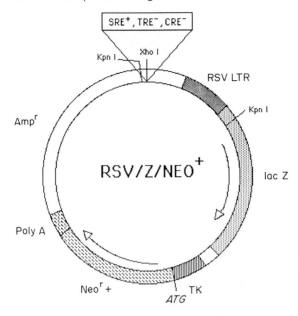

FIG. 1.   The plasmid construct used for generation of mammalian indicator cell lines. The *lacZ* gene is under the control of the Rous sarcoma virus promoter (RSV) and an enhancer containing either serum response elements (SREs), cAMP response elements (CREs) or tumour promoter response elements (TREs). The neomycin resistance gene (Neo[r]) driven by the thymidine kinase promoter allows for selection in mammalian cells; transcriptional activation is in the same orientation as the *lacZ* gene. *Kpn*I, *Xho*I, restriction enzyme sites. Amp[r], ampicillin resistance gene.

Features of cell lines transfected with indicator plasmids

Biological analysis of Rat-2 fibroblasts transfected with indicator plasmids similar to those described above gives the following results (Meinkoth et al 1990). Cells containing constructs with CRE elements upstream from the *lacZ* gene express β-galactosidase (as judged by staining with the chromogenic agent, X-gal) when the endogenous cAMP-dependent protein kinase (A-kinase) is activated or when the concentration of catalytic subunit of A-kinase is artificially raised by microinjection (Riabowol et al 1988). In the former case, treatment of the cells with 8bromocAMP, dibutyryl cAMP or forskolin can be used to increase intracellular cAMP and activate the endogenous A-kinase. In contrast, treatment of these cells with serum or tumour promoters does not induce expression of β-galactosidase (Fig. 2).

Cell lines containing TRE elements regulating β-galactosidase expression show distinctly different properties. Treatment of these cells with serum, tumour promoters or agents which act via phosphoinositide turnover or the activation

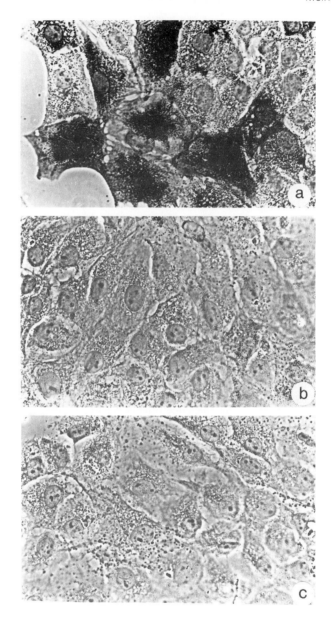

FIG. 2.   Rat-2 fibroblast cell lines were constructed containing the *lacZ* gene under the control of CRE enhancer sequences. Cells were treated with (a) 8bromocAMP, (b) 10% serum, or (c) TPA (12-*O*-tetradecanoylphorbol 13-acetate). After treatment with the drug for four hours the cells were fixed and stained with X-gal. Cells that express β-galactosidase appear blackened.

of protein kinase C induces expression of β-galactosidase. However, these cells do not express β-galactosidase in response to agents that activate endogenous A-kinase. In microinjection experiments, we found that the *ras* oncogene protein induces expression of β-galactosidase in the TRE cell lines but not in the CRE cell lines. We are in the process of comparing the responses of constructs containing SREs, TREs and CREs to various signalling agents.

Conclusion

We think that the combined approach of microinjection and indicator cell lines will provide useful information regarding the signal transduction mechanisms that regulate gene expression. Cell lines that respond to factors by differentiation or growth can be used in these analyses. The rapidity with which the responses can be detected will allow for more certain conclusions concerning the specific involvement of a signalling factor in the transcriptional event, and should complement transient co-transfection studies.

Acknowledgements

This work was supported by grants from the NIH and the Council for Tobacco Research. We wish to thank S. Fink and R. Goodman for providing various plasmids.

References

Bar-Sagi D, Feramisco JR 1986 Induction of membrane ruffling and fluid-phase pinocytosis in quiescent fibroblasts by *ras* proteins. Science (Wash DC) 233:1061–1068

Cold Spring Harbor Symposium on Quantitative Biology 1988 Molecular biology of signal transduction (vol 53)

Meinkoth JL, Montminy MR, Fink JS, Feramisco JR 1990 Expression of a cAMP-responsive gene in living cells requires the nuclear factor CREB. (Submitted)

Riabowol K, Fink S, Gilman M, Walsh D, Goodman R, Feramisco JR 1988 The catalytic subunit of cAMP-dependent protein kinase induces expression of genes containing cAMP-responsive enhancer elements. Nature (Lond) 336:83–86

Schonthal A, Herrlich P, Rahmsdorf H, Ponta H 1988 Requirement for c-*fos* expression in the transcriptional activation of collagenase by other oncogenes. Cell 54:325–334

Stacey DW, Watson T, Kung HS, Curran T 1987 Microinjection of transforming *ras* protein induces c-*fos* expression. Mol Cell Biol 7:523–527

DISCUSSION

Verma: Mark Montminy has isolated and cloned a CRE binding protein (CREB). I call it the wild-type because it can transactivate on phosphorylation. When an antibody against this protein is put into cells containing the 5X TRE-TK-CAT construct, which carries five copies of a TRE, there is an increase in chloramphenicol acetyl transferase (CAT) activity. From work done in our lab by Bill Lamph and others, we think CREB suppresses transcription from the AP1 promoter. If you co-transfect the AP1 promoter and CREB, there is complete suppression of AP1 transcription. If you then phosphorylate CREB by putting in the catalytic subunit of protein kinase A, there is complete relief of this suppression. So we think one of the ways to separate the two pathways is phosphorylation of the CREB.

I think that's an interesting way to look at separation of the signal transduction pathways, because one can suppress the transcription of the other.

Feramisco: I think Mark Montminy's clone represents probably the only protein that's been implicated in CRE transcription, so far.

Harlow: Jim, in your experiments using the Rat-2 lines that contain the inducible clones, it looked like whatever reagents you added, the response was always heterogeneous. Is there an explanation for that? If that's true, how are you ever going to get the microinjection experiments to work?

Feramisco: The experiments clearly will work. These responses are always heterogeneous. Even in yeast populations isolated from clonal lines (from Mike Wigler's lab) that express, for example, various *ras* mutants, the individual cells are very heterogeneous in their responses to a variety of things. These particular cells, either glioma C6 cell lines or Rat-2 fibroblasts—we don't have any results yet from PC12 cells—always give a good population response.

If you look at individual cells, in the TRE lines 40–60% of the cells in the population respond to a given signal. For the cAMP-driven promoters the percentage of cells responding is less than 30%. That's also true if you inject the catalytic subunit of A-kinase into those cells. So if you are looking at the activation, going from no response to a positive response, and 200 of 500 cells injected over a few hours show an activation, that's statistically significant compared to no activated cells when you inject a control substance. When we did blocking experiments with antibodies, we had to inject 500 cells to be convinced that we were depressing the number of cells that responded to cAMP. We count only the injected cells that turn blue, we ignore all the others. The cells that turn blue no matter what are covered by the control microinjections of the antigen—only 20–30% of those cells respond by synthesizing β-galacto-sidase. This is the same as in a population of cells into which you injected nothing, so the injection itself doesn't affect the results. You just have to do sufficient numbers to get a meaningful result.

With the CRE-dependent expression there appears to be a cell cycle component. If the cells are serum starved, then re-fed, at the starved stage that response rate of 30% can be increased to 60%. So the optimum conditions need to be determined for those particular cells. But in the TRE lines and the SRE lines one can get 60–70% response, which we feel is sufficient.

Harlow: Other than starving and re-feeding, is there any explanation for the heterogeneity?

Feramisco: We have tried cloning the cells and re-plating, the population is immediately heterogeneous. Similar experiments with other promoters being done by Steve Fink and Richard Goodman are giving a very heterogeneous response. Even if a cell sorter is used with the β-galactosidase substrate to select the high expressers or good responders, when those are put back into a culture and expanded, you get exactly the same heterogeneity. These are very minimal promoters. With the endogenous genes, like c-*fos*, if one treats populations of cells with cAMP, there is a much higher response, so there may be more complex factors that this minimal promoter is lacking.

Wagner: Have you looked at a minimal *fos* promoter and found expression of *lacZ*?

Feramisco: It's being done, we haven't put that construct into the cells yet.

Hunter: Jim, could the variability be due to something like the number of molecules per cell?

Feramisco: It's not due to the number of molecules that get into the cell. When you inject the pure catalytic subunit of A-kinase into 100 cells, you get about 30% blue cells, which is the same as in a population that has been treated with cAMP.

Independently of that, Ying Yeh and Wes Yamamoto from Susan Taylor's lab have made a variety of fluorescent modifications of catalytic subunits of A-kinase and regulatory subunits of A-kinase, which they have characterized biochemically. We have used those to localize in living cells where the subunits go during proliferation and during stimulation of signal transduction by cAMP after injection of the subunits into the cell. The catalytic subunit activates about 30% of the cells. In independent experiments, we see that in about 30% of the cells there is discrete nuclear localization. Within the other cells the distribution is very diffuse. Experiments are underway to see whether these two sub-populations are the same.

Hunter: But you haven't tried nuclear injections?

Feramisco: No. It would be a good way to get around the heterogeneity. If this nuclear localization is the reason for the heterogeneity of the response, we will try nuclear injection.

Verma: If you put the *fos* CRE into A126 cells, then add the A-kinase subunit, you can initiate transcription.

Feramisco: We did one interesting experiment we never published, which relates to the PC12 cells. As a control for *ras*, the logical choice was the catalytic subunit of A-kinase. We fully expected this to promote neurite outgrowth because cAMP is traditionally a promoter of neurite outgrowth in PC12 cells. The catalytic subunit had no effect of the morphology of the PC12 cells.

Brugge: Mike Greenberg thinks that depolarization-induced *fos* expression is mediated by a transacting factor related to the one that is regulated by the A-kinase (personal communication). Have you addressed that?

Feramisco: fos expression, for example induced by A-kinase in fibroblasts, is not sufficient to promote growth. It may mean that the kinase is blocking growth and *fos* is trying to stimulate growth or that *fos* expression is necessary for proliferation but not sufficient in our experiment.

Brugge: But in the experiments with the catalytic subunit of A-kinase in PC12 cells, did you look at *fos* induction or only at neurite extension?

Feramisco: Only at neurite extension.

Verma: That's been published and there is plenty of *fos* expression after treatment of PC12 cells with K^+, cAMP or nerve growth factor.

Brugge: But is the catalytic activity of the A-kinase alone sufficient for induction of *fos*?

Feramisco: We didn't look in our experiments whether that was driving *fos* expression.

Verma: Nerve growth factor will induce *fos* without the requirement for CRE.

Brugge: Basically, my question was related to whether depolarization-induced *fos* expression itself was mediated by A-kinase. One could study that, as Mike Greenberg has done, by using Don Walsh's inhibitor and John Wagner's mutants that are defective in the A-kinase.

Méchali: Do you know anything about the stability of the complex? If you remove the inducer at different times during the cell cycle, what happens?

Feramisco: If you take these cells in which β-galactosidase expression is driven by the TRE, for example, consider them as exponentially growing cells and stain at any time with X-gal, 50–60% of the cells will be expressing β-galactosidase. If you then put those cells into serum-free conditions and stain them every four hours, by about 12–24 hours the bulk of the β-galactosidase staining has disappeared. So we usually keep the cells in serum-free conditions for 24 hours before we do an induction experiment to remove the residual β-galactosidase that is present after growing up the cells.

We don't have similar information about the SRE yet. The CRE doesn't drive expression unless we specifically agonize the cAMP response, so background expression is not a problem.

Wagner: In the experiments where *fos* is induced via these elements, is the *fos* product always localized in the nucleus?

Feramisco: Whenever we have done immunostaining, the protein has been in the nucleus. If you stain populations of cells, there's often cytoplasmic staining in addition to nuclear staining. It is hard to discern whether that is just an artifact of immunofluorescence or a real cytoplasmic staining.

Wagner: There are the reports by the groups of J. M. Blanchard and Philippe Jeanteur in France of differential localization of *fos* between the cytoplasm and the nucleus.

Feramisco: When one injects the catalytic subunit of A-kinase, the nucleus becomes kidney shaped. This is seen naturally in some virally infected cells. If you inject *ras*, which like A-kinase induces expression of c-*fos*, the nucleus stays perfectly round. What that has to do with the biology of the cells, I don't know.

Verma: The only time we ever saw cytoplasmic *fos* protein was in CHO cells, where the gene had been amplified 50-fold and the capacity of the cytoplasm was not enough to cope with all of the protein, so there was morphological distortion.

Feramisco: We microinjected *myc* proteins and E1A proteins made in *E. coli* into fibroblasts at levels that exceeded the bounds of discretion; the nucleus just filled up and there were incredible clusters of staining inside the nucleus. The nucleus can accommodate enormous amounts of *myc* based on molecular calculation of how much we injected.

Hunter: Did you co-inject the protein kinase inhibitor peptide and the catalytic subunit of A-kinase?

Feramisco: Ned Lamb in my lab did that when we tried to confirm that we were studying the effects of the catalytic subunit of A-kinase on regulation of the cytoskeleton.

The inhibitor peptide came from Don Walsh and Dave Glass. This 20 amino acid peptide has a K_d of less than nanomolar for blocking the activity of the catalytic subunit of A-kinase. When this was injected into cells which were then challenged with cAMP analogues, it prevented the effects of cAMP analogues on the cytoskeleton. Endogenous A-kinase activated by cAMP analogues shows the same thing.

The problem with those experiments is that the peptides can only block this effect for a couple of hours, even though we inject large amounts. They are either being degraded or stimulating something that overcomes their inhibitory action. Dave Glass and Don Walsh are now making another set of analogues that have D amino acids at their N termini, which may prevent their degradation.

Hunter: How quickly does the cell respond by increasing the level of the regulatory subunit when you inject the catalytic subunit?

Feramisco: The only way we have addressed that is by following changes in the phosphorylation pattern by two-dimensional gel electrophoresis in patches of cells on chips of glass. The increase in the specific activity of these proteins when pulse labelled over 30 minute periods goes away about six hours after injection of catalytic subunit of A-kinase. Its effects on the cytoskeleton last for about 20 hours.

There are two possible conclusions. One is that within those six hours the catalytic subunit causes the expression of the regulatory subunit, as Pam Mellon and Stan McKnight and others have indicated by transfection experiments, to such a degree that it complexes with the catalytic subunit that we inject and inhibits this activity or induces the inhibitor protein. The second is that the catalytic subunits are just being degraded or lost by some other mechanism. We can't distinguish those possibilities yet, but we will be able to look with the fluorescently tagged catalytic subunit from Susan Taylor's lab to see if it's degraded very rapidly and put into lysosomes or if it's just in the cytoplasm.

Heath: You use multiple copies of the elements in all your constructs, why?

Feramisco: Only for historical reasons at the moment. There probably is a difference between single and multiple enhancer sequences. We wanted something that would work in the first set of experiments and constructs with four or five copies were easy to make. We are making constructs containing single enhancer sequences and other variations.

Verma: We use mostly one TRE for all our work and those constructs work just as well. The multiples may be a good idea to increase the signal.

Feramisco: According to Steve Fink and Richard Goodman, using their constructs of CRE promoters with β- galactosidase they find that single copies give good responses but not as nice as the lines carrying five copies of the element.

Hunter: Jim, I know that you are a great proponent of microinjection; can you compare its virtues with scrape loading as a means of getting proteins into cells?

Feramisco: There are a variety of techniques that we all use to change the composition of cells. Ideally, there will be a nice promoter system some day, in which in the absence of an inducer there is no expression, no leakiness, and when you add an otherwise inert modifier you get very rapid expression of a marker gene. In the absence of that, we have been relying on needle micro-injection because, as opposed to the old techniques of fusion and erythrocyte loadings, there is less impact on the cell in terms of changing the membrane composition.

So now comes the technique of scrape loading, which we're trying in the lab. It seems to require massive amounts of proteins. These amounts are certainly out of the question for affinity-purified antibodies. The technique is useful but you are bathing the cells in these antibodies or proteins that may have some extracellular effects; and scraping the cells to me seems like injecting them with 20 needles at the same time. Just making holes in the cell with a single needle causes a rapid rise in ion flux—although not in *fos*!

Verma: Something does not induce *fos*!

Feramisco: Needle injection is one of the few things that doesn't induce *fos*. I would like to see whether scrape loading induces *fos*.

Land: It does a little bit.

Verma: A technical note: everybody is looking for this totally inducible promoter. We have been trying lipofection of mRNA into cells. With a protein like *fos*, which is only made for a short period of time, by three hours there is enough protein in the nucleus that you can detect it. One can lipofect *fos* RNA and then make a single cell or multicell cDNA library, because the only thing that was introduced into the cell was the cDNA or mRNA to *fos*.

Feramisco: That might be useful, but when you say simple lipofection, it is not that easy!

Tyrosine phosphorylation of membrane-associated tubulin in nerve growth cones enriched in pp60$^{c\text{-}src}$

Patricia F. Maness and Wayne T. Matten

Department of Biochemistry, University of North Carolina School of Medicine, Chapel Hill, North Carolina 27599, USA

Abstract. The product of the c-*src* proto-oncogene (pp60$^{c\text{-}src}$) is a tyrosine-specific protein kinase that is expressed in two phases of neural development. In post-mitotic neuronal cells undergoing terminal differentiation, pp60$^{c\text{-}src}$ is present at high levels in the membrane of nerve growth cones and proximal axon shafts. Membrane-associated forms of α- and β-tubulin are the major phosphotyrosine-modified proteins in growth cone membranes *in vivo*. pp60$^{c\text{-}src}$ phosphorylates purified, unassembled tubulin subunits *in vitro*, inhibiting their ability to polymerize into microtubules. It is conceivable that tubulin phosphorylation by pp60$^{c\text{-}src}$ in the growth cone may regulate neurite extension by altering adhesion of cells to the substratum.

1990 Proto-oncogenes in cell development. Wiley, Chichester (Ciba Foundation Symposium 150) p 57–78

The normal cellular c-*src* gene encodes a 60 kDa tyrosine-specific protein kinase (pp60$^{c\text{-}src}$) that is homologous to the transforming protein of Rous sarcoma virus (Collett et al 1978, Opperman et al 1979, Hunter & Sefton 1980, Takeya & Hanafusa 1983). The c-*src* proto-oncogene exhibits a high degree of evolutionary conservation among metazoa, and it is structurally and functionally similar to the tyrosine kinase domains of certain growth factor receptors. These features imply a critical role for pp60$^{c\text{-}src}$ in animal cell growth or differentiation. Yet, despite intensive scrutiny, the biological role of pp60$^{c\text{-}src}$ remains elusive.

Analysis of pp60$^{c\text{-}src}$ expression in the developing nervous system has indicated that a form of pp60$^{c\text{-}src}$ is probably important in neuronal differentiation. High levels of a novel form of pp60$^{c\text{-}src}$ are expressed in embryonic nervous tissues (Cotton & Brugge 1983, Levy et al 1984, Sorge et al 1984, Fults et al 1985, Maness et al 1986, Cartwright et al 1988). pp60$^{c\text{-}src}$ is also present in other cell types, including platelets (Golden et al 1986) and erythrocytes (Hillsgrove et al 1987). Immunocytochemical studies in the developing chick embryo have identified two phases of pp60$^{c\text{-}src}$ expression in the neural lineage. Early neuroectodermal cells of gastrulating chick embryos express pp60$^{c\text{-}src}$ coincident with their commitment to neuronal or glial pathways (Maness et al 1986). After neurulation, immunoreactive pp60$^{c\text{-}src}$ is not detectable in the neural tube. Later, when committed neuroepithelial cells initiate terminal neuronal differentiation, cease proliferating and extend processes (neurites), pp60$^{c\text{-}src}$ is expressed at high levels (Sorge et al 1984, Fults et al 1985). During this second phase of *src* expression, pp60$^{c\text{-}src}$ is preferentially localized in the growth cone and regions of the developing nervous system that contain a high density of neurites (Maness et al 1988). In *Drosophila*, c-*src* transcripts also accumulate at two stages: in early embryos and post-mitotic neural tissues of pupae (Simon et al 1985). Interestingly, tyrosine-specific protein kinases encoded by the c-*yes* (Sudol et al 1988), c-*fyn* (Yamanashi et al 1987) and c-*abl* (Gertler et al 1989) proto-oncogenes are also expressed at elevated levels in the nervous system. In the cerebellum, c-*yes* and c-*src* are expressed in at least some of the same neurons but at different stages of development (Sudol et al 1988).

pp60$^{c\text{-}src}$ is produced by neuronal cells in both the central (Brugge et al 1985, Cartwright et al 1988) and peripheral nervous systems (Maness 1986, Le Beau et al 1987). A specialized role for pp60$^{c\text{-}src}$ in neuronal cells is underscored by the finding that many neurons and neuroblastoma cells contain a form distinct from that found in non-neuronal cells, which exhibits increased protein kinase activity (Brugge et al 1985, Cartwright et al 1988, Bolen et al 1985, Matten & Maness 1987). This neuronal form of pp60$^{c\text{-}src}$ contains additional amino acids in its N-terminal effector domain and arises as a consequence of alternative mRNA splicing (Martinez et al 1987, Levy et al 1987). The higher molecular weight form of pp60$^{c\text{-}src}$ appears to be preferentially expressed in the central nervous system, whereas the other form is predominantly in the peripheral nervous system (LeBeau et al 1987).

Several lines of evidence provide support for a role for neuronal pp60$^{c\text{-}src}$ in the extension of neuronal processes. (1) During development of the chick retina and cerebellum, the expression of pp60$^{c\text{-}src}$ in neuronal cells correlates temporally and spatially with neurite extension (Sorge et al 1984, Fults et al 1985). (2) Retinoic acid-induced differentiation of the mouse embryonal carcinoma cell line P19 into neuronal cells is accompanied by a shift to the synthesis of the neuronal form of pp60$^{c\text{-}src}$ concomitant with process extension (Lynch et al 1986). (3) Rat PC12 pheochromocytoma cells infected by Rous sarcoma virus are induced to extend neurites by the action of the retroviral form of pp60^{src} (Alemà et al 1985).

In this paper we discuss our recent results demonstrating that the activated, neuronal form of pp60$^{c\text{-}src}$ is highly concentrated in membranes of a subcellular fraction enriched in nerve growth cones from fetal rat brain (Maness et al 1988). The growth cone is a dynamic structure located at the tip of the growing axon, which is rich in cytoskeletal elements and responsible for contact-mediated guidance of the neurite as it migrates through the developing nervous system prior to forming a synapse. pp60[c-src] antibodies preferentially bind the growth cone and proximal axon shaft of retinal neurons in culture and, *in vivo*, label regions of the optic tract where growth cones are abundant. Other studies have shown that growth cone membranes contain high levels of tyrosine kinase activity, resulting in the phosphorylation of several endogenous proteins at tyrosine residues (Aubry & Maness 1988). Two proteins of 53–59 kDa are the most prominent tyrosine-phosphorylated proteins in the subcellular fraction enriched for growth cones. These proteins are not prominent in synaptosomal membranes from adult brain, indicating that their synthesis or phosphorylation is developmentally regulated, and that they may mediate a function either specific to, or amplified in, nerve growth cones. We describe new findings that identify the major tyrosine-phosphorylated proteins in growth cone membranes as membrane-associated α- and β-tubulin subunits, cytoskeletal proteins whose modification may be important in regulation of neurite extension at the growth cone, where new tubulin subunits are added to growing microtubules (W. T. Matten et al, unpublished observations).

Nerve growth cone membranes from fetal rat brain contain large amounts of pp60[c-src]

The high levels of pp60[c-src] in regions of the developing retina and cerebellum rich in neurites led us to investigate the hypothesis that pp60[c-src] is preferentially localized in growth cones (Maness et al 1988). Pfenninger et al (1983) have described the preparation of a subcellular 'A-fraction' from fetal rat brain, of which at least 70% of the components have been identified by electron microscopy as growth cone particles. pp60[c-src] was analysed in this and other subcellular fractions from the brain of fetal rats at day 18 of gestation (E18) by quantitative immunoblotting with a monoclonal antibody that recognizes the mammalian form of pp60[c-src] (Lipsich et al 1983). pp60[c-src] was enriched approximately ninefold in growth cone membranes compared to membranes prepared from fetal brain homogenates (Table 1). All of the pp60[c-src] detected was associated with membranes; none was in the soluble fraction. Growth cone membrane-associated pp60[c-src] consists primarily of the neuronal form of the enzyme that has increased activity (Maness et al 1988). Levels of pp60[c-src] in synaptosomal membranes from adult rat brain were 15-fold lower than in fetal growth cone membranes (Table 1). In adult brain, pp60[c-src] was not present at significantly higher concentrations in synaptosomal membranes compared to membranes prepared from adult brain homogenate, suggesting that the relatively low levels of pp60[c-src] in mature neurons may not be concentrated at synaptic terminals. Comparison of the levels of pp60[c-src] in the low-speed supernatant

TABLE 1   Relative levels of pp60$^{c\text{-}src}$ in fetal and adult rat brain membrane fractions

Cellular fraction	pp60$^{c\text{-}src}$ (per mg protein)
Early Neural Tube (E11)	
Low-speed supernatant membranes	1.3
Fetal Brain (E18)	
Homogenate membranes	1.0
Low-speed supernatant membranes	2.3
Growth cone membranes	9.0
Adult Brain	
Homogenate membranes	0.4
Low-speed supernatant membranes	0.3
Synaptosomal membranes	0.6

pp60$^{c\text{-}src}$ was quantitated in fractions (100 µg) by scanning densitometrically the 60 kDa bands on autoradiograms obtained from immunoblots. Absorbance values were normalized to those measured in homogenate membranes, as described in Maness et al (1988).

membrane fraction (Pfenninger et al 1983) from early (E11) neural tube, which contains primarily proliferating precursors, fetal brain (E18), and adult rat brain revealed that pp60$^{c\text{-}src}$ increased from fetal days 11 to 18, concomitant with neurite extension, then decreased approximately eightfold during the fetal to adult transition. The developmental regulation of pp60$^{c\text{-}src}$ underlines a primary role for this proto-oncogene in nervous system development, rather than mature brain function.

To confirm that growth cones are enriched for pp60$^{c\text{-}src}$, indirect immunofluorescence staining of primary cultures of chick retinal neurons was performed (Maness et al 1988). Neurofilament-positive neuronal cells with long processes exhibited more intense pp60$^{c\text{-}src}$ immunofluorescence in the growth cone and adjacent region of the neurite than in the older region of the neurite, closer to the cell body (Fig. 1b), consistent with the subcellular fractionation results. The cell body was also stained, but at least some of the staining was non-specific, as indicated by failure to abolish fluorescence completely when the antibody was preincubated with purified antigen (Fig. 1c).

These results demonstrate that the amount of the neuronal form of pp60$^{c\text{-}src}$ is increased in membranes of the growth cone and proximal axon shaft of developing neurons. It is not known if pp60$^{c\text{-}src}$ resides exclusively in the plasma membrane, or is present in membrane-bounded vesicles in the growth cone cytoplasm.

Localization of pp60$^{c\text{-}src}$ in the optic tract during the retinotectal projection

During embryogenesis, growth cones of retinal ganglion cell axons migrate from the neural retina along the optic tract to the optic tectum, where they form synapses with target neurons. Immunoperoxidase staining of the optic tract of

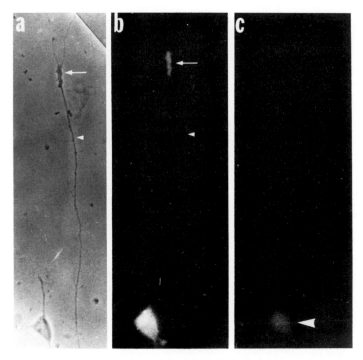

FIG. 1.   Indirect immunofluorescence staining of retinal neurons with anti-pp60$^{c\text{-}src}$ antibodies. (a) Phase contrast. (b) Growth cones, contiguous region of neurite and cell body labelled with anti-pp60$^{c\text{-}src}$ antibodies. (c) Competition control photographed under conditions identical to (b), showing a low level of staining only in the cell body when the primary antibody was preincubated with purified antigen. Arrow, growth cone; small arrowhead, neurite; large arrowhead, cell body. Magnification × 860 (× 80% at printing).

the Stage 36 chick embryo during this retinotectal projection revealed pp60$^{c\text{-}src}$ immunoreactivity to be highly concentrated at the surface of the optic tectum, the optic nerve, optic chiasm, and in the layers of the retina with a high density of neurites (Fig. 2a). The pattern of staining was similar to the distribution of the 200 kDa subunit of neurofilament protein that is a component of developing neuronal processes (Fig. 2b). No pp60$^{c\text{-}src}$ immunoreactivity was observed in the proliferating, undifferentiated neuroblasts of the retina. The highest levels of pp60$^{c\text{-}src}$ immunoreactivity were seen at the inner surface of the retina (the nerve fibre layer, which comprises primarily retinal ganglion cell axons) and at the surface of the optic tectum, where growth cones of arriving ganglion cell axons accumulate (Fig. 2c). These results provide evidence that pp60$^{c\text{-}src}$ is localized in processes of post-mitotic, differentiating neurons *in vivo* and is especially abundant at sites of accumulation of growth cones and proximal neurites, consistent with subcellular fractionation studies and localization of pp60$^{c\text{-}src}$ in neuronal cell cultures.

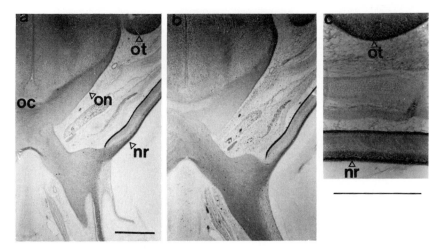

FIG. 2. Localization of pp60$^{c\text{-}src}$ in the retinotectal projection of the chick embryo at Stage 36. Paraffin sections (5 µm) were subjected to immunoperoxidase staining as described previously (Sorge et al 1984) using as primary antibodies: (a,c) a rabbit polyclonal antibody raised against pp60$^{v\text{-}src}$ (Gilmer & Erikson 1983), and (b) a mouse monoclonal antibody against the 200 kDa subunit of neurofilament protein (Wood & Anderton 1981). oc, optic chiasm; on, optic nerve; nr, neural retina; ot, optic tectum. Bar in (a) and below (c) = 25 µm.

Tubulin is the major tyrosine phosphorylated protein in growth cone membranes

To identify potential targets of pp60$^{c\text{-}src}$ kinase activity in growth cones, membrane proteins from the growth cone-rich A-fraction from fetal rat brain were analysed for tyrosine phosphorylation by immunoblotting with a polyclonal antibody specific for phosphotyrosine residues (Wang 1985). The major tyrosine-phosphorylated proteins in growth cone membranes migrated between the 53 and 59 kDa standards (Fig. 3, lane 2) (Aubry & Maness 1988). Phosphotyrosine-modified proteins of 34, 42, 95 and more than 205 kDa were also observed. Staining of the same immunoblot with antibodies specific for either α- or β-tubulin revealed co-localization with the majority of the 53–59 kDa protein recognized by phosphotyrosine antibodies (lane 3). This form of tubulin was tightly bound to membranes as it could not be removed by washing in high salt (0.3 M Na$_2$SO$_4$). The 53–59 kDa protein did not correspond to autophosphory-lated pp60$^{c\text{-}src}$, which migrated above the 59 kDa marker, as shown by immuno-blotting with *src* monoclonal antibody 327 (lane 5). The fetal rat brain homogenate (lane 1), containing both membrane and soluble components, showed much less prominent labelling of proteins of 53–59 kDa than did growth cone membranes, despite containing a large amount of tubulin. This suggests

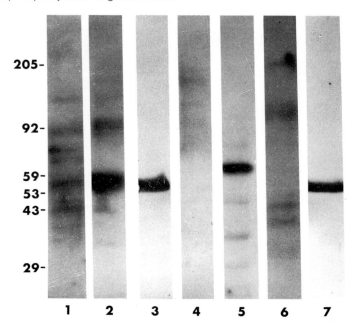

FIG. 3.   Tyrosine-phosphorylated proteins in nerve growth cone and synaptosome membranes detected by immunoblotting. Fetal rat brain homogenate (lane 1), growth cone membranes (lanes 2–5) and synaptosomal membranes (lanes 6, 7) were immunoblotted with phosphotyrosine antibodies (lanes 1, 2, 6), phosphotyrosine antibodies preincubated with 10 mM o-phospho-L-tyrosine (lane 4); *src* monoclonal antibody 327 (lane 5). After autoradiography, the immunoblots of growth cone membranes (lane 3) and synaptosomal membranes (lane 7) were immunostained with α- and β-tubulin antibodies and alkaline phosphatase-conjugated secondary antibodies (W. T. Matten et al, unpublished).

that the tyrosine-phosphorylated protein is highly concentrated in growth cone membranes. Preincubation of the antibody with 10 mM phosphotyrosine blocked binding to the 53–59 kDa protein in growth cone membranes, indicating that antibody recognition was specific (lane 4). Immunoblotting of synaptosomal membrane protein (30 μg) from adult rat brain with phosphotyrosine antibodies revealed proteins of 95, 42 and 35 kDa, but not 53–59 kDa (lane 6). The lack of tyrosine phosphorylation of the 53–59 kDa proteins was not due to the absence of tubulin from synaptosomal membranes, because tubulin was readily detected on immunoblots stained with tubulin antibodies (lane 7).

To determine whether the 53–59 kDa phosphotyrosine-modified proteins in membranes from the growth cone-rich fraction were tubulin subunits, phosphotyrosine-containing proteins were immunoprecipitated from growth cone membrane extracts with a monoclonal phosphotyrosine antibody PY20 (Glenney et al 1989), separated by two-dimensional gel electrophoresis, transferred

(A)   −   +

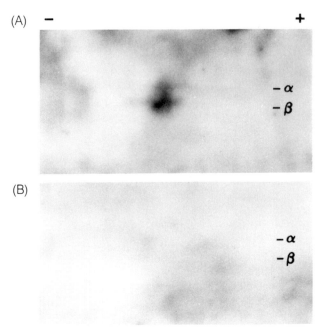

(B)

FIG. 4.   Immunoprecipitation of tubulin from growth cone membranes by phospho-tyrosine antibodies. Phosphotyrosine-containing proteins in an extract of A-fraction membranes enriched for growth cones (50 μg) were immunoprecipitated with (A) a mouse monoclonal antibody specific for phosphotyrosine (PY20, ICN Biochemicals) or (B) non-immune mouse immunoglobulin G, then separated by two-dimensional gel electrophoresis, transferred to Immobilon membranes (Millipore, Inc.) and immuno-blotted with monoclonal antibodies specific for α- and β-tubulin and [^{125}I] secondary antibodies (W. T. Matten et al, unpublished).

to an Immobilon membrane, and immunoblotted with mouse monoclonal antibodies specific for α- and β-tubulin and [^{125}I] goat anti-mouse immuno-globulin. Two proteins with the same electrophoretic behaviour as α- and β-tubulin were immunoprecipitated from A-fraction membranes by the phospho-tyrosine antibodies and were recognized by tubulin antibodies (Fig. 4). Thus, tubulin in growth cone membranes is phosphorylated at tyrosine residues *in vivo*. Whether one or both subunits are phosphorylated is not certain, however, because α- and β-tubulin subunits can form dimers during immuno-precipitation.

These studies suggest that a minor isotype or other form of tubulin that is associated with growth cone membranes is phosphorylated *in vivo* at tyrosine residues. Tyrosine-phosphorylated tubulin appears to be developmentally regulated in either its synthesis or phosphorylation, as it was not detectable in synaptosomal membranes from adult brain.

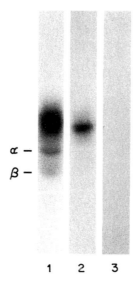

α –
β –

                                    1   2   3

FIG. 5.   Direct phosphorylation of purified tubulin by pp60[c-src] *in vitro*. pp60[c-src] was immunoprecipitated from growth cone-rich A-fraction membranes with anti-*src* monoclonal antibody 327 and secondary antibodies coupled to agarose, then incubated with 40 µM [γ-^{32}P] ATP in an immune complex protein kinase assay (Matten & Maness 1987). Lane 1, reaction with unassembled bovine brain tubulin; lane 2, reaction with pp60[c-src], without tubulin; lane 3, reaction with tubulin, without pp60[c-src]. The products of the reaction were subjected to SDS polyacrylamide gel electrophoresis at pH 9.2 and analysed by autoradiography.

Tubulin is a direct target of pp60[c-src]

To determine whether pp60[c-src] from the growth cone-rich fraction could phosphorylate purified tubulin directly, it was immunoprecipitated from an extract of A-fraction membranes and incubated with [γ-^{32}P] ATP (40 µM) in an immune complex protein kinase assay (Matten & Maness 1987) with native, unassembled tubulin purified from adult bovine brain. In the reaction, both α- and β-tubulin subunits were phosphorylated (Fig. 5). Phosphoamino acid analysis revealed only phosphotyrosine in each tubulin subunit. Autophosphorylation of pp60[c-src] appeared to be stimulated in the presence of tubulin. There was no detectable phosphorylation of tubulin by endogenous protein kinase activity present in purified tubulin in reactions without pp60[c-src]. The extent of phosphorylation of unassembled bovine brain tubulin by pp60[c-src] from membranes of the growth cone-rich fraction was 1.8×10^{-4} mole P_i/mole α-tubulin and 4.5×10^{-4} mole P_i/mole β-tubulin, when maximal incorporation of [^{32}P] into each subunit was reached after 40 minutes. This stoichiometry was much lower than the phosphorylation of α-tubulin

$(6.8 \times 10^{-2}$ mole P_i/mole) and β-tubulin $(4.5 \times 10^{-2}$ mole P_i/mole) that occurred when A-fraction membranes were incubated with $[\gamma\text{-}^{32}P]$ATP in an endogenous phosphorylation reaction *in vitro* (W. T. Matten et al, unpublished). This suggests that the particular isotype or other form of tubulin that is associated with growth cone membranes may be the preferred substrate of pp60[c-src] *in vivo*.

To determine if phosphorylation of tubulin by pp60[c-src] affected the ability of tubulin to polymerize, bovine brain tubulin was phosphorylated by pp60[c-src] from A-fraction membranes with $[\gamma\text{-}^{32}P]$ATP in the immune complex kinase assay, then polymerized in the presence of excess unlabelled tubulin. Polymerized microtubules were separated from unpolymerized tubulin by centrifugation, and equal amounts of tubulin in the pellet and supernatant were subjected to two-dimensional gel electrophoresis. Autoradiography showed that a form of phosphorylated α-tubulin that co-migrated with the most basic region of α-tubulin protein staining was found preferentially in the supernatant containing unassembled tubulin, and at a much lower level in the microtubule-containing pellet. Phosphorylated β-tubulin that co-migrated with the most basic region of β-tubulin staining was distributed preferentially in the unassembled tubulin fraction, too, although to a much lesser extent than basic α-tubulin. Acidic forms of phosphorylated α- and β-tubulin were present at similar levels in both the polymerized and unpolymerized tubulin fractions, and accounted for approximately half of the tubulin phosphorylated by pp60[c-src]. These results can be interpreted as indicating that tyrosine phosphorylation of α- and β-tubulin from bovine brain by pp60[c-src] can prevent assembly into microtubules *in vitro*.

The endogenous tubulin that was phosphorylated by pp60[c-src] *in vitro* may represent a subpopulation associated with growth cone membranes. It is possible that such a form does not normally polymerize, but instead serves some other cellular function. Membrane-associated tubulin has been identified in other systems, where it has been postulated to play a role in vesicle transport, tethering of cytoskeletal elements to the plasma membrane, or signal transduction (Stephens 1986). The effect of tyrosine phosphorylation of tubulin in growth cone membranes on its precise cellular function there remains to be elucidated.

It is tempting to speculate that phosphorylation of tubulin at the growth cone membrane by pp60[c-src] may block polymerization locally, and that such an event may regulate pathfinding by the growth cone. Although the majority of microtubules in the growing neurite terminate at the junction of the axon shaft with the growth cone (Forscher & Smith 1988), some microtubules are found near sites of adhesion, where they may stabilize cell–substratum contacts (Rinnerthaler et al 1988). Tubulin phosphorylation might be triggered by pp60[c-src] in response to extracellular signals and destabilize adhesion sites in the growth cone, which would affect the direction of migration. A role for pp60[v-src] in cell adhesion is supported by its localization in adhesion plaques of transformed cells (Willingham et al 1979, Shriver & Rohrschneider 1981).

If pp60^{v-src} participated in a similar cellular mechanism, disruption of cell–substratum contact by pp60^{v-src} could lead to anchorage-independent cell growth, the parameter of transformation most closely linked with tumorigenicity (Maness 1981).

Acknowledgements

This work was supported by NIH grant NS26620. PFM is a recipient of an NIH Research Career Development Award. We thank Dr Muriel Aubry, Janet West, Dr Carol G. Shores, Dr Michael Ignelzi, Michael E. Cox, Dr Christine A. Ingraham, Jacqueline S. Biscardi, and Julie Atashi for contributing to this work.

References

Alemà S, Casalbore P, Agostini E, Tato F 1985 Differentiation of PC12 phaeochromocytoma cells induced by v-src oncogene. Nature (Lond) 316:557–559

Aubry M, Maness PF 1988 Developmental regulation of protein tyrosine phosphorylation in rat brain. J Neurosci Res 21:473–479

Bolen JB, Rosen N, Israel MA 1985 Increased pp60^{c-src} tyrosyl kinase activity in human neuroblastomas is associated with amino-terminal tyrosine phosphorylation of the src gene product. Proc Natl Acad Sci USA 82:7275–7279

Brugge JS, Cotton PC, Queral AE, Barrett JN, Nonner D, Keane RW 1985 Neurons express high levels of a structurally modified activated form of pp60^{c-src}. Nature (Lond) 316:554–557

Cartwright CA, Simantov R, Cowan MW, Hunter T, Eckhart W 1988 pp60^{c-src} expression in the developing rat brain. Proc Natl Acad Sci USA 85:3348–3352

Collett MS, Brugge JS, Erikson RL 1978 Characterization of a normal avian cell protein related to the avian sarcoma virus transforming gene product. Cell 15:1363–1369

Cotton P, Brugge J 1983 Neural tissues express high levels of the cellular src gene product pp60^{c-src}. Mol Cell Biol 3:1157–1162

Forscher P, Smith SJ 1988 Actions of cytochalasins on the organization of actin filaments and microtubules in a neuronal growth cone. J Cell Biol 107:1505–1516

Fults DW, Towle AC, Lauder JM, Maness PF 1985 pp60^{c-src} in the developing cerebellum. Mol Cell Biol 5:27–32

Gertler FB, Bennett RL, Clark MJ, Hoffmann FM 1989 Drosophila abl tyrosine kinase in embryonic CNS axons: a role in axonogenesis is revealed through dosage-sensitive interactions with disabled. Cell 58:103–113

Gilmer TM, Erikson RL 1983 Development of anti-pp60^{c-src} serum antigen produced in *Escherichia coli*. J Virol 45:462–465

Glenney JR, Zokas L, ad Kamps MP 1988 Monoclonal antibodies to phosphotyrosine. J Immunol Meth 109:277–285

Golden A, Nemeth SP, Brugge JS 1986 Blood platelets express high levels of the pp60^{c-src}-specific tyrosine kinase activity. Proc Natl Acad Sci USA 83:852–856

Hillsgrove D, Shores CG, Parker JC, Maness PF 1987 Band 3 tyrosine kinase in avian erythrocyte plasma membrane is immunologically related to pp60^{c-src}. Am J Physiol 253:C286–C295

Hunter T, Sefton BM 1980 Transforming gene product of Rous sarcoma virus phosphorylates tyrosine. Proc Natl Acad Sci USA 77:1311–1315

Le Beau JM, Wiestler OD, Walter G 1987 An altered form of pp60$^{c\text{-}src}$ is expressed primarily in the central nervous system. Mol Cell Biol 7:4115–4117

Levy BJ, Dorai T, Wang L-H, Brugge JS 1987 The structurally distinct form of pp60$^{c\text{-}src}$ detected in neuronal cells is encoded by a unique c-src mRNA. Mol Cell Biol 7:4142–4145

Levy BT, Sorge LK, Meymandi A, Maness PF 1984 pp60$^{c\text{-}src}$ kinase is in embryonic tissues of chick and human. Dev Biol 104:9–17

Lipsich LA, Lewis AJ, Brugge JS 1983 Isolation of monoclonal antibodies that recognize the transforming proteins of avian sarcoma virus. J Virol 48:352–360

Lynch SA, Brugge JS, Levine JM 1986 Induction of altered c-src product during neural differentiation of embryonal carcinoma cells. Science (Wash DC) 234:873–876

Maness PF 1981 Actin structure in fibroblasts: its possible role in transformation and tumorigenesis. In: Dowben RM, Shay JW (eds) Cell and muscle motility, vol 1. Plenum, New York

Maness PF 1986 pp60$^{c\text{-}src}$ encoded by the proto-oncogene c-src is a product of sensory neurons. J Neurosci Res 16:127–139

Maness PF, Sorge LK, Fults DW 1986 An early developmental phase of pp60$^{c\text{-}src}$ expression in the neural ectoderm. Dev Biol 117:83–89

Maness PF, Aubry M, Shores CG, Frame L, Pfenninger KH 1988 c-src gene product in developing rat brain is enriched in nerve growth cone membranes. Proc Natl Acad Sci USA 85:5001–5005

Martinez R, Mathey-Provot B, Bernards A, Baltimore D 1987 Neuronal pp60$^{c\text{-}src}$ contains a six-amino acid insertion relative to its non-neuronal counterpart. Science (Wash DC) 237:411–415

Matten WT, Maness PF 1987 Vmax activation of pp60$^{c\text{-}src}$ tyrosine kinase activity from neuroblastoma Neuro-2A. Biochem J 248:691–696

Opperman H, Levinson AD, Varmus HE, Levintow L, Bishop JM 1979 Uninfected vertebrate cells contain a protein that is closely related to the product of the avian sarcoma virus transforming gene src. Proc Natl Acad Sci USA 76:1804–1808

Pfenninger KH, Ellis L, Johnson MP, Friedman LB, Somlo S 1983 Nerve growth cones isolated from fetal rat brain. I: subcellular fractionation and characterization. Cell 35:573–584

Rinnerthaler G, Geiger B, Small JV 1988 Contact formation during fibroblast locomotion: involvement of membrane ruffles and microtubules. J Cell Biol 106:747–760

Shriver K, Rohrschneider L 1981 Organization of pp60$^{c\text{-}src}$ and selected cytoskeletal proteins within adhesion plaques and junctions of Rous sarcoma virus-transformed rat cells. J Cell Biol 89:525–535

Simon MA, Drees B, Kornberg T, Bishop JM 1985 The nucleotide sequences and the tissue-specific expression of Drosophila c-src. Cell 42:831–840

Sorge LK, Levy BT, Maness PF 1984 pp60$^{c\text{-}src}$ is developmentally regulated in the neural retina. Cell 36:249–257

Stephens RE 1986 Membrane tubulin. Biol Cell 57:95–109

Sudol M, Alvarez-Buylla A, Hanafusa H 1988 Differential developmental expression of c-yes and c-src protein in cerebellum. Oncogene Res 2:345–355

Takeya T, Hanafusa H 1983 Structure and sequence of the cellular gene homologous to the RSV src gene and the mechanism for generating the transforming virus. Cell 32:881–890

Wang JYJ 1985 Isolation of antibodies for phosphotyrosine by immunization with a v-abl oncogene-encoded protein. Mol Cell Biol 5:3640–3643

Willingham MC, Jay G, Pasta I 1979 Localization of the avian sarcoma virus src gene product to the plasma membrane of transformed cells by electron microscopic immunocytochemistry. Cell 18:125–134

Wood JN, Anderton BH 1981 Monoclonal antibodies to mammalian neurofilaments. Biosci Rep 1:263–268

Yamanashi Y, Fukushige S-I, Semba K et al 1987 The yes-related cellular gene lyn encodes a possible tyrosine kinase similar to p56[lck]. Mol Cell Biol 7:237–243

DISCUSSION

Feramisco: Do you know any more about the 92 kDa protein? It looked like a very acidic protein. Was it localized in the growth cones?

Maness: It contains phosphotyrosine; it is localized in the growth cone membranes and in synaptosome membranes. It doesn't seem to be developmentally regulated. We don't know what it is, except that it's not synapsin I.

Feramisco: It could be the heat shock 90 kDa protein.

Hunter: There is no evidence that the 90 kDa heat shock protein is phosphorylated on tyrosine in cells that have activated tyrosine kinases.

Feramisco: But tubulin is not necessarily seen as a protein that contains phosphotyrosine.

Alemà: Is tubulin phosphorylated by pp60[v-src] in RSV-infected PC12 cells or fibroblasts?

Maness: We haven't yet looked in fibroblasts. In the PC12 cells there are two bands in the 55 kDa region that do contain phosphotyrosine, but we are not yet sure if they are tubulin subunits.

Regarding the question of whether phosphorylation of tubulin is observed in transformed cells, we have just begun to look at that seriously and our initial efforts indicate that it may not be. We have immunoprecipitated phosphotyrosine-containing proteins with the monoclonal anti-phosphotyrosine antibody, PY20, from 2mg of proteins from the rat tumour cell line, RR1022, which expresses pp60[v-src], but have not detected tubulin in the immunoprecipitate. However, compared to the amount of growth cone membranes that we use for a similar analysis, we do not have quite the same amount of protein, so we want to repeat the immunoprecipitation with more protein from the transformed cells, or with membranes isolated from the RR1022 cells, which may be enriched in tubulin containing phosphotyrosine.

Hunter: We reported this experiment in 1982 (Sefton et al). In transformed chick cells total tubulin shows a very low level of tyrosine phosphorylation; the membrane-associated form of tubulin, however, might be preferentially phosphorylated on tyrosine.

Maness: We looked for tyrosine phosphorylated tubulin in [^{32}P]labelled RR1022 cells and couldn't find any, but that was before we began to use vanadate, an inhibitor of tyrosine phosphatases.

Hunter: We have not done any experiments pretreating with vanadate. I could imagine that tyrosine-phosphorylated tubulin might accumulate under those conditions.

Feramisco: Ray Erickson has also shown very active phosphorylation of tubulins *in vitro.*

Brugge: It was necessary for Marc Collett to boil the tubulin for it to be a good substrate for pp60^{v-src} (personal communication).

Maness: We have shown that heat-inactivated tubulin is a substrate for purified pp60^{v-src}, but it is inefficiently phosphorylated (Maness & Levy 1983). In contrast, the experiments I described today, in which pp60^{c-src} phosphorylated tubulin, were done with phosphocellulose-purified tubulin that had not been heat denatured.

Brugge: Patricia, did you take precautions to avoid phosphorylation of tyrosine during the isolation of the growth cones? Tyrosine phosphorylation can occur after cell lysis, even though you are diluting the endogenous ATP considerably (Hamaguchi & Hanafusa 1989).

Maness: Yes, we have paid a lot of attention to the possibility of post-lysis modification. We have isolated growth cone membranes in the presence of N-ethyl maleimide, which inhibits pp60^{c-src} and other protein kinases; AMP-P-C-P, a non-hydrolysable ATP analogue; and GTP, which might prevent the denaturation of tubulin. In each case, tubulin is recognized by anti-phosphotyrosine antibodies, indicating that it is phosphorylated on tyrosine *in vivo.* We also included [γ^{32}P]ATP in the homogenization buffer, and found that there was minimal incorporation into tubulin. Because we were still concerned about modification after cell lysis, we labelled primary cultures of cortical neurons with [^{32}P], and found that specific isoforms of tubulin contained phosphotyrosine. Thus, we are confident that a small fraction of tubulin is phosphorylated at tyrosine residues in neuronal cells *in vivo.*

Verma: Your model is that phosphorylation might prevent polymerization of the tubulin, but the amount of phosphorylation that you showed was extremely small, only 5–10% of tubulin was phosphorylated.

Maness: In subpopulations of tubulin at local sites within the cell, for example the membrane-associated tubulin, the stoichiometry of phosphorylation might be much higher than the average found in the cell extracts. We know that at least 80% of tubulin in fetal rat brain is not phosphorylated. Most of this is in the soluble fraction. The majority of tubulin in neurons is in the axoplasm; only a small amount is associated with membranes. We are now investigating whether a particular isoform of tubulin in growth cone membranes is preferentially phosphorylated at tyrosine residues with high stoichiometry. Such a form may have a unique biological role in developing neurons.

Verma: Can you study the polymerization *in vitro*? Can you introduce phosphorylated tubulin and see how that affects polymerization?

Maness: We would like to isolate the tyrosine-phosphorylated tubulin from the membrane and do polymerization analysis, but that may be tricky. Triton X114 may provide a means of extracting tubulin from the membrane, but we haven't tried it yet.

Feramisco: The ratio of end tubulin molecules to protofilament molecules can be one to several thousands. So the phosphorylation could be like capping the end of the microtubule and you wouldn't require a high concentration of phosphorylated tubulin molecules to see a dramatic effect on polymerization.

It is like actin polymerization; there are capping molecules present in miniscule amounts relative to the amount of actin within the cell but they are very effective at blocking growth of filaments. Along that line, Patricia, have you looked at the effect of the tyrosinylation reaction that can occur on tubulin?

Maness: That residue does not appear to be phosphorylated. We treated phosphotyrosine-modified tubulin with carboxypeptidase A, which removes only the terminal tyrosine. We then used a monoclonal antibody (γL½) specific for the terminal tyrosine-containing peptide and immunoblotting showed that all of the terminal tyrosine residues had been removed from α-tubulin. We still saw the same amount of phosphotyrosine in the tubulin.

Feramisco: What if phosphorylation of the terminal tyrosine blocked the peptidase and also blocked binding by the antibody?

Maness: That would be hard to rule out.

Brugge: Do colchicine and other agents that disrupt microtubule polymerization prevent neurite extension in model systems?

Maness: Colchicine blocks neurite extension, indicating that polymerization of microtubules is required.

Noble: I don't understand why when you put v-*src* into a cell you see neurite extension and then you are talking about blocking polymerization.

Maness: Our model implies that *src* kinase is regulated locally at different regions of the growth cone membrane. In regions where it is not affected by a local adhesion signal, polymerization occurs and neurites grow in that direction. Alternatively, it could be that viral *src* protein has a different substrate from pp60[c-src], and is substituting for another tyrosine kinase that affects differentiation rather than neurite extension.

Verma: When you say *src*, do you mean a generic *src* or could it be c-*yes* or c-*fyn*? c-*yes* is certainly present in the brain.

Maness: Sudol et al (1988) have studied the expression of c-*yes* and c-*src* in developing cerebellum. *src* is expressed principally during embryogenesis and then declines to lower levels in the adult. Expression of c-*yes* is highest in the mature cerebellum. We looked for c-*yes* protein in the growth cone membrane fraction, but did not detect it. We also looked for c-*fyn* protein, another *src*-related proto-oncogene product that seems to be preferentially expressed in the adult.

Regarding what other tyrosine kinases could be present in the growth cone membranes, we have not seen increased tubulin phosphorylation in growth cone membranes treated with: insulin-like growth factor 1, insulin, bombesin, PDGF, EGF or basic FGF. Nor do we see stimulation of phosphorylation of proteins of the expected sizes of the receptors for those ligands.

Alemà: I would like to point out that there is no firm evidence that c-*src* is involved in the differentiation of nerve cells, especially PC12 cells. The action of v-*src* in PC12 cells might be due to the acquisition of a new function which is not shared by c-*src*. Many groups have been looking at variations in the kinase activity of pp60^{c-src} in PC12 cells after treatment with nerve growth factor (NGF), but they have found no significant changes. We have treated PC12 cells with vanadate, a specific inhibitor of phosphotyrosine phosphatases, and observed that the levels of phosphotyrosine increase, presumably as the result of the action of a whole range of kinases. This leads to an abortive morphological and biochemical differentiation, which is not as complete as that triggered by v-*src* or NGF.

Maness: Pam Maher found that treatment of PC12 cells with NGF induces a rapid, transient phosphorylation of tyrosine residues in a number of different proteins (Maher 1988). She observed some 55 kDa proteins migrating on gels in the same region as tubulin that are rapidly phosphorylated after stimulation. We intend to see if those proteins represent α- and β-tubulin subunits.

McMahon: As we are going back to development, is there any experimental evidence as to what pp60^{c-src} might be doing in the presumptive neural ectoderm and also what might be responsible for its induction?

Maness: We haven't focused on the early stage of pp60^{c-src} expression biochemically, because there is so little tissue at that stage. I am not aware that anyone else has worked on that either. The early stage of *src* expression might be that of the non-neuronal form; the second stage of expression might be the neuronal form that results from alternative splicing of the mRNA. By *in situ* hybridization or using antibodies to the neuronal insert one might be able to look at that. Joan, when do you find expression of the neuronal insert (pp60^{c-src+}) being switched on in development?

Brugge: We have looked only at a specific region, the neural retina. There we didn't see pp60^{c-src+} at Day 4, we did see it at Day 6. This is exactly the time when differentiation of the neurons in the neural retina begins. So expression of the neuronal form of the c-*src* protein was correlated with neuronal differentiation, but it is difficult to distinguish whether the conversion to c-*src*$^+$ occurred just before or just after the final differentiation step. In cultured embryonal carcinoma cells that can be induced to differentiate *in vitro*, expression of the neuronal form of c-*src* comes on at the time when the cells start to extend neurites.

Two different labs, those of Harold Varmus and Philippe Soriano, are doing gene replacement experiments where they are replacing the normal c-*src* gene with constructs that will not allow production of the c-*src*+ protein. Hopefully they will get an answer regarding the unique function of c-*src*+.

Maness: The *Drosophila* studies of Mike Simon and Mike Bishop (1985) are also relevant. They showed that there are two phases of *src* expression in *Drosophila*. The first stage of expression is in gastrulating embryos. In larvae and pupae *src* expression falls to much lower levels. In addition they saw expression of c-*src* in the intestine, which we have not looked at.

Brugge: Mary Sugrue, a student in my lab, has been working with Eric Gustafson in Paul Greengard's lab to examine the expression of c-*src*+ in the adult rat brain. We have made an antibody against the six amino acid insert within the neuronal form of pp60$^{c\text{-}src}$. We are interested to know whether this form of the *src* protein is localized to a unique place in the neuronal cell.

The ideal result would have been that pp60$^{c\text{-}src\,+}$ has a unique localization compared to the form of the c-*src* protein that doesn't contain the insert, thus suggesting that the neuronal insert confers specificity on the interaction of pp60$^{c\text{-}src\,+}$ with a unique membrane protein, for example within an axon terminal or a growth cone. This was not what we found. pp60$^{c\text{-}src\,+}$ is expressed throughout the central nervous system; however, it is not expressed in all neurons. From the localization studies, there are specific types of cells, for instance the Purkinje cells in the cerebellum, that express very high levels of *src*+. The entire dendritic tree is immunoreactive, as well as the membrane around the cell body. Therefore, pp60$^{c\text{-}src\,+}$ seems to be localized throughout the membranes, both in the dendritic arbors and in the cell body. We haven't seen any immunolocalization in the axonal outgrowth that emanates from the Purkinje cell.

In other parts of the brain *src*+ shows really nice immunoreactivity with regions that are rich in axon terminals or axonal projections. For instance, within the inferior and superior olivary complex, pp60$^{c\text{-}src\,+}$ is localized in the axon-rich regions with a distinct axon-like staining. This evidence, with neurochemical studies which we have previously done with Paul Greengard's lab, suggests that pp60$^{c\text{-}src\,+}$ is not localized exclusively in cell bodies and in dendritic processes, but also is present somewhere in axonal membranes. We are trying to look at this by electron microscopy, but so far have nothing to report.

Another point is that although it's very likely that non-receptor tyrosine kinases can be involved in events that occur in the growth cone, we also have to consider that these non-receptor tyrosine kinases might be involved in synaptic transmission or might at least be associated with synaptic vesicles. The c-*src* protein has been found associated with secretory vesicles in three different cell types. Parsons & Creutz (1986) and Grandori & Hanafusa (1988) showed that pp60$^{c\text{-}src}$ is associated with chromaffin granules. In collaboration with Dennis Pang and Flavia Valtorta in Paul Greengard's lab, we have found that synaptic vesicles are rich in pp60$^{c\text{-}src\,+}$. In addition, pp60$^{c\text{-}src\,+}$ is associated with platelet secretory vesicles (Rendu et al 1989, W. Isenberg, J. Fox, A. Golden, J. Brugge, unpublished results).

We have to consider the possibility that *src*-related kinases are very flexible molecules that might be involved in events that take place in intracellular membranes as well as in the plasma membrane.

Hunter: How does the distribution of pp60$^{c\text{-}src+}$ compare with that of normal pp60$^{c\text{-}src}$?

Brugge: The problem is that we don't have an antibody that specifically recognizes the form of pp60$^{c\text{-}src}$ that lacks the insert. If you stain the cerebellum or cerebral cortex with an antibody that recognizes both pp60$^{c\text{-}src+}$ and pp60$^{c\text{-}src}$, you see a staining pattern similar to that observed using anti-pp60$^{c\text{-}src+}$ serum. pp60$^{c\text{-}src+}$ contains every amino acid that's contained in pp60$^{c\text{-}src}$, so it is likely that it localizes to all the same sites. The question is whether pp60$^{c\text{-}src+}$ is localized in a unique place or can interact with a specific protein that the other protein doesn't recognize.

Hunter: Did you test whether the anti-pp60$^{c\text{-}src+}$ antibody blocks binding of monoclonal antibody 327? If so, you might be able to pre-stain with anti-pp60$^{c\text{-}src+}$ then use MAb 327 to localize pp60$^{c\text{-}src}$.

Brugge: No, we haven't checked that.

Feramisco: With immunofluorescence using a small target peptide, there is always the inherent danger that that region is involved in some complex formation in the cell so that when you fix the cell and stain, you are actually seeing everywhere where binding of that region of the protein is not important.

Brugge: That's definitely a concern. Maybe we should denature the tissue sections before staining.

Feramisco: You could do the experiment with a control piece of neuronal tissue or cells that express *src*$^+$. If the antibodies bound to *src*$^+$ in intact membranes and you could recover all that when you denatured, you could deduce that the antibody was able to find its target in the native membrane.

Hunter: How does your pattern of expression of *src*$^+$ compare to Chris Ross' *in situ* hybridization studies with a *src*$^+$-specific oligonucleotide probe (Ross et al 1988)?

Brugge: It's very similar, but we can see individual cells at higher resolution using immunohistochemical staining. However, we are not as interested in which cell types express this protein as in where in the cell it is expressed.

Maness: In terms of the synaptic vesicle localization, how good are those data? Have you assayed synaptic plasma membranes obtained from the fractionation?

Brugge: In every experiment, we have looked at each membrane and soluble cell fraction through the entire isolation procedure. As you purify the vesicles, you continue to enrich for pp60$^{c\text{-}src}$, however, the concentration of pp60$^{c\text{-}src}$ (per mg total protein) in the synaptic vesicle preparation is similar to that in the sucrose gradient fraction that's enriched for synaptic vesicles; it's about eight times greater than the concentration of pp60$^{c\text{-}src}$ in the total plasma membrane fraction.

Maness: We don't see any significant enrichment for pp60$^{c\text{-src}}$ in synaptosomal membranes relative to homogenate membranes.

Brugge: We agree, the enrichment in synaptosomes is only approximately 1.4-fold.

Maness: Sally Parsons has found a significant amount of pp60$^{c\text{-src}}$ in the plasma membrane of chromaffin cells, in addition to that in the chromaffin granule membranes (Parsons & Creutz 1986).

Brugge: I think pp60$^{c\text{-src}}$ is probably in the plasma membrane in growth cones and in synaptic vesicles. The question is whether it has a function in each of these places. In the platelets, pp60$^{c\text{-src}}$ is concentrated in the intracellular vesicles; after thrombin-induced secretion, the protein is all at the plasma membrane (W. Eisenberg, J. Fox, A. Golden, J. Brugge, unpublished).

Maness: We are also interested in pp60$^{c\text{-src}}$ and other tyrosine kinases in synaptic vesicles. In synaptic vesicle fractions from adult chicken retina, we have examined phosphotyrosine-containing proteins and found that three phosphotyrosine-containing proteins of 80, 42 and 35 kDa increased dramatically during the transition from embryo to adult, in contrast to other parts of the brain where phosphotyrosine-containing proteins decrease during development. We have now completed fractionation from adult retina and those three proteins appear to be associated with synaptic vesicle membranes. So I think there may be something quite interesting about tyrosine phosphorylation in synaptic vesicles.

Brugge: One must also consider the possibility that tyrosine kinases might be involved in transmitting signals that are mediated by extracellular matrix proteins or cell adhesion molecules. From studies in platelets, Jim Ferrell in Steve Martin's lab (Ferrell & Martin 1989) and Andy Golden in my lab have found that treatments that block the interaction between fibrinogen and its receptor, prevent thrombin-induced activation of tyrosine phosphorylation. In addition, Andy has found that incubation of platelets with fibrinogen and ADP alone causes a stimulation of tyrosine phosphorylation. These results suggest that an extracellular matrix/cell adhesion type molecule can stimulate tyrosine phosphorylation.

Going back to growth cones, growth cone extension *in vivo* involves communication between cell adhesion/extracellular matrix molecules and their receptors. It is possible that interaction between the cell adhesion molecules is responsible for activating tyrosine phosphorylation.

Hunter: I just want to address the question of whether p60$^{c\text{-src}}$ is normally active in the cell. When you assay *in vitro* it's clearly active as a tyrosine kinase, and perhaps that would imply that it is active constitutively in the cell. However, what we know about p60$^{c\text{-src}}$ in fibroblasts suggests that it's probably not constitutively active because the majority of p60$^{c\text{-src}}$ is phosphorylated at tyrosine 527, which regulates kinase activity negatively. Have you looked to see whether in the growth cones p60$^{c\text{-src}+}$ is phosphorylated at tyrosine 416, which might indicate that it is active?

Maness: We haven't completed phosphopeptide mapping of pp60$^{c\text{-src}}$ in the growth cone-enriched membrane fraction. However, it is clear that pp60$^{c\text{-src}+}$ does seem to be more active than the non-neuronal form of *src*, when assayed in an immune complex kinase assay *in vitro*.

Feramisco: Joan, you saw with your anti-peptide series of experiments an enrichment like that Patricia sees biochemically with growth cone enriched preparations. Patricia, there is enrichment of the neuronal form of *src* in growth cones, is that right?

Maness: The majority of pp60$^{c\text{-src}}$ in the growth cone membrane fraction seems to be the neuronal form, as judged by electrophoretic mobility and specific activity, but on Western blots with MAb 327, the band is broad, so we think one or more minor forms of pp60$^{c\text{-src}}$ may also be present.

Feramisco: In staining of neurite processes, do you see any selective staining of growth cones with your anti-peptide serum?

Brugge: In adult brain, growth cones are not detectable. In embryonal carcinoma P19 cells that have differentiated *in vitro*, Mary Sugrue from my lab has observed pp60$^{c\text{-src}+}$ staining of the growth cone and proximal neuritic shaft similar to that Patricia Maness showed for cultured retina neurons (Maness et al 1988).

Alemà: Concerning the regulation of gene expression, what are the effects of overexpression of c-*src*$^+$ in fibroblasts or PC12 cells?

Brugge: In fibroblasts the higher activity of c-*src*$^+$ just seems to tickle the cells to a greater extent than does c-*src*, but we don't see any unique c-*src*$^+$-induced events (Levy & Brugge 1989). We haven't looked at specific gene expression.

Hunter: So we are really at a loss to know what that regulator of pp60$^{c\text{-src}}$ is.

Brugge: If pp60$^{c\text{-src}}$ behaves like the protein encoded by the *lck* oncogene, pp56^{lck} (Veillette et al 1988), and interacts with a membrane receptor, like the T cell CD4 and CD8 proteins, then one would assume that pp60$^{c\text{-src}}$ and pp60$^{c\text{-src}+}$ communicate with cell surface receptors.

Hunter: No one has found anything with which it is specifically associated, except Carla Grandori and Saburo Hanafusa (1988) found a 38 kDa protein associated with pp60$^{c\text{-src}}$ in neuronal tissues. Have you seen anything like that?

Maness: We see a 38 kDa protein that contains phosphotyrosine on prolonged autoradiographic exposure of phosphotyrosine immunoblots of proteins from growth cone membranes, but we have never investigated whether pp60$^{c\text{-src}}$ is specifically associated with this protein, or whether it phosphorylates it.

Hunter: Your protein is not synaptophysin, which has also been reported to be a substrate for pp60^{src} (A. Barnekow, M. Schartl, personal communication)?

Maness: No, nor is Carla's.

Feramisco: Does the *src*$^+$ show up on PC12 cells in any conditions?

Brugge: No.

Feramisco: Do PC12 cells show growth cones?

Maness: Yes, but PC12 cells have smaller club-like growth cones than central nervous system neurons, which have very broad, flat growth cones.

Feramisco: You could put the neuronal *src*+ into PC12 cells and see if the morphology of the growth cones changes.

Brugge: We have done this experiment in collaboration with Simon Halegrena at SUNY, Stonybrook. We used a retroviral vector that expressed the *neo* and *src*+ genes to infect PC12 cells. In contrast to cells infected with a retrovirus carrying v-*src*, where every *neo*^{R} colony has elaborate neurite extensions, overexpression of c-*src* or c-*src*+ caused mixed phenotypic changes in the PC12 cells. In about 10% of the colonies you see neurite extension; however, 90% of the colonies behaved like normal PC12 cells. We believe that the cells which show neurite extension might express the highest levels of *src*+ or *src*. We are currently examining this possibility.

Hunter: Did you try NGF in conjunction with *src* or *src*+?

Brugge: When we treat the c-*src*+-expressing cells with NGF, they appear to be more sensitive to NGF but it's very difficult to quantitate. The cells have the same morphology as normal PC12 cells.

Maness: We have an interesting result with a PC12 cell transfected with a plasmid containing the v-*src* gene driven by the metallothionein I promoter, which was developed by Michael Cox in my laboratory. When we induce v-*src* gene expression with heavy metals and then treat the cells with NGF, we get increased neurite extension relative to that seen with NGF stimulation alone. Thus, pp60$^{v\text{-}src}$ appears to synergize with NGF in inducing neurite extension in PC12 cells.

Verma: So there is a difference in the two proteins.

Brugge: Yes, there is clearly a difference between the c-*src*+ and v-*src* proteins.

References

Ferrell JE Jr, Martin GS 1989 Tyrosine-specific protein phosphorylation is regulated by glycoprotein IIb-IIIa in platelets. Proc Natl Acad Sci USA 86:2234–2238

Grandori C, Hanafusa H 1988 p60$^{c\text{-}src}$ is complexed with a cellular protein in subcellular compartments involved in exocytosis. J Cell Biol 107:2125–2135

Hamaguchi M, Hanafusa H 1989 Localization of major potential substrates of p60$^{v\text{-}src}$ kinase in the plasma membrane matrix fraction. Oncogene Res 4:29–37

Levy J, Brugge JS 1989 Biological and biochemical properties of the c-*src*+ gene product overexpressed in chicken embryo fibroblasts. Mol Cell Biol 9:3332–3341

Maher PA 1988 Nerve growth factor induces protein tyrosine phosphorylation. Proc Natl Acad Sci USA 85:6788–6791

Maness PF, Levy BT 1983 Highly purified pp60-src induces the actin transformation in microinjected cells and phosphorylates selected cytoskeletal proteins in vitro. Mol Cell Biol 3:102–112

Maness PF, Aubry M, Shores CG, Frame L, Pfenninger KH 1988 c-*src* gene product in developing rat brain is enriched in nerve growth cone membranes. Proc Natl Acad Sci USA 85:5001–5005

Parsons SJ, Creutz CE 1986 p60[c-*src*] activity detected in the chromaffin granule membrane. Biochem Biophys Res Commun 134:736–742

Rendu F, Lebret M, Danielian S, Sagard R, Levy-Toledano S, Fischer S 1989 High pp60[c-*src*] level in human platelet dense bodies. Blood 73:1545–1551

Ross CA, Wright GE, Resh MD, Pearson RCA, Snyder SH 1988 Brain specific src oncogene messenger RNA mapped in rat brain by in situ hybridization. Proc Natl Acad Sci USA 85:9831–9835

Simon MA, Drees B, Kornberg T, Bishop JM 1985 The nucleotide sequences and the tissue-specific expression of Drosophila c-src. Cell 42:831–840

Sudol M, Alvarez-Buylla A, Hanafusa H 1988 Differential development expression of c-yes and c-src protein in cerebellum. Oncogene Res 2:345–355

Veillette A, Bookman MA, Horak EM, Bolen JB 1988 The CD4 and CD8 T cell surface antigens are associated with the internal membrane tyrosine- protein kinase p56[lck]. Cell 55:301–308

Molecular biology of the *hst-1* gene

Takashi Sugimura, Teruhiko Yoshida, Hiromi Sakamoto, Osamu Katoh, Yutaka Hattori and Masaaki Terada

Genetics Division, National Cancer Center Research Institute, 1-1 Tsukiji 5-chome, Chuo-ku, Tokyo 104, Japan

Abstract. The *hst-1* gene (or *HSTF1* by human gene nomenclature) was originally identified in our laboratory by an NIH/3T3 focus formation assay using DNA from a human gastric cancer. Sequence analysis predicted the *hst-1* product to be a novel growth factor with 30–50% homology with six other heparin-binding growth factors: basic and acidic fibroblast growth factors (FGFs), the *int-2* protein, FGF5, the *hst-2*/FGF6 protein and keratinocyte growth factor (KGF). A recombinant *hst-1* protein was synthesized in silkworm cells and found to be a potent heparin-binding mitogen for murine fibroblasts and human vascular endothelial cells. Although *hst-1* expression cannot be detected in most cancer cells, including gastric cancers, it is expressed in mouse embryos and in some germ cell tumours. Both *hst-1* and *int-2* are located on band q13.3 of human chromosome 11 within a distance of 35 kbp; in the mouse genome these two genes are separated by less than 20 kbp. They are differentially transcribed in the F9 mouse terato-carcinoma cell line; *hst-1* is expressed in undifferentiated stem cells and *int-2* in differentiated endodermal cells. The *hst-1* and *int-2* genes were coamplified in a variety of cancer cells, most notably in more than 50% of oesophageal cancers.

1990 Proto-oncogenes in cell development. Wiley, Chichester (Ciba Foundation Symposium 150) p 79–98

Gastric cancer is the leading cause of cancer death among Japanese people; the reasons for this are not known. Our initial purpose in the study that resulted in the discovery of the *hst-1* oncogene was to identify an essential genetic aberration of gastric cancer. A test of the transforming activity of DNA samples prepared from cancerous and non-cancerous portions of gastric mucosae obtained at the time of surgery identified a novel transforming gene called *hst-1*, an acronym for *human stomach*. It soon became obvious that the *hst-1* protein belongs to a potentially large family of heparin-binding growth factors and oncogene products; however, we do not yet have concrete evidence for specific involvement of *hst-1* in any human cancers, including gastric cancers.

Hybridization analyses and molecular cloning indicated the presence of a close homologue of *hst*, designated *hst-2* (Sakamoto et al 1988); accordingly, *hst* was renamed *hst-1*. In human gene nomenclature, this gene was registered officially as *HSTF1* (*h*eparin-binding *s*ecretory *t*ransforming *f*actor *1*) (Yoshida MC et al 1988).

79

In this paper, we describe the homology of the *hst-1* protein with other heparin-binding growth factors, the mitogenic activities of a recombinant *hst-1* protein, the expression of *hst-1* in embryonic tissues and in germ cell tumours, the chromosomal localization of *hst-1* to band 11q13.3 and coamplification of *hst-1* and *int-2* in human cancers, most notably in oesophageal cancers.

Materials and Methods

Cloning of the transforming gene, hst-1

High molecular weight DNA was extracted from 37 cancerous tissues of patients with gastric cancer and also from 21 non-cancerous portions of adjacent gastric mucosae. The NIH/3T3 focus formation assay revealed unequivocal transforming activity in two cancer-derived DNAs and also in DNA from a non-cancerous gastric mucosa. Human genomic sequences present in the DNA of an NIH/3T3 secondary transformant, T361-2nd-1, were cloned using a human specific repetitive sequence, *Alu*, as a probe (Sakamoto et al 1986). Four DNA fragments which hybridized to mRNA from the T361-2nd-1 transformant, but not to that from parental NIH/3T3 cells, were isolated. These fragments did not hybridize with a series of known oncogene probes and were considered to represent a novel transforming gene, termed *hst* (Sakamoto et al 1986). A corresponding cDNA clone was obtained by screening a cDNA library made from mRNA of T361-2nd-1 cells (Taira et al 1987). The cDNA fragment was then used to screen human and mouse genomic libraries to clone proto-*hst* genomic fragments (Yoshida et al 1987a, Sakamoto et al 1988); both the cDNA clone and genomic fragments of *hst* were sequenced in full by the dideoxy chain termination method using m13mp18 phages.

Identification of an open reading frame

Two possible open reading frames longer than 150 amino acids were deduced by sequence analysis and designated ORF1 and ORF2. An *hst-1* cDNA was placed downstream of the SV40 early promoter to make a eukaryotic expression vector, pKOc1 (Taira et al 1987). A series of derivatives of this were constructed, each of which has ORF1 or ORF2 inactivated by frame-shift mutations; these plasmids were transfected into NIH/3T3 cells, and a coding region necessary and sufficient for transform... activity was determined.

Synthesis of the hst-1 protein and its biological activity

The coding region of *hst-1* was cloned into vector pBM030 and transfected into *Bombyx mori* silkworm-derived *Bm*N cells to generate a recombinant virus by homologous recombination (Maeda et al 1985). Culture medium from *Bm*N cells infected with the virus was collected and purified by Affi-Gel heparin column and reversed-phase high performance liquid chromatography (Miyagawa et al 1988). The recombinant *hst-1* protein was added to NIH/3T3 cells or human umbilical endothelial cells and subjected to [^{3}H]thymidine uptake assay. A soft agar colony formation assay was performed using NIH/3T3 cells and NRK-49F (normal rat kidney fibroblasts) cells as described in Miyagawa et al (1988).

Southern and Northern blot analyses

Hybridization and washing were carried out at the stringent conditions given in Yoshida et al (1988a). Probes used were: AA (Sakamoto et al 1988) and LGC (Yoshida et al 1988a) for human *hst-1*, M1.8 (Yoshida et al 1988a) for mouse *hst-1*, BB4 and SS6 for human *int-2* (Casey et al 1986).

Results and discussion

Identification of the coding region of the hst-1 cDNA

Evaluation of the transforming activity of various frame-shift mutants in pKOc1 has led us conclude that ORF1 is the coding region of the transforming protein (Taira et al 1987). However, we needed to know that this coding region was not affected by artificial gene rearrangements which might have occurred during transfection into NIH/3T3 cells. Proto-*hst-1* genomic fragments were thus cloned directly from two human genomic libraries. The coding region of the cDNA matched completely the corresponding sequences of the proto-*hst-1* genomic fragment (Yoshida et al 1987b); this means that the amino acid sequence encoded by ORF1 (Fig. 1) is that of a normal human protein. The deduced sequence consists of 206 amino acids and has a stretch of hydrophobic residues at the N-terminus, a characteristic of a secretory protein.

Transforming activity of proto-hst-1 genomic fragments

A *Bam*HI-*Sal*I 6.2 kbp fragment of proto-*hst-1* was cloned from a normal human genomic library and found to transform NIH/3T3 cells without any additional promoter/enhancer element. Furthermore, a mouse proto-*hst-1* gene was cloned from the genome of NIH/3T3 cells, and the cosmid clone containing this gene could transform NIH/3T3 cells. The transforming activities of the human and mouse proto-*hst-1* genomic fragments were comparable (about 100–300 foci/pmol) to that of the *hst-1* genomic fragment cloned from the original transformant, T361-2nd-1 (Sakamoto et al 1988). These observations were not unexpected, because of the identity of the coding sequences in the *hst-1* cDNA derived from T361-2nd-1 cells and the proto-*hst-1* genomic fragment

```
ATG.TCG.GGG.CCC.GGG.ACG.GCC.GCG.GTA.GCG.CTG.CTC.CCG.GCG.GTC.CTG.CTG.GCC.TTG.CTG
Met-Ser-Gly-Pro-Gly-Thr-Ala-Ala-Val-Ala-Leu-Leu-Pro-Ala-Val-Leu-Leu-Ala-Leu-Leu    20

GCG.CCC.TGG.GCG.GGC.CGA.GGG.GGC.GCC.GCC.GCA.CCC.ACT.GCA.CCC.AAC.GGC.ACG.CTG.GAG
Ala-Pro-Trp-Ala-Gly-Arg-Gly-Gly-Ala-Ala-Ala-Pro-Thr-Ala-Pro-Asn-Gly-Thr-Leu-Glu    40

GCC.GAG.CTG.GAG.CGC.CGC.TGG.GAG.AGC.CTG.GTG.GCG.CTC.TCG.TTG.GCG.CGC.CTG.CCG.GTG
Ala-Glu-Leu-Glu-Arg-Arg-Trp-Glu-Ser-Leu-Val-Ala-Leu-Ser-Leu-Ala-Arg-Leu-Pro-Val    60

GCA.GCG.CAG.CCC.AAG.GAG.GCG.GCC.GTC.CAG.AGC.GGC.GCC.GGC.GAC.TAC.CTG.CTG.GGC.ATC
Ala-Ala-Gln-Pro-Lys-Glu-Ala-Ala-Val-Gln-Ser-Gly-Ala-Gly-Asp-Tyr-Leu-Leu-Gly-Ile    80

AAG.CGG.CTG.CGG.CGG.CTC.TAC.TGC.AAC.GTG.GGC.ATC.GGC.TTC.CAC.CTC.CAG.GCG.CTC.CCC
Lys-Arg-Leu-Arg-Arg-Leu-Tyr-Cys-Asn-Val-Gly-Ile-Gly-Phe-His-Leu-Gln-Ala-Leu-Pro   100

GAC.GGC.CGC.ATC.GGC.GGC.GCG.CAC.GCG.GAC.ACC.CGC.GAC.AGC.CTG.CTG.GAG.CTC.TCG.CCC
Asp-Gly-Arg-Ile-Gly-Gly-Ala-His-Ala-Asp-Thr-Arg-Asp-Ser-Leu-Leu-Glu-Leu-Ser-Pro   120

GTG.GAG.CGG.GGC.GTG.GTG.AGC.ATC.TTC.GGC.GTG.GCC.AGC.CGG.TTC.TTC.GTG.GCC.ATG.AGC
Val-Glu-Arg-Gly-Val-Val-Ser-Ile-Phe-Gly-Val-Ala-Ser-Arg-Phe-Phe-Val-Ala-Met-Ser   140

AGC.AAG.GGC.AAG.CTC.TAT.GGC.TCG.CCC.TTC.TTC.ACC.GAT.GAG.TGC.ACG.TTC.AAG.GAG.ATT
Ser-Lys-Gly-Lys-Leu-Tyr-Gly-Ser-Pro-Phe-Phe-Thr-Asp-Glu-Cys-Thr-Phe-Lys-Glu-Ile   160

CTC.CTT.CCC.AAC.AAC.TAC.AAC.GCC.TAC.GAG.TCC.TAC.AAG.TAC.CCC.GGC.ATG.TTC.ATC.GCC
Leu-Leu-Pro-Asn-Asn-Tyr-Asn-Ala-Tyr-Glu-Ser-Tyr-Lys-Tyr-Pro-Gly-Met-Phe-Ile-Ala   180

CTG.AGC.AAG.AAT.GGG.AAG.ACC.AAG.AAG.GGG.AAC.CGA.GTG.TCG.CCC.ACC.ATG.AAG.GTC.ACC
Leu-Ser-Lys-Asn-Gly-Lys-Thr-Lys-Lys-Gly-Asn-Arg-Val-Ser-Pro-Thr-Met-Lys-Val-Thr   200

CAC.TTC.CTC.CCC.AGG.CTG
His-Phe-Leu-Pro-Arg-Leu
```

206 amino acids
22,000 dalton

FIG. 1. Nucleotide sequence and deduced amino acid sequence of the coding region (ORF1) of the *hst-1* cDNA. A hydrophobic core sequence for a signal peptide is underlined.

cloned directly from human genomic libraries. We suppose that the normal *hst-1* protein could transform NIH/3T3 cells, if the normal regulatory machinery for gene expression were disrupted.

Homology with heparin-binding growth factors and oncogene products

There are no genes identical to *hst-1* in current databases, but striking homologies with acidic and basic fibroblast growth factors (FGFs), the *int-2* protein, FGF5, *hst-2*/FGF6 and keratinocyte growth factor (KGF) are noted (Fig. 2).

Acidic and basic FGFs are well-known heparin-binding growth factors (reviewed by Baird et al 1986). *int-2* is a cellular gene which is activated by proviral insertion of mouse mammary tumour virus in murine mammary carcinogenesis (Peters et al 1983), and its human counterpart has been sequenced (Brookes et al 1989). Recently, it was shown that the *in vitro* translation product of the mouse *int-2* gene is mitogenic for fibroblast and epithelial cell lines, and that the *int-2* cDNA could effectively support anchorage-independent growth of cultured cells, when its weakly hydrophobic N-terminus was replaced by the immunoglobulin signal peptide (R. Deed et al and G. R. Merlo et al, unpublished papers, 5th Annual Meeting on Oncogenes, 30 June 1989). FGF5 was identified by defined medium transformation assay (Zhan et al 1988). *hst-2*/FGF6, a close homologue of *hst-1* (Sakamoto et al 1988), has been partially sequenced (Marics et al 1989). KGF is the newest member of this family (Finch et al 1989).

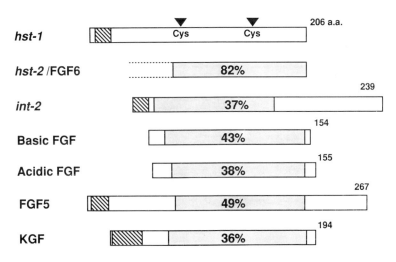

FIG. 2. Topological comparison of the human *hst-1* gene product and related human proteins. A region with strong similarity to the *hst-1* protein is shown as a shaded box with the percentage amino acid identity. Two conserved cysteine residues are marked by arrowheads. Hydrophobic core residues for a signal peptide are hatched. The N-terminal sequence for *hst-2*/FGF6 has not been determined.

N-terminal hydrophobic core sequences are present in all these proteins except acidic and basic FGFs. Two cysteine residues (Cys 88 and 155 in *hst-1*) are conserved. Acidic and basic FGFs are mitogens for vascular endothelial and epithelial cells, as well as mesodermal cells. KGF is specific to epithelial cells (Finch et al 1989). Biological functions of the other members of this family are not yet clear.

Biological activity of the recombinant hst-1 protein

The recombinant *hst-1* protein synthesized in silkworm cells markedly stimulated thymidine uptake by NIH/3T3 cells at a concentration as low as 0.1–1.0 ng/ml (Fig. 3). The mitogenic response was augmented by the presence of 50 µg/ml heparin; this enhancement was also reported for acidic FGF (Schreiber et al 1985). Similar stimulation of DNA synthesis by the recombinant *hst-1* protein was observed in a primary culture of human umbilical endothelial cells (Miyagawa et al 1988). Since FGFs are known to be potent angiogenic substances both *in vitro* and *in vivo*, it is possible that *hst-1* is also involved in certain angiogenic events, such as organogenesis in embryos. Soft agar colony formation by NIH/3T3 and NRK-49F cells was stimulated by the *hst-1* protein in a dose-dependent manner (Miyagawa et al 1988).

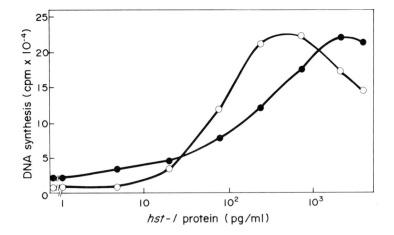

FIG. 3.   Dose–response curve of the stimulation of DNA synthesis in NIH/3T3 cells by purified recombinant *hst-1* protein. The [^{3}H] thymidine uptake induced in quiescent NIH/3T3 cells by *hst-1* protein was measured in the absence (filled circles) and presence (open circles) of 50 µg/ml of heparin.

Expression of hst-1

Northern blot analysis failed to detect the *hst-1* transcript in most cancer cells, including gastric cancers (Yoshida et al 1988a). The only exception known today is germ cell tumours; as shown in Fig. 4A, *hst-1* was expressed in five of nine surgical specimens of testicular germ cell tumours and in a cell line derived from a human immature teratoma, NCC-IT (Teshima et al 1988). The major transcripts were 3.0 and 1.7 kb. The expression of *hst-1* in normal mouse tissues was investigated using probe M1.8, a mouse *hst-1* genomic fragment. A single band of 3.0 kb was detected in mid-stage embryos at 11 and 14 days post-coitum but not at a later stage of development nor after birth (Yoshida et al 1988a). A mouse teratocarcinoma cell line, F9, can be induced to differentiate into endodermal cells *in vitro*; the process is considered to simulate an early stage of embryonic development (Jakobovits et al 1986). We found that *hst-1* is abundantly transcribed in undifferentiated F9 cells and dramatically down-regulated on differentiation into parietal endodermal cells caused by treatment with retinoic acid and dibutyryl cAMP (Fig. 4B); in contrast, *int-2* is preferentially expressed in differentiated F9 cells (Jakobovits et al 1986). These two closely related heparin-binding growth factors may have distinct roles in early embryogenesis.

Proximity of hst-1 and int-2 and their coamplification in cancer

We mapped the *hst-1* gene to band q13.3 of human chromosome 11 (Yoshida MC et al 1988), the same locus at which *int-2* is localized (Casey et al 1986).

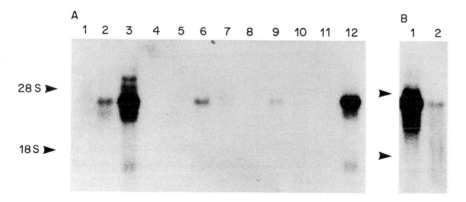

FIG. 4.   RNA blot analysis of germ cell tumours for the expression of *hst-1*. (A) Lanes 1 to 9 are surgical specimens of human testicular germ cell tumours; 10 and 11 are non-cancerous portions of the testes. Lane 12 is a human immature teratoma cell line, NCC-IT. (B) Lane 1 is a mouse teratocarcinoma cell line, F9. Lane 2 is F9 cells that have differentiated to parietal endodermal cells *in vitro*. Each lane contains 3 µg of poly(A)$^+$ RNA. Arrowheads indicate the positions of the 28S and 18S ribosomal RNAs.

TABLE 1 Coamplification of *hst-1* and *int-2* in human cancers

Cancer	Incidence of coamplification
Oesophagus	
Surgical specimens	22/41 (54%)
Cultured cells	5/13 (38%)
Stomach	1/32
Lung	1/53
Bladder	1/18
Kidney	0/23
Breast[a]	21/176 (12%)

[a]DNA was extracted from formalin-paraffin blocks of breast cancers and analysed by dot-blot analysis. Other cancers were analysed by standard Southern blot analysis.

Physical mapping of this region showed that the distance between *hst-1* and *int-2* is about 35 kbp in the human (Fig. 5) and less than 20 kbp in the mouse genome (Wada et al 1988, Yoshida et al 1988b). They are arranged in the same transcriptional orientation with *int-2* upstream of *hst-1*. This *int-2–hst-1* region is amplified in various types of human cancers as shown in Table 1 (Tsutsumi et al 1988, Yoshida MC et al 1988, Tsuda H et al 1989, Tsuda T et al 1989). The degree of amplification is the same for both genes, ranging from three- to 20-fold. The incidence of the amplification is especially high in oesophageal cancers. We have looked at the expression of these genes in 13 oesophageal cancer cell lines, among which five have coamplifed *hst-1* and *int-2* genes; Northern blot analysis failed to detect either message (O. Katoh, unpublished work 1989). There are several possible explanations for the absence of measurable expression in the presence of amplification; the level of expression may be very low, even when the genes are amplified, but sufficient to endow some growth advantage on the cells; the genes may be actively transcribed at a certain early stage of oesophageal carcinogenesis and then turned off; the genes may be down-regulated in standard culture conditions; another gene essential for oesophageal carcinogenesis and/or its progression may be present in the

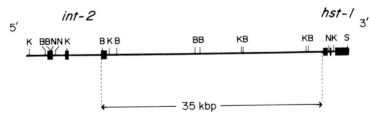

FIG. 5. A physical map of the *int-2–hst-1* region. The *int-2* gene is located 5′ of the *hst-1* gene on the same DNA strand. Each of the two oncogenes has three exons shown as boxes. The distance between the 5′ end of the last exon of *int-2* and the TATA box of *hst-1* is 35 kbp. K, *Kpn*I; B, *Bam*HI; N, *Not*I; S, *Sal*I.

int-2–hst-1 amplification unit. We favour the last possibility, and chromosome walking in this region is in progress.

Future research

The transforming gene *hst-1* was found to encode a heparin-binding growth factor homologous to acidic and basic FGFs, the *int-2* protein, FGF5, *hst-2*/FGF6 and KGF, and there is no evidence that this is the end of the list. Except for acidic and basic FGFs, we should say that the study of *in vivo* biological functions has just started for members of this family. FGFs are potent angiogenic factors and also mitogens for epithelial as well as mesodermal cells. It may be possible to develop clinical applications for some members of this family or their derivatives for treatment of nerve and other injuries, burns, cardiovascular and cerebrovascular disorders and organ grafts, for instance. Antagonists of these heparin-binding growth factors may be useful for interfering with abnormal angiogenesis in cancer and diabetic retinopathy. The *hst-1* and *int-2* genes are unique, because they may be involved specifically in embryogenesis at distinct stages; it should be an interesting target of research to study how a regulatory system activates these neighbouring genes differentially in the course of embryonic development and also suppresses expression of these potentially transforming genes in normal cells after birth. A receptor molecule that binds basic FGF was recently cloned (Lee et al 1989), but receptors specific for each of the other members have not yet been identified. Analysis of the amplification unit at chromosome 11q13.3 may be another fruitful subject in the study of cancer; two important questions we would like to answer are (1) which genes are amplified and overexpressed, and (2) by what mechanism is this region frequently amplified in certain types of human cancers?

Increased DNA content per cell is often associated with more malignant types of cancers. We used the DNA renaturation in gel method to demonstrate the frequent presence of amplified DNA sequences in various cell lines (Nakatani et al 1985, 1986). One of the genes on these amplified DNA fragments in a gastric cancer cell line was identified, and designated the *sam* gene. Sequence analysis indicated that the *sam* gene encodes a novel receptor type tyrosine kinase (H. Nakatani et al, unpublished data).

Isolation and characterization of the amplified sequence containing *hst-1* and *int-2* on chromosome 11q13.3 will lead to better understanding of the biological significance of amplified genes in cancer.

Acknowledgements

I would like to thank our collaborators, especially Drs M. Furusawa, Y. Mitsui, K. Miyagawa, T. Muramatsu and E. Tahara. This work was supported in part by the Grant-in-Aid for a Comprehensive 10-Year Strategy for Cancer Control from the Ministry of Health and Welfare, Japan.

References

Baird A, Esch F, Mormede P et al 1986 Molecular characterization of fibroblast growth factor: distribution and biological activities in various tissues. Recent Prog Horm Res 42:143–205

Brookes S, Smith R, Casey G, Dickson C, Peters G 1989 Sequence organization of the human *int-2* gene and its expression in teratocarcinoma cells. Oncogene 4:429–436

Casey G, Smith R, McGillivray D, Peters G, Dickson C 1986 Characterization and chromosome assignment of the human homolog of *int-2*, a potential proto-oncogene. Mol Cell Biol 6:502–510

Finch PW, Rubin JS, Miki T, Ron D, Aaronson SA 1989 Human KGF is FGF-related with properties of a paracrine effector of epithelial cell growth. Science (Wash DC) 245:752–755

Jakobovits A, Shackleford GM, Varmus HE, Martin GR 1986 Two proto-oncogenes implicated in mammary carcinogenesis, *int-1* and *int-2*, are independently regulated during mouse development. Proc Natl Acad Sci USA 83:7806–7810

Lee PL, Johnson DE, Cousens LS, Fried VA, Williams LT 1989 Purification and complementary DNA cloning of a receptor for basic fibroblast growth factor. Science (Wash DC) 245:57–60

Maeda S, Kawai T, Obinata M et al 1985 Production of human α-interferon in silkworm using a baculovirus vector. Nature (Lond) 315:592–594

Maniatis T, Fritsch EF, Sambrook J 1982 Molecular Cloning: a laboratory manual. Cold Spring Harbor Laboratory, Cold Spring Harbor, New York p 150–162, p 187–206, p 280–281, p 382–389

Marics I, Adelaide J, Raybaud F et al 1989 Characterization of the *HST*-related *FGF.6* gene, a new member of the fibroblast growth factor gene family. Oncogene 4:335–340

Miyagawa K, Sakamoto H, Yoshida T et al 1988 *hst-1* transforming protein: expression in silkworm cells and characterization as a novel heparin-binding growth factor. Oncogene 3:383–389

Nakatani H, Tahara E, Sakamoto H, Terada M, Sugimura T 1985 Amplified DNA sequences in cancers. Biochem Biophys Res Commun 130:508–514

Nakatani H, Tahara E, Yoshida T, Sakamoto H et al 1986 Detection of amplified DNA sequences in gastric cancers by a DNA renaturation method in gel. Jpn J Cancer Res (Gann) 77:849–853

Peters G, Brookes S, Smith R, Dickson C 1983 Tumorigenesis by mouse mammary tumor virus: evidence for a common region for proviral integration in mammary tumors. Cell 33:369–377

Sakamoto H, Mori M, Taira M et al 1986 Transforming gene from human stomach cancers and a noncancerous portion of stomach mucosa. Proc Natl Acad Sci USA 83:3997–4001

Sakamoto H, Yoshida T, Nakakuki M et al 1988 Cloned *hst* gene from normal human leukocyte DNA transforms NIH/3T3 cells. Biochem Biophys Res Commun 151:965–972

Schreiber AB, Kenney J, Kowalski WJ, Friesel R, Mehlman T, Maciag T 1985 Interaction of endothelial cell growth factor with heparin: characterization by receptor and antibody recognition. Proc Natl Acad Sci USA 82:6138–6142

Taira M, Yoshida T, Miyagawa K, Sakamoto H, Terada M, Sugimura T 1987 cDNA sequence of human transforming gene, *hst*, and identification of the coding sequence required for transforming activity. Proc Natl Acad Sci USA 84:2980–2984

Teshima S, Shimosato Y, Hirohashi S et al 1988 Four new human germ cell tumor cell lines. Lab Invest 59:328–336

Tsuda H, Hirohashi S, Shimosato Y et al 1989 Correlation between long-term survival in breast cancer patients and amplification of two putative oncogene-coamplification units: *hst-1/int-2* and c-*erb*B-2/*ear*-1. Cancer Res 49:3104–3108

Tsuda T, Tahara E, Kajiyama G, Sakamoto H, Terada M, Sugimura T 1989 High incidence of coamplification of *HST1* and *INT-2* genes in human esophageal carcinomas. Cancer Res 49:5505–5508

Tsutsumi M, Sakamoto H, Yoshida T et al 1988 Coamplification of the *hst-1* and *int-2* genes in human cancers. Jpn J Cancer Res (Gann) 79:428–432

Wada A, Sakamoto H, Katoh O et al 1988 Two homologous oncogenes, *HST1* and *INT2*, are closely located in human genome. Biochem Biophys Res Commun 157:828–835

Yoshida MC, Wada M, Satoh H et al 1988 Human *HST1* (*HSTF1*) gene maps to chromosome band 11q13 and coamplified with the *INT2* gene in human cancer. Proc Natl Acad Sci USA 85:4861–4864

Yoshida T, Sakamoto H, Miyagawa K, Little P, Terada M, Sugimura T 1987a Genomic clone of *hst* with transforming activity from a patient with acute leukemia. Biochem Biophys Res Commun 142:1019–1024

Yoshida T, Miyagawa K, Odagiri H et al 1987b Genomic sequence of *hst*, a transforming gene encoding a protein homologous to fibroblast growth factors and the *int-2*-encoded protein. Proc Natl Acad Sci USA 84:7305–7309

Yoshida T, Tsutsumi M, Sakamoto H et al 1988a Expression of the *HST1* oncogene in human germ cell tumors. Biochem Biophys Res Commun 3:1324–1329

Yoshida T, Muramatsu H, Muramatsu T et al 1988b Differential expression of two homologous and clustered oncogenes, *Hst1* and *Int-2*, during differentiation of F9 cells. Biochem Biophys Res Commun 157:618–625

Zhan X, Bates B, Hu X, Goldfarb M 1988 The human FGF-5 oncogene encodes a novel protein related to fibroblast growth factors. Mol Cell Biol 8:3487–3495

DISCUSSION

Westermark: You alluded to the use of the *hst* gene product to treat stomach ulcers. Have you shown that the mucosal cells actually respond to this growth factor?

Sugimura: The *hst* gene product shows growth-promoting activity on various cell types. Therefore it may be useful in wound healing, particularly for stomach ulcers. The NIH/3T3 cells and NRK-49F cells responded well. Endothelial cells from the human umbilical vein also respond to *hst*. We have not yet tested the epithelial cells that line the stomach.

Feramisco: In the samples of human stomach cancer, were there any where you couldn't find amplification or evidence of a dominant-acting oncogene? Are there any chromosomal deletions that are correlated with some of those samples?

Sugimura: I don't think there is any firm evidence yet. There are several reported oncogene amplifications in gastric cancer and also point mutations in oncogenes, for example K-*ras*. We have failed to detect any chromosomal loci showing a high incidence of loss of heterozygosity (Wada et al 1988). However, there was a report of increased loss of heterozygosity at chromosomes

1q and 12q in stomach cancer (Fey et al 1989). *ras* gene activation is not frequently observed in stomach cancer, it is in pancreatic cancer. Amplification of the *sam* oncogene was found in some histological types of gastric cancer. Alterations of c-*erbB2* were found in 40% of tubular adenocarcinomas of the stomach, but not in other types of stomach cancer (Yokota et al 1988).

Verma: Does *sam* transform cells?

Sugimura: We have cloned *sam* into an expression vector with the MLV-LTR promoter and introduced it into NIH/3T3 cells. There is some indication that expression of *sam* causes transformation.

Hunter: Do you know that the amplified *sam* gene you have cloned is normal? Have you compared its sequence to that of the normal *sam* cDNA?

Sugimura: There is no definite answer on that point yet.

Westermark: How does your *sam* gene product relate to the FGF receptor that was cloned by Rusty Williams?

Sugimura: It's hard to say at the moment. We have isolated two *sam* genes. They are closely related to each other; one of them may be related to the FGF receptor.

Hunter: How was the second *sam* gene isolated?

Sugimura: By screening a cDNA library made from mRNA from a human immature teratoma with the *sam-1* cDNA.

Méchali: Is *hst-1* gene expressed during mesoderm induction in the early stages of development in mice?

Sugimura: We are trying to do *in situ* hybridization to see in which tissues in mice *hst* is expressed during embryogenesis.

Heath: We have shown that keratinocyte growth factor (KGF) is a very potent mesoderm-inducing agent. There appear to be *Xenopus* homologues of KGF expressed at the time of normal mesoderm induction.

Wagner: It has also been shown by R. Basilico's group and others that it is expressed very highly in embryonic stem cells. In those cells it is also shut off rapidly when the cells differentiate.

Heath: It's only shut off down certain lineages. It is shut off if the cells differentiate in the direction of parietal endoderm. Our results suggest that if the cells differentiate into mesodermal cell types then expression of KGF is maintained for a few days. It appears to be not strictly correlated with the state of the embryonic stem cell.

Wagner: Dr Sugimura, do you know whether expression of *hst-1* is shut off in the F9 cells?

Sugimura: Yes, expression of *hst* is shut off during differentiation of the F9 cells.

Wagner: What's responsible for that shut-off?

Sugimura: The cells are induced to differentiate by retinoic acid or dibutyryl cAMP, but I don't know how the *hst-1* gene is turned off.

Sherr: Do you see cases of coamplification of *erbB2* with *hst* and *int-2* or are they separate?

Sugimura: They are separate. *erbB2* is amplified in a differentiated type of gastric cancer. *hst* and *int-2* are coamplified in oesophageal cancer. In breast cancer, *hst* and *int-2* are coamplified; sometimes *erbB2* is also amplified but this is an independent event.

Sherr: If you look at Kaplan–Meier survival curves for cases that have amplification of both genes, is the predicted survival of those patients worse or the same when both *hst* and *erbB2* are amplified versus only one set?

Sugimura: We found that amplification of c-*erbB2* alone and coamplification of *hst-1* and *int-2* are both correlated with a poor prognosis of breast cancer (Tsuda et al 1989).

Sherr: So you think the genes independently contribute to survival?

Sugimura: Probably, yes.

Verma: *int-2* and *hst-1* are coamplified and the prognosis is bad but there is no expression of the two genes?

Sugimura: That's right. Although *hst-1* and *int-2* are amplified in about 50% of oesophageal cancers, Northern blot analysis failed to detect expression of these genes. This suggests that another gene is important in the malignant phenotype of oesophageal cancers. We are looking for such genes in the region between *hst-1* and *int-2*.

Jessell: *hst*, as opposed to some other members of this family, has an Arg-Gly-Asp cell adhesion domain. Do you have any evidence that this is functional, that some of the properties of the *hst* gene product are mediated by that domain?

Sugimura: We have no information on this.

McMahon: Recently, novel forms of FGF have been described which utilize an upstream CTG as an initiation codon (Florkiewicz & Sommer 1989); I believe the same is true for *int-2* (Gordon Peters & Clive Dickson, personal communication). That has rather profound consequences for the protein, because if it has a secretory signal sequence, that will become buried inside the protein. Do you have any data as to whether *hst* uses upstream CTG sequences as initiation codons?

Sugimura: No, we don't have that information.

Heath: In reticulocyte lysate *in vitro* translation systems, full length KGF protein is secreted across the microsomal membrane, so it behaves as if it was a secreted protein. This also occurs in COS cells where the protein is secreted into the medium.

Sugimura: *hst* protein is secreted in the baculovirus expression system, which requires a proper signal sequence.

Verma: Dr Sugimura, what about the second ORF, is there a protein or an mRNA being made from that sequence?

Sugimura: We have no evidence that the protein encoded by the sequence ORF2 is made *in vivo*. From the study with the *hst-1* cDNA expression vector, we know that ORF2 does not encode a transforming protein.

Hunter: There are now seven known members of the FGF family, and there are additional genes related to the cloned FGF receptor, which suggests that there are multiple FGF receptors.

Waterfield: Rusty Williams has evidence that multiple FGF receptors exist (personal communication).

Hunter: The question is do these different receptors have higher affinities for certain members of the FGF family?

Heath: We have just published that the *int-2* protein is significantly less potent in both mitogenic and mesoderm-inducing assays than KGF or acidic or basic FGF (Paterno et al 1989). Our interpretation is that we are observing cross-reactivity of *int-2* with the receptors which mediate those responses. That suggests that there is a class of receptors which have a higher affinity for *int-2*-like members of the FGF family. It's entirely consistent with the idea that there are multiple families of receptors with overlapping specificities.

Hunter: Would the *int-2* receptor be related or would it be a totally separate type of receptor?

Heath: I think the exciting prediction is that it would be one of the related receptors that Mike Waterfield mentioned.

Hunter: If that's the case, it raises the issue of whether one could form hetero-dimers between these different FGF receptors, like those seen with the two PDGF receptors, or does one require a dimeric growth factor like PDGF for this?

Heath: You have to have a different mechanism to the one that Bengt Westermark (this volume) described, because, as I understand it, that relies on the fact that PDGF is a polymeric molecule.

Hunter: There's no direct evidence that it relies on PDGF being a dimer. It is a nice model but I think there's no evidence that one growth factor dimer binds independently the two members of the receptor dimer. It is clear that EGF, another dimeric growth factor, causes dimerization, which is presumably mediated through receptor–receptor interactions.

Waterfield: It is also very easy to oligomerize EGF.

Verma: But doesn't the EGF receptor dimerize before EGF binds to it?

Waterfield: The current model suggested by Jossi Schlessinger and colleagues is that dimerization is induced by the ligand (Schlessinger 1988). There is also evidence that one molecule of EGF binds to one molecule of receptor. The EGF receptor is a multifunctional allosteric protein.

The EGF receptor external domain expressed in cells will not dimerize on ligand binding. This domain has only low affinity binding sites, however. We don't really understand the exact details of the dimerization mechanism; it may be that some part of the receptor other than the external domain is required for dimerization (Greenfield et al 1989).

Verma: So what is the evidence that EGF receptor needs to be dimerized in order for EGF to relay a signal? When I hear Schlessinger talk, it sounds as though dimerization is essential.

Sherr: There are a series of published reports from people other than Schlessinger that say the same thing (Böni-Schnetzler & Pilch 1987, Fanger et al 1989, Heldin et al 1989).

Hunter: There are also contradictory papers, for example from Roger Davis, who says that signal transduction from the EGF receptor does not need dimerization (Northwood & Davis 1988). I think it's not cast in stone, but personally, I believe that the dimer is the active signalling form of the receptor.

Sherr: We have done some experiments with some unusual chimaeras composed of the CD2 extracellular domain fused to the CSF-1 receptor kinase domain. Monoclonal antibodies to CD2, which activate T cell responsiveness, will activate the *fms* kinase in the these constructs. So one would argue that activation of the kinase is simply the result of aggregation by an antibody.

Hunter: But that doesn't always work. There are many anti-EGF receptor monoclonals that do not activate the kinase.

Sherr: There are many anti-*fms* monoclonals that don't activate the kinase, but it may simply be that they don't effectively cross-link the receptors.

Westermark: May I clarify one point about the PDGF receptor that relates to the paper by Heldin and collaborators (1989): they very clearly show that purified β-receptors for PDGF form dimers in solution when exposed to PDGF-BB and that dimerization occurs perfectly in parallel with the activation of the kinase.

Heath: Does the AB form of PDGF cause dimerization of receptor in those conditions?

Westermark: It does at higher concentrations.

Heath: What is your explanation for that?

Westermark: I think dimerization of the PDGF receptor may be caused both by the bivalence of the ligand and also by other changes within the receptor, for example conformational changes within the extracellular domain, perhaps also coupled to the specificity of the transmembrane domain.

Waterfield: The same experiment with the EGF receptor has shown that you can activate the receptor kinase in solution as a monomer, as judged by the kinetic studies. What happens in the membrane is another matter.

Sherr: How is one confident that one has a monomer in solution?

Waterfield: We just measure enzyme concentration and activity and extrapolate.

Verma: Couldn't you allow the ligand to bind to the receptor, freeze it and then do immunogold labelling, for example?

Hunter: I think it's clear that dimers form in the intact cell, the question is whether dimerization is essential for signal transduction, or is it for instance, part of the mechanism for down-regulation.

Waterfield: What would be the definitive experiment to show that formation of the dimer is necessary for signal transduction?

Sherr: I am rather attracted by the experiments that Bengt Westermark presented (this volume). There you have a correlation between dimerization and predicted biological effects of heterodimeric and homodimeric growth factors. I don't see how you can interpret his data easily in another way. Bengt has converted an agonist to an antagonist.

Hunter: That says that the dimers can form and that this can sequester the other subunit, but it doesn't say that's the way it has to work to get the signal.

Sherr: You would accept that biological activity is critically dependent on kinase activity in all of these systems. I find it hard to argue that the purpose of dimerization is simply down-regulation. The most economical model based on the results with different PDGF isoforms is that it has the ability to form α- and β-receptor homodimers and heterodimers, which generates the spectrum of biological specificity that was discussed. Otherwise how would you get it?

Verma: You are right, but that's PDGF where a BB dimer has already formed. With EGF the situation is different.

Sherr: So you would accept that dimerization may be essential for signal transduction of the PDGF receptor but you would like to leave the EGF receptor open?

Hunter: It's a little teleological. The only reason you would want to postulate the existence of PDGF receptor heterodimers seems to me to be because of the AB form of PDGF. We don't really know that AB PDGF has a biological function. It's present in human platelets but porcine platelets don't contain AB, is that right?

Waterfield: We don't know about AB, we only know that they have BB.

Westermark: A little while ago, we didn't know that human platelets had the BB form of PDGF. Now we know that they have BB and I have just learned from Russell Ross in Seattle (personal communication) that human platelets also have the AA form. So my guess is that pig platelets have all three isoforms. We know that cultured porcine endothelial cells have the AA form.

Waterfield: I am sure there are differences between pig and human platelets; however, these may be due to the methodology. It was very easy to get fresh pig platelets and the purification of the BB form of PDGF may relate to this or to the conditions used during purification, which are slightly different to those used to purify human PDGF (Stroobant & Waterfield 1984).

Mechanistically we already know that, for example, the association of other proteins with the PDGF receptor dimer may be very different from with the EGF receptor. We can't see all the same proteins associated in the EGF receptor complex; presumably there are different functional pathways used by these two different receptors (unpublished work).

Sherr: I agree that there are different pathways used by the receptors, but I don't think that speaks to the point. I think the specificity is built into the kinase domain. The domain swapping experiments, where you take an extracellular domain of receptor X and put it on the kinase domain of receptor Y and the product has the specificity of receptor Y, tell you that. The specificity is a function of the receptor domain inside the cell. The issue here is what kind of physical change is necessary to activate the kinase and whether that can occur through some push-pull mechanism involving monomers or whether you actually need to generate some form of leverage by forming dimers. I had a distinct impression that over the last few years the emerging evidence is coming more and more to favour the formation of dimers in all the systems that have been examined. I don't know what people would accept as proof, but the data with the EGF receptor, in spite of some conflicting reports in the literature, are generally consistent with what we heard from Bengt Westermark.

Hunter: It may not be possible to answer this until we know the contact sites in the dimer. Then if you mutate those residues and concomitantly block activation, with total concordance between the two, to me that would prove dimerization is essential for activation.

Sherr: There are other points that may speak to the issue of receptor aggregation. For example, the bovine papillomavirus (BPV) E5 protein has been reported to exert its transforming function indirectly by having an effect on the EGF receptor or the CSF-1 receptor (Martin et al 1989). It will be interesting to see whether or not the indirect effects of proteins like BPV E5 or the converse effects of the adenovirus E3 protein (Gooding et al 1988) are mediated by either enhancing or interfering with receptor aggregation. It's another way to approach the problem.

Harlow: I don't think they know that BPV E5 works indirectly. Doug Lowy's group from NIH have not been able to show that there is an interaction between E5 and receptors. The question almost can't be asked: there are few antibodies against E5 and E5 is very small. All the antibodies recognize parts that would be needed for interaction with a receptor.

Sherr: What I am trying to suggest is that the biological systems that are emerging may provide alternative ways of getting at the same question. In other words, a study of BPV E5 might show that receptor aggregation is important for kinase activation. Similarly, in a study of c-*fms* activating mutants, where the activating mutation is in the extracellular domain, one could investigate by cross-linking experiments whether the mutations affect the monomer–dimer equilibrium.

Hunter: It has been done for *erbB2/neu*, where the activating mutation is in the transmembrane domain.

Sherr: And the result claimed is that the activating mutation favours dimer formation.

Hunter: That is the published result from Mark Greene's lab (Weiner et al 1989), but several other groups have tried very hard and failed to see an increase in dimer formation. I don't know that that case is solved. For *neu* it certainly is not a very dramatic effect; it isn't as though 100% becomes dimer, only a very small fraction of the receptor is dimeric in those experiments.

Brugge: Does the CSF-1 receptor dimerize?

Sherr: We don't know. Chemical cross-linking of these receptors is very difficult. It requires a high receptor number, and may require solubilization of the purified receptor at high concentration in order to do the experiment. My view is that the cross-linking reagents aren't really powerful enough to analyse this critically in membranes where there are not two million receptors per cell. They won't work if you have 50 000 receptors, which, biologically speaking, is a good healthy number. Those are the experiments that provide chemical proof; everything else is indirect.

Brugge: Can you use non-denaturing gels, glycerol gradients or sucrose gradients to assay for dimerization?

Sherr: In glycerol gradients, the receptors sediment as high molecular weight forms. I remind you that others are claiming that the receptors associate with other molecules—*raf* protein, PI-3′ kinase, phospholipase Cγ—and these associations may also involve equilibria which give high molecular weight forms of receptor. Until you can identify what's in the complex, show that the ligand is present, show the stoichiometry and demonstrate by chemical cross-linking under denaturing conditions what you have, you can't really interpret the experiment.

Hunter: There are better cross-linking reagents than people use conventionally, which would help.

Verma: Tony mentioned the binding site of EGF receptor. I thought Jossi Schlessinger had made a chicken–human chimaeric receptor and found that the human EGF bound much better to the human receptor than to the chicken receptor, so they know which sequences constitute the binding site.

Hunter: Yes, Jossi has partially defined the EGF binding site as what he would call subdomain 3, which lies between the two cysteine-rich boxes (Lax et al 1988, 1989).

Waterfield: The problem is how to define a binding site. Until we have a three-dimensional structure for the protein, we can't fully understand the binding: that's true for immunoglobulins and other enzymes. The results Jossi Schlessinger and colleagues have from nice experiments are on the difference between EGF and TGF-α binding to human and chicken receptors, and from domain swapping experiments which show that the high affinity binding follows a particular domain (Lax et al 1989). That domain is outside the cysteine-rich domains.

Shall I just clarify the structure of the human receptor? From the N-terminus, there's a large domain called L1 followed by three small cysteine-rich domains, then a second large domain, L2, which is cysteine free, followed by three more cysteine-rich domains. L2 is the domain associated with the acquisition of high affinity binding of human EGF by the chicken receptor. Schlessinger's group have also done cross-linking experiments which show that within that same domain you can cross-link the ligand to the receptor (Lax et al 1988).

Hunter: Also antibodies that block EGF binding regulate that same region. Jossi Schlessinger has some data on N-terminal deletions that affect EGF binding, which suggest that L1 is involved (personal communication).

Verma: Just a small twist to this. In these domain swapping experiments, how important is the transmembrane domain?

Sherr: There was a claim for some specificity in the transmembrane region (Escobedo et al 1988). Rusty now admits that those results are probably not correct and the experiments are being repeated.

Hunter: For the EGF receptor there are quite extensive deletion and mutation analyses, which suggest that you can alter the transmembrane region considerably and still have a functional receptor—you can shorten it, put inserts into it, put some charged residues into it and retain a functional receptor (Kashles et al 1988). That needs to be done for other receptors.

Verma: I thought that in the oncogenic *neu*-encoded receptor the only change from the normal receptor was in the transmembrane domain.

Hanley: Several people have tested whether the mutated *neu* transmembrane sequence can function as a sort of universal activator of single transmembrane segment-type receptors, but it appears to activate only *neu* (Williams et al 1988).

Hunter: A similar mutation doesn't activate the EGF receptor (Kashles et al 1988).

Waterfield: In the PDGF receptor there is remarkable conservation, which is almost never seen amongst families of transmembrane sequences, between species, whereas within transmembrane domains from the two PDGF receptors from the same species there is not so much conservation (Claesson-Welsh et al 1988).

Sherr: Could the homology reflect, for example, the fact that the exon which encodes the transmembrane domain also encodes the juxtamembrane portion of the receptor? There is a high degree of conservation of the N-terminal juxtamembrane portion between the α and β subunits of the receptor.

Westermark: We haven't studied the gene.

References

Böni-Schnetzler M, Pilch PF 1987 Mechanism of epidermal growth factor receptor autophosphorylation and high affinity binding. Proc Natl Acad Sci USA 84:7832–7836

Claesson-Welsh L, Eriksson A, Morén A et al 1988 cDNA cloning and expressing human platelet-derived growth factor (PDGF) receptor specific for B-chain containing molecules. Mol Cell Biol 8:3476–3486

Escobedo JA, Barr PJ, Williams LT 1988 Role of tyrosine kinase and membrane spanning domains in signal transduction by the platelet derived growth factor receptor. Mol Cell Biol 8:5126–5131

Fanger BO, Stephens JE, Staros JV 1989 High-yield trapping of EGF- induced receptor dimers by chemical cross-linking. FASEB (Fed Am Soc Exp Biol) J 3:71–75

Fey MF, Hesketh C, Wainscoat JS, Gendler S, Thein SL 1989 Clonal allele loss in gastrointestinal cancers. Br J Cancer 59:750–754

Florkiewicz RZ, Sommer A 1989 Human basic fibroblast growth factor gene encodes four polypeptides: three initiate translation from non-AUG codons. Proc Natl Acad Sci USA 86:3978–3981

Gooding LR, Elmore LW, Tollefson AE, Brady HA, Wold WSM 1988 A 14,700 MW protein from the E3 region of adenovirus inhibits cytolysis by tumour necrosis factor. Cell 53:341–346

Greenfield C, Miles I, Waterfield MD et al 1989 Epidermal growth factor binding induces a conformational change in the external domain of its receptor. EMBO (Eur Mol Biol Organ) J 8:4115–4125

Heldin CH, Ernlund A, Rorsman C, Rönnstrand L 1989 Dimerization of B-type PDGF receptors occurs after ligand binding and is closely associated with receptor kinase activation. J Biol Chem 264:8905–8912

Kashles O, Szapary D, Bellot F, Ullrich A, Schlessinger J, Schmidt A 1988 Ligand induced stimulation of epidermal growth factor receptor mutants with altered transmembrane regions. Proc Natl Acad Sci USA 85:9567–9571

Lax I, Burgess WH, Bellot F, Ullrich A, Schlessinger J, Givol D 1988 Localization of a major receptor-binding domain for epidermal growth factor by affinity labelling. Mol Cell Biol 8:1831–1834

Lax I, Bellot F, Howk R, Ullrich A, Givol D, Schlessinger J 1989 Functional analysis of the ligand binding site of EGF receptor utilizing chimeric chicken/human receptor molecules. EMBO (Eur Mol Biol Organ) J 8:421–427

Martin P, Vass WC, Schiller JT, Lowy DR, Velu TJ 1989 The bovine papillomavirus E5 transforming protein can stimulate the transforming activity of EGF and CSF-1 receptors. Cell 59:21–32

Northwood IC, Davis RJ 1988 Activation of the epidermal growth factor receptor tyrosine protein-kinase in the absence of receptor oligomerization. J Biol Chem 263:7450–7453

Paterno GD, Gillespie L, Dixon M, Slack JMW, Heath JK 1989 Mesoderm inducing potency of oncogene encoded growth factors K-FGF and int-2. Development 106:79–84

Schlessinger J 1988 The epidermal growth factor receptor is a multifunctional allosterioprotein. Biochemistry 27:3119–3123

Stroobant P, Waterfield MD 1984 Purification and properties of porcine platelet-derived growth factor. EMBO (Eur Mol Biol Organ) J 3:2963–2967

Tsuda H, Hirohashi S, Shimosato Y et al 1989 Correlation between long-term survival in breast cancer patients and amplification of two putative oncogene coamplification units: *hst*-1/*int*-2 and c-*erb*B-2/*ear*-1. Cancer Res 49:3104–3108

Wada M, Yokota J, Mizoguchi H, Sugimura T, Terada M 1988 Infrequent loss of chromosomal heterozygosity in human stomach cancer. Cancer Res 48:2988–2992

Weiner DB, Liu J, Cohen JA, Williams WV, Greene MI 1989 A point mutation in the neu oncogene mimics ligand induction of receptor aggregation. Nature (Lond) 339:230–231

Westermark B, Claesson-Welsh L, Heldin C-H 1990 Structural and functional aspects of platelet-derived growth factor and its receptors. In: Proto-oncogenes in cell development. Wiley, Chichester (Ciba Found Symp 150) p 6–22

Williams LT, Escobedo JA, Keating MT, Coughlin SR 1988 Signal transduction by the platelet-derived growth factor receptor. Cold Spring Harbor Symp Quant Biol 53:455–465

Yokota J, Yamamoto T, Miyajima N et al 1988 Generic alterations of the c-*erb*B-2 oncogene occur frequently in tubular adenocarcinoma of the stomach and are often accompanied by amplification of the v-*erb*A homologue. Oncogene 2:283–287

Growth factor-like action of lysophosphatidic acid: mitogenic signalling mediated by G proteins

W. H. Moolenaar and E. J. van Corven

Division of Cellular Biochemistry, The Netherlands Cancer Institute, Plesmanlaan 121, 1066 CX Amsterdam, The Netherlands

Abstract. Several classes of growth factors can be distinguished that act through different signal transduction pathways. One class is constituted by the peptide growth factors that bind to receptors with ligand-dependent protein tyrosine kinase activity. Another class of mitogens activates a phosphoinositide-specific phospholipase C via a receptor-linked G protein. An intriguing member of this class is lysophosphatidic acid (LPA). LPA mitogenicity is not dependent on other mitogens and is blocked by pertussis toxin. LPA evokes at least three separate signalling cascades: (i) activation of a pertussis toxin-insensitive G protein mediating phosphoinositide hydrolysis; (ii) release of arachidonic acid in a GTP-dependent manner, but independent of prior phosphoinositide hydrolysis; and (iii) activation of a pertussis toxin-sensitive G_i protein mediating inhibition of adenylate cyclase. The peptide bradykinin mimics LPA in inducing responses (i) and (ii), but fails to activate G_i and to stimulate DNA synthesis. Our results suggest that the mitogenic action of LPA occurs through G_i or a related pertussis toxin substrate and that, unexpectedly, the phosphoinositide hydrolysis pathway is neither required nor sufficient, by itself, for mitogenesis.

1990 Proto-oncogenes in cell development. Wiley, Chichester (Ciba Foundation Symposium 150) p 99–111

The molecular mechanisms by which growth factors simulate cell proliferation are poorly understood. At least two classes of mitogens can be distinguished that utilize different signalling pathways. A major class is represented by the family of peptide growth factors that bind to receptors with intrinsic protein tyrosine kinase activity, the prototype of which is epidermal growth factor (EGF). Ligand-stimulated receptor tyrosine kinase activity is essential, but not necessarily sufficient, for the multiple cellular responses to EGF (Chen et al 1987, Moolenaar et al 1988). Another class of mitogens activates a phosphoinositide-specific phospholipase C, through the participation of a GTP-binding protein (Cockcroft 1987), to mobilize Ca^{2+} from intracellular stores and to activate protein kinase C. Although a direct cause–effect relation between the phospholipase C pathway and the eventual initiation of DNA synthesis has not been established, it is widely assumed that receptor-mediated phospho-inositide hydrolysis is necessary and sufficient, by itself, for this class of mitogens to elicit a proliferative response.

An intriguing growth stimulus, apparently belonging to the class of Ca^{2+}-mobilizing mitogens, is the simple phospholipid, phosphatidic acid (PA) (Moolenaar et al 1986). PA is normally produced in stimulated cells, but it is also capable of exerting its own biological effects (Murayama & Ui 1987). In particular, exogenous PA stimulates DNA synthesis in fibroblasts, an effect attributed to its ability to activate phospholipase C (Moolenaar et al 1986). We recently found that lysophosphatidic acid (LPA) is even more mitogenic than PA (van Corven et al 1989). At least three separate signal transduction cascades in the action of LPA have been identified. This chapter will briefly describe the nature of these signalling pathways.

Signalling pathways in the action of LPA

Our previous work has raised the question of how PA exerts its growth promoting effects when added to quiescent fibroblasts (Moolenaar et al 1986). From studies on A431 epidermoid carcinoma cells, where PA triggers the formation of inositol phosphates and the release of intracellularly stored Ca^{2+}, it was proposed that exogenous PA acts by increasing phosphoinositide turnover with consequent production of second messengers which stimulate growth. Whatever the precise mechanism underlying PA stimulation of the phospho-inositide pathway, this working hypothesis of PA action would seem plausible in view of the popular belief that activation of the phosphoinositide signalling cascade is sufficient to lead to enhanced DNA synthesis in responsive cells.

Several striking observations concerning PA mitogenic activity were made by van Corven et al (1989). First, removal of one fatty acid chain at the 2-position, which yields LPA, results in a considerable enhancement of mitogenic potency when tested on quiescent fibroblasts (Fig. 1). The effect of LPA is specific in that other lysophospholipids tested, including those with weak detergent-like properties such as lysophosphatidyl choline, were found to lack any mitogenic activity.

Second, LPA mitogenicity is selectively blocked by pertussis toxin (half-maximal inhibition at $0.2\,ng/ml$), whereas the mitogenic response to EGF is unaffected; pertussis toxin ADP-ribosylates a 40–41 kDa protein substrate in the fibroblasts tested (van Corven et al 1989).

Third, LPA activates a subclass of G proteins to initiate at least three independent second messenger cascades. In summary: (i) using a permeabilized cell system it is possible to show that LPA activates phospholipase C in a GTP-dependent manner and that this effect of LPA is blocked by GDPβS, indicating the involvement of a G protein (putative G_p). As a consequence, LPA stimulates

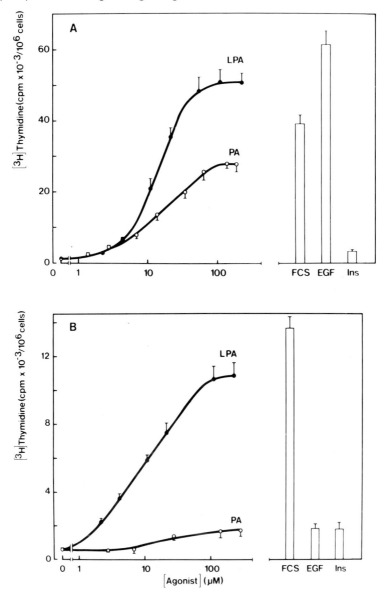

FIG. 1. DNA synthesis induced by phosphatidic acid and lysophosphatidic acid. Confluent cultures of Rat-1 (A) or diploid human skin fibroblasts (B) were stimulated with the indicated concentrations of LPA (1-oleoyl) or PA (1,2-dioleoyl) obtained from commercial sources (purity >95%). 24 hours after addition of phospholipid, cultures were assayed for [^{3}H] thymidine incorporation. Right part shows DNA synthesis induced by fetal calf serum (10%), epidermal growth factor (10 ng/ml) and insulin (5 µg/ml).

mobilization of Ca^{2+} from intracellular stores and causes activation of protein kinase C. (ii) LPA stimulates the formation of free arachidonic acid in a GTP-dependent manner, presumably through the action of phospholipase A_2; and (iii) LPA inhibits the accumulation of cAMP. Only the latter response is abolished by pertussis toxin, indicating that the decrease in cAMP is mediated by the inhibitory G protein, G_i, which inhibits adenylate cyclase. The phosphoinositide response to LPA is completely blocked by phorbol ester, whereas the arachidonic acid response is inhibited by only 30%. We therefore conclude that arachidonic acid release is not simply a result of phosphoinositide metabolism. Results obtained from permeabilized cells suggest that LPA-induced liberation of arachidonic acid may be mediated, at least in part, by a pertussis toxin-insensitive G protein that activates phospholipase A_2. Activation of phospholipase A_2 via a pertussis toxin-insensitive G protein has previously been reported to occur in bradykinin-stimulated Swiss 3T3 cells (Burch & Axelrod 1987). Although this hypothesis is the simplest one, our data do not eliminate the formal possibility that LPA could liberate arachidonate from diacylglycerol, produced by hydrolysis of phospholipids other than phosphoinositide. In this regard, there is preliminary evidence suggesting that LPA, like other Ca^{2+}-mobilizing agonists, is able to promote hydrolysis of phosphatidyl choline in its target cells (W. van Blitterswijk et al, personal communication).

The important question arises as to which, if any, of these biochemical responses accounts for the mitogenic effect of LPA. Quite unexpectedly, and contrary to our previous hypothesis, it seems unlikely that activation of the phosphoinositide hydrolysis–Ca^{2+} protein kinase C pathway is of major importance in mitogenesis. When phosphoinositide hydrolysis is blocked by phorbol ester there is no effect on LPA-induced DNA synthesis. Furthermore, when protein kinase C activity is functionally removed by long-term treatment with TPA (12-O-tetradecanoylphorbol 13-acetate), LPA is still fully mitogenic. It thus appears that LPA does not rely on the phosphoinositide hydrolysis–protein kinase C cascade to stimulate cell proliferation. Furthermore, we found that bradykinin mimics LPA in stimulating phosphoinositide hydrolysis and arachidonic acid release, at least qualitatively, but fails to elicit a detectable mitogenic response (van Corven et al 1989). The latter finding further argues against the view that activation of the phospholipase C cascade is sufficient for mitogenesis, at least in previously quiescent fibroblasts.

It is also unlikely that LPA-stimulated arachidonate release is important for the mitogenic action of LPA. First, addition of exogenous arachidonate or prostaglandins to quiescent cells does not stimulate DNA synthesis. Second, the estimated dose of LPA for half-maximal arachidonate release is at least an order of magnitude greater than that for mitogenesis. In this regard, low concentrations of LPA (approximately 5 µM) can elicit a significant proliferative response, while arachidonate release is barely detectable (van Corven et al 1989).

human fibroblasts

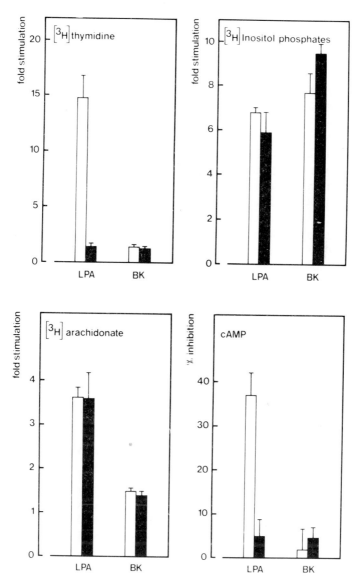

FIG. 2.   Second messenger pathways generated by lysophosphatidic acid (70 μM) and bradykinin (1 μM) in human skin fibroblasts in the presence (black bars) or absence (open bars) of 100 ng/ml pertussis toxin. For measurements of inhibition of cAMP accumulation, cells were first stimulated with prostaglandin E_1 (1 μM) for 10 minutes to raise intracellular cAMP levels severalfold above control values. For further experimental details see van Corven et al (1989).

The results obtained with pertussis toxin strongly suggest that G_i, the inhibitory G protein of adenylate cyclase, is critical for LPA-induced mitogenesis. This notion is further supported by the finding that the non-mitogenic peptide bradykinin, which activates G_p but not G_i, fails to stimulate DNA synthesis. Interestingly, Seuwen et al (1988) recently inferred a similar important role for G_i from their studies on serotonin-induced mitogenesis. Fig. 2 illustrates the various signalling pathways in the action of LPA that have been identified.

How could activation of the G_i pathway with subsequent inhibition of adenylate cyclase lead to DNA synthesis? cAMP has long been known as a modulator of cell growth in many cell systems (Pastan et al 1975, Rozengurt 1986). In normal fibroblasts, including Rat-1 and human fibroblast cells, agents that raise intracellular cAMP levels exert a marked growth-inhibitory effect (Heldin et al 1989, A. Ladoux and W. H. Moolenaar, unpublished observations). Conversely, it seems reasonable to assume that a reduction in cellular cAMP may have a growth-promoting effect in these cells. Alternatively, G_i or a related member of the pertussis toxin-sensitive G_i family could interact with an as yet unidentified effector system that is critical for mitogenesis. Finally, the intriguing possibility exists that in addition to G_p and G_i, the GTP-binding p21 *ras* protein may be activated by LPA. *ras* proteins are related to the α subunit of signal-transducing G proteins and play an important role in the action of growth factors; in particular, PA-induced mitogenesis is completely blocked by a neutralizing anti-*ras* antibody (Yu et al 1988). Recently, Tsai et al (1989) reported that the GTPase activity of cellular p21 *in vitro* could be specifically inhibited by some, but not all, phosphatidic acids. Whether LPA may have a similar action, not only *in vitro* but also in intact cells, remains to be investigated. Fig. 3 illustrates a hypothetical scheme of action of LPA.

Site of action and possible biological function of LPA

A major question that emerges concerns the mechanism by which extracellular LPA acutely activates certain G proteins at the inner face of the plasma membrane. One possibility is that LPA dissolves in and perturbs the lipid bilayer in such a way that G proteins are selectively activated. So far, however, there are no data to indicate how the structure and composition of the lipid bilayer may affect G protein function. An alternative and perhaps more attractive hypothesis is that LPA binds to and directly activates a specific cell surface receptor that couples to G proteins. We are currently testing this hypothesis.

Whatever the precise site of action of LPA may be, the results presented here indicate that this simple phospholipid may serve as a potent tool to dissect the role of G protein-mediated signalling pathways in different cell systems, particularly those involved in mitogenesis. Some biological effects of LPA other than cell proliferation have been reported previously, namely platelet

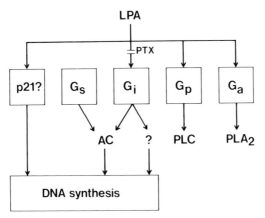

FIG. 3. Hypothetical scheme of signal transduction pathways in the action of lysophosphatidic acid. PTX, pertussis toxin; p21, GTP-binding *ras* proteins; G_s and G_i, stimulatory and inhibitory G proteins regulating adenylate cyclase (AC) activity; G_p, G protein controlling phospholipase C (PLC) activity; G_a, putative G protein controlling phospholipase A_2 (PLA$_2$) activity.

aggregation and smooth muscle contraction, but the underlying mechanisms have not been investigated (Vogt 1963, Benton et al 1982).

Finally, these novel results, together with the fact that LPA and PA are rapidly generated in stimulated cells, raise the intriguing possibility that these low-abundance phospholipids may have a function in signal transduction as agonists or as second messengers. A critical test of this hypothesis requires elucidation of the precise site of action of LPA.

Acknowledgements

We thank Paulien Sobels for preparing the manuscript. This work was supported by the Netherlands Cancer Foundation (Koningin Wilhelmina Fonds).

References

Benton AM, Gerrard JM, Michiel T, Kindom SE 1982 Are lysophosphatidic acids involved in stimulus activation coupling in platelets? Blood 60:642–649

Burch RM, Axelrod J 1987 Dissociation of bradykinin-induced prostaglandin formation from PI turnover in Swiss 3T3 fibroblasts: evidence for G protein regulation of phospholipase A$_2$. Proc Natl Acad Sci USA 84:6374–6378

Chen WS, Lazar CS, Poenie M, Tsien RY, Gill GN, Rosenfeld M 1987 Requirement for intrinsic protein tyrosine kinase in the immediate and late actions of the EGF receptor. Nature (Lond) 328:820–823

Cockroft S 1987 Polyphosphoinositide phosphodiesterase: regulation by a novel guanine nucleotide binding protein. Trends Biochem Sci 12:75–78

Heldin NE, Paulsson Y, Forsberg K, Heldin CH, Westermark B 1989 Induction of cAMP synthesis by forskolin is followed by inhibition of ^{3}H-thymidine incorporation in human fibroblasts. J Cell Physiol 138:17–23

Moolenaar WH, Kruijer W, Tilly BC, Verlaan I, Bierman AJ, de Laat SW 1986 Growth factor-like action of phosphatidic acid. Nature (Lond) 323:171–173

Moolenaar WH, Bierman AJ, Tilly BC et al 1988 A point mutation at the ATP-binding site of the EGF-receptor abolishes signal transduction. EMBO (Eur Mol Biol Organ) J 7:707–710

Murayama T, Ui M 1987 Phosphatidic acid may stimulate membrane receptors mediating adenylate cyclase inhibition and phospholipid breakdown in 3T3 fibroblasts. J Biol Chem 262:5522–5529

Pastan I, Johnson GS, Anderson WB 1975 Role of cyclic nucleotides in cell growth. Annu Rev Biochem 44: 491–522

Rozengurt E 1986 Early signals in the mitogenic response. Science (Wash DC) 234:161–166

Seuwen K, Magnaldo I, Pouysségur J 1988 Serotonin stimulates DNA synthesis in fibroblasts acting through 5-HT$_{1B}$ receptors coupled to a G$_i$ protein. Nature (Lond) 335:254–256

Tsai M-H, Yu C-H, Wei F-S, Stacey DW 1989 The effect of GTPase activating protein upon ras is inhibited by mitogenically responsive lipids. Science (Wash DC) 243:522–526

van Corven EJ, Groenink A, Jalink K, Eichholtz T, Moolenaar WH 1989 Lysophosphatidate-induced cell proliferation; identification and dissection of signalling pathways mediated by G proteins. Cell 59:45–54

Vogt W 1963 Pharmacologically active acidic phospholipids and glycolipids. Biochem Pharmacol 12:415–420

Yu C-L, Tsai M-H, Stacey DW 1988 Cellular *ras* activity and phospholipid metabolism. Cell 52:63–71

DISCUSSION

Knopf: Is there any concentration of bradykinin that is mitogenic?

Moolenaar: No. Bradykinin is not very stable so we tried re-adding it at short intervals but it had no effect. This is not an exclusive property of bradykinin. For instance, histamine is also a potent inducer of Ca^{2+} mobilization and protein kinase C activation in fibroblasts and it is not mitogenic, unless you add a sub-optimal concentration of a growth factor that acts through the tyrosine kinase pathway, such as EGF.

Hanley: Angiotensin (Hanley et al, this volume) clearly does 'something else', besides stimulate inositol lipid hydrolysis through phospholipase C. It is very comparable to your LPA story. Phospholipase C may be necessary but not sufficient for mitogenesis.

In relation to the question of where PA comes from, there is a lot of interest among people working on signal transduction mechanisms in phosphatidyl choline hydrolysis.

Moolenaar: I thought the evidence was that the highly purified phospholipase C that breaks down phosphoinositides does not hydrolyse phosphatidyl choline.

Hanley: I don't think there is evidence for a ubiquitous enzyme with phospholipase C activity specific for phosphatidyl choline, such as that accepted for inositol lipids. The difficulty in concluding that such an enzyme exists in a given cell type comes from the lipid interconversions you describe. For example, diacylglycerol, the product of phospholipase C attack on phosphatidyl choline, is rapidly converted to PA, the product of phospholipase D, by diacylglycerol kinase. Similarly, PA can be converted to diacylglycerol by phosphatidic acid phosphohydrolase.

There is a unique marker activity for phospholipase D, an interesting transphosphatidylation activity, which can use alcohols rather than water. If you add ethanol, phospholipase D action will make the unnatural lipid, phosphatidyl ethanol (Liskovitch 1989). In our hands, in several cell types, including NG115-401L cells, the only form of receptor-regulated phosphatidyl choline metabolism is a phospholipase D, which appears to be externally orientated. It produces PA at the cell surface after activation of more than one intracellular pathway, but particularly protein kinase C (Hanley et al 1988).

What we appear to be observing is an inside-out transduction system which takes cytosolic messages, for example increased activity of protein kinase C, and communicates them to the cell surface. Why should there be such a system? It produces PA in the outer leaflet of the membrane, so there is a potential physiological system for producing PA, which you have shown to have growth factor activity (Moolenaar et al 1986). I don't think PA comes from serum, I think it is produced *in situ*; the cells have an autocrine mechanism to generate this lipid regulator as part of a self-amplification system. This may not seem to make much biological sense, but neutrophil activation by platelet-activating factor appears to require an autocrine loop, involving the production of thromboxane A2, which amplifies the initial response through its own cell surface receptors acting on the common inositol lipid signalling machinery (Hanley & Downes 1987).

Moolenaar: We tested several cell systems to see whether this is a universal phenomenon. Neutrophils are not activated by LPA; mast cells do not secrete their vesicle contents in response to LPA. Also, peripheral blood lymphocytes do not show a mitogenic response to LPA. So, perhaps cells of the immune system do not respond to LPA, whereas other cell types do. These observations support the idea of a receptor-mediated signalling cascade, found specifically in a limited number of cell types.

Knopf: Your LPA is similar to the precursor for plasminogen activating factor (PAF), have you looked at that?

Moolenaar: That was our first candidate. We tested PAF and PAF antagonists with negative results.

Sherr: What was the basis for the hypothesis that the G_i protein couples to another effector?

Moolenaar: When we add a well defined hormone, adrenaline, that couples through G_i and lowers cAMP by itself, we cannot mimic the response to LPA. So, activation of G_i alone (the G_i that couples to adenyl cyclase) cannot achieve the same effect as LPA. I am still ready to believe that lowering cAMP by itself could be a mitogenic signal; for instance, in the induction of oocyte maturation by progesterone, cAMP is a sufficient stimulus for germinal vesicle breakdown.

Vande Woude: Have you investigated whether LPA will induce maturation of *Xenopus* oocytes?

Moolenaar: First we wanted to see whether the oocytes have endogenous LPA receptors. They have, when you add LPA to oocytes, there is a significant calcium-activated chloride current, so oocytes cannot be used as a transient expression system for a putative LPA receptor.

We then added LPA to immature oocytes in an attempt to induce maturation, but our preliminary results are inconclusive.

Sherr: Could you comment on effects described in the literature that cAMP can interfere with mitogenic effects in several systems.

Moolenaar: In some cells, for example Swiss 3T3 cells, cAMP is a mitogen. But in most cells, including diploid normal fibroblasts, raising cAMP concentrations completely blocks mitogenesis induced by all growth factors, including EGF and LPA. The site of action is not known, but it must be downstream of all these effects that I have described.

Land: In Swiss 3T3 cells where cAMP is a mitogen, is LPA mitogenic?

Moolenaar: Yes, it is. In our hands it is more potent than bombesin.

Feramisco: Would you comment on the reports from Dennis Stacey's lab concerning the effect of PA on GTP activating protein (GAP) and *ras*, and what you mentioned about the sensitivity of *ras*-transformed cells to LPA.

Moolenaar: They published that PA and related lipids inactivate GAP and thereby activate normal *ras* proteins. We used Rat-1 cells transfected with normal H-*ras*. Although you cannot make them fully quiescent, we found that they are more sensitive to LPA, in that the dose–response curve is shifted to the left, but not by an order of magnitude, it is perhaps a factor of two. Whether that is significant I don't know, because these cells produce their own growth factors and all kinds of synergies are possible. We are going to look at GAP activity more directly to see whether lysophospholipids can bring about the same effect.

Feramisco: We have tried numerous experiments injecting the catalytic subunit of cAMP-dependent protein kinase (A- kinase) and *ras* protein into cells to see if the subunit suppresses growth activation by *ras*. In Rat cells at least, there doesn't seem to be any effect of loading a cell with catalytic subunit of A-kinase, *ras* still activates growth.

Hunter: The fact that LPA appears to affect several different G proteins might imply that it interacts with the receptors that couple to these proteins and in some way potentiates their activity, perhaps even in the absence of ligand.

Moolenaar: It also could be that with LPA we activate a receptor for which

LPA is not the natural ligand. Perhaps an LPA-like structure remains to be discovered. The concentrations of LPA that we are using are fairly high, half maximal concentrations are in the micromolar range and that is not the usual affinity for hormone receptors.

Hunter: What is the likely concentration of LPA in the serum?

Moolenaar: I don't know. There is a lot of phosphatidyl choline in the serum.

Hanley: The precursor lipids are generally thought to be bound to protective carrier molecules and are not really accessible. The active molecules like LPA are turned over rapidly and there are only very low amounts in serum.

Hunter: Presumably there are non-hydrolysable analogues of LPA?

Moolenaar: Yes, we have tried ether lipids and our preliminary results suggest that non-hydrolysable ether-linked LPA exerts a significant mitogenic effect.

Hunter: Have you tested whether LPA affects protein kinase C directly *in vitro*?

Moolenaar: Yes, we have and it doesn't. That's not a novel result. J. F. Kuo and colleagues have looked at the direct effects of lysophospholipids on the activity of protein kinase C (Oishi et al 1988). The lysolipid that has some effect is lysophosphatidyl choline.

Hunter: How many binding sites for LPA are there per cell?

Moolenaar: We don't know yet.

Knopf: In your binding experiments, can you see some cell types that have fewer receptors?

Moolenaar: Yes, we use K562 leukaemic cells as negative controls, because they do not respond mitogenically, and do not give a calcium signal. They do not have specific binding sites, all the binding is non- specific.

Brugge: Does LPA affect the GTP binding of $G_{i\alpha}$ itself?

Moolenaar: We looked at GTPγS binding. The effects are very small, we have not yet found the right conditions to measure GTP binding.

Hunter: To come back to the potential G_i that might be involved, pertussis toxin ADP-ribosylated at least two bands. Have you analysed those on two-dimensional gels to see how many species there are?

Moolenaar: Not yet.

Hunter: I suppose it's a question of whether LPA activates a novel $G_{i\alpha}$ or whether it is a conventional $G_{i\alpha}$ simply coupled to a novel effector.

Moolenaar: My personal bias would be that it is a novel effector.

Hunter: Did you investigate whether LPA can induce phosphorylation of phosphatidylinositol 3-phosphate (PI-3-P)?

Moolenaar: No, we didn't. I don't see how PI-3-kinase could ever give rise to an interesting second messenger.

Hanley: I agree. There are suggestions that other inositol lipids, such as phosphatidylinositol 3,4-bisphosphate (PI-3,4-P_2) or phosphatidylinositol 3,4,5-trisphosphate (PI-3,4,5-P_3), are the real informational products (Traynor-Kaplan et al 1989).

Moolenaar: Michael Berridge has proposed that the situation is very similar to what happens in glucose metabolism, where there are two types of phosphofructokinase. The second, phosphofructokinase II, acts by fine regulating the activity of phosphofructokinase I.

Hanley: The idea is that it is a regulatory shunt, PI-3-P acts as an allosteric regulator of other enzymes rather than as a second messenger. Making the analogy with glucose metabolism, PI-3 kinase activity could be coincident with mitogenic events, not causal. However, the role of PI-3 kinase has to be evaluated from several perspectives; that the enzyme activity is informational, that the lipid products are informational or that the breakdown products are informational. No phospholipase C isoform has been shown to hydrolyse PI-3-P, PI-3,4-P_2 or PI-3,4,5-P_3, so the last notion is unlikely. Also, the resting levels of PI-3-P are very low, suggesting that it does not act as a signal generator. As Lewis Cantley has said, the low levels of this lipid may mean that it itself is the signal.

Hunter: It depends a little on how fast it turns over.

Hanley: In our hands it is extremely stable and we find neither the bisphosphate nor the trisphosphate.

Hunter: In vivo?

Hanley: Yes, in living 401L-C3 cells. The data that we are discussing on PI-3 kinase stimulation in PDGF receptor complexes occur rather distant in time from acute cell regulation. You immunoprecipitate a receptor complex and measure the regulation of PI-3-P production some hours after addition of the PDGF. Remember, that stable complex has been extracted from the cell with detergent, immunoprecipitated, taken through several washes and then assayed.

Hunter: But PI-3 kinase is activated very rapidly by growth factors such as PDGF (Auger et al 1989) and insulin. You can precipitate PI-3 kinase activity together with the PDGF receptor five minutes after treatment with PDGF (Coughlin et al 1989).

Hanley: That has been reported for only one or two cell types, such as PDGF stimulation of intact smooth muscle cells (Auger et al 1989), but the amount of PI- 3-P does not change much in intact cells, whereas PI- 3,4,5-P_3 changed dramatically and rapidly after stimulation with PDGF. We have not assayed rigorously whether PI-3 kinase is an inositol lipid-3 kinase or whether other enzymes are involved. In neutrophils stimulated with the chemotactic peptide, f-Met-Leu-Phe, new PI-3,4,5-P_3 can be detected within 45 seconds (Traynor-Kaplan et al 1989). The relationship of this to mitogenic peptide growth factors is unknown, although f-Met-Leu-Phe is thought to work through a G protein coupled to phosphoinositide hydrolysis, as with the *mas* gene product.

Hunter: This may be special to PDGF. Personally, I don't think PI-3-P has to be a precursor for a second messenger, because PI-3-P itself could be a second messenger. We are beginning to find more and more proteins that interact with phospholipids, such as GAP. Therefore, I don't think PI-3-P has to be hydrolysed in order to be an interesting molecule. No other system generates PI-3-P or the 3,4 or 3,4,5 phosphates, as far as I know.

Hanley: There is obviously a very interesting case that, so far, is uniquely associated with the PDGF receptor.

Hunter: I am not sure there is going to be any universal feature of mitogenesis.

References

Auger KR, Serunian LA, Soltoff SP, Libby P, Cantley LC 1989 PDGF-dependent tyrosine phosphorylation stimulates production of novel polyphosphoinositides in intact cells. Cell 57:167–175

Coughlin SR, Escobedo JA, Williams LT 1989 Role of phosphatidylinositol kinase in PDGF receptor signal transduction. Science (Wash DC) 243:1191–1194

Hanley MR, Downes CP 1987 Second messengers or second agonists? Nature (Lond) 330:319

Hanley MR, Cheung WT, Hawkins P et al 1990 The *mas* oncogene as a neural peptide receptor: expression, regulation and mechanism of action. In: Proto-oncogenes in cell development. Wiley, Chichester (Ciba Found Symp 150) p 23–46

Liskovitch M 1989 Phosphatidylethanol biosynthesis in ethanol-exposed NG108-15 neuroblastoma X glioma hybrid cells. J Biol Chem 264:1450–1456

Moolenaar W, Kruijer W, Tilly BC, Verlaan I, Bierman AJ, DeLaat SW 1986 Growth factor-like action of phosphatidic acid. Nature (Lond) 323:171–173

Oishi K, Raynor RL, Charp PA, Kuo JF 1988 Regulation of protein kinase C by lysophospholipids. Potential role in signal transduction. J Biol Chem 263:6865–6871

Traynor-Kaplan AF, Thompson BL, Harris AL, Taylor P, Omann G, Sklar LA 1989 Transient increase in phosphatidylinositol trisphosphate during activation of human neutrophils. J Biol Chem 264:15668–15673

Phospholipase C isozymes:
structural and functional similarities

Ronald Kriz*, Lih-Ling Lin*, Lisa Sultzman*, Christine Ellis†, Carl-Henrik Heldin°, Tony Pawson†, and John Knopf*

*Genetics Institute, 87 Cambridge Park Drive, Cambridge MA 02140, USA; †Division of Molecular and Developmental Biology, Mt. Sinai Hospital Research Institute, 600 University Avenue, Toronto, Ontario M5G 1X, Canada and °Ludwig Institute for Cancer Research, Uppsala Branch, Box 595, S-751 23 Uppsala, Sweden

Abstract. Phospholipase C (PLC) is shown to comprise at least nine isoforms. These isoforms can be separated into three structurally related classes. Within a class the isozymes have similar enzymological properties. In the case of the PLCγ class, both isoforms may be regulated by tyrosine phosphorylation. For $PLC\gamma_1$ we show that the tyrosine phosphorylation sites are contained within the SH2/SH3 region or 'modulatory domain'. The overexpression of $PLC\gamma_1$ in Rat-2 cells results in increased phosphatidylinositol breakdown in response to PDGF treatment, demonstrating that $PLC\gamma_1$ mediates this response. We note that thrombin activates $PLC\gamma_1$ in addition to other PLC isoforms.

1990 Proto-oncogenes in cell development. Wiley, Chichester (Ciba Foundation Symposium 150) p 112–127

The binding of various ligands to their specific cell surface receptors rapidly induces the formation of two second messenger molecules derived from phosphatidylinositol 4,5-bisphosphate: diacylglycerol and inositol 1,4,5-trisphosphate. The production of these second messenger molecules is mediated by activated phosphatidylinositol (PI)-specific phospholipase C (PLC) enzymes (Berridge 1987). Evidence has accumulated for the existence of a number of immunologically distinct enzymes with PI-specific PLC activities (Hofmann & Majerus 1982, Bennett & Crooke 1987, Katan & Parker 1987, Rebecchi & Rosen 1987, Stahl et al 1988, Suh et al 1988).

Recently three cDNAs encoding PI-specific PLCs have been isolated; these are referred to as $PLC\beta_1$, $PLC\gamma_1$ and $PLC\delta_1$. A diagrammatic representation of these enzymes is shown in Fig. 1. This figure indicates that two separate regions of sequence similarity are shared among all the isozymes; they are referred to as domains I and II. The N-terminal 300 amino acids of each of these enzymes display a sequence similarity of only approximately 20%.

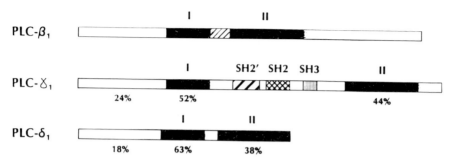

FIG. 1.   Diagrammatic representation of three classes of phospholipase C isozymes. Percentages reflect the degree of sequence identity with the corresponding region of PLCβ₁. SH2′, SH2 and SH3 refer to *src*-homology regions 2′, 2 and 3, respectively.

Interestingly, domain I begins 300 amino acid residues from the N-terminus in all the isozymes, which may suggest some common function for this domain.

PLCγ₁ is unusual in that domains I and II are separated by a region referred to as the *src*-homology or SH region. This region may serve a regulatory function, because of the sequence similarity with the modulatory domain of *src* (Jove & Hanafusa 1987) and because it is dispensable for *in vitro* catalytic activity (Bristol et al 1988). Both domains I and II of PLCγ₁ are required for catalytic activity, as shown previously by deletion mapping experiments (Bristol et al 1988). It is likely therefore that domains I and II interact with one another to form a catalytic core and that the SH2/SH3 region serves as a hinge. Both domains I and II are also likely to be required for the catalytic activity in PLCβ and PLCδ, and the short non-conserved region between the domains may act as a hinge in these isozymes.

As our primary interest is the role of PLC in inflammation, we wanted to examine the distribution of PLC in the cells which participate in this process, such as neutrophils and macrophages. We have previously shown that, at high stringency, neither PLCγ₁ nor PLCβ₁-related messages could be detected in HL-60 cells, precursors of neutrophils and macrophages, despite the fact that these cells have significant levels of PLC activity. Therefore, we felt that additional isotypes of these enzymes must exist. Based on the sequence similarity we observed, it was likely that we could isolate cDNAs encoding these other isotypes by cross-hybridization with a probe made from the conserved domain I of PLCβ₁. Low stringency Northern blots revealed the existence of novel RNA transcripts in HL-60 cells, as well as in fibroblasts. We isolated full length cDNA clones encoding these transcripts and deduced their amino acid sequences. A diagrammatic representation of these new sequences, PLCβ₂, PLCβ₃, PLCγ₂ and PLCδ₃, is included with the representation of other known cloned PLC isozymes in Fig. 2. Each of the new related genes is most similar to, and is structurally related to, one of the three previously isolated PLCs. This has led to the arrangement of PLC enzymes into classes as discussed below.

Kriz et al

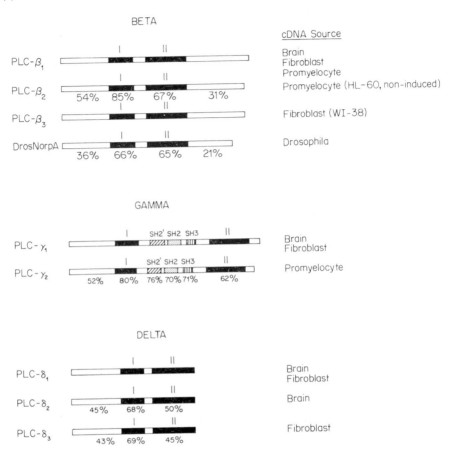

FIG. 2.   Diagrammatic representation of the three classes of PLC isoforms, percentages reflect degree of sequence identity with PLCβ₁, PLCγ₁ or PLCδ₁ for each member of a class.

PLCγ class

One cDNA clone isolated from an HL-60 library encodes a PLC gene most closely related to PLCγ₁ and is therefore referred to as PLCγ₂. PLCγ₁ and PLCγ₂ have a relatively high sequence identity in domains I and domains II (80 and 62%, respectively). Additionally, the N-terminal 300 amino acids display significant sequence identity (52%), which is not shared with the PLCβ or PLCδ classes. The distinguishing structural feature of the PLCγ class is a region of 400 amino acids inserted between domains I and II which has sequence similarity with the regulatory region of *src* proto-oncogene (Stahl et al 1988). The amino acid alignment of PLCγ₁ and γ₂ shows much higher sequence identity in the

SH2 and SH3 regions than in the neighbouring spacer regions. This again implies an important regulatory function for these regions.

PLCβ class

Two other putative PLC cDNA clones encode enzymes most closely related to $PLC\beta_1$, and are referred to as $PLC\beta_2$ and $PLC\beta_3$. $PLC\beta_2$ was isolated from the HL-60 library. A sequence alignment of $PLC\beta_2$ with $PLC\beta_1$ shows a high degree of identity throughout the N-terminal 800 amino acids, and most notably over domains I and II. This class of PLC has a C-terminal 400 amino acid domain which distinguishes it from the PLCγ and PLCδ classes. It had been previously noted that this C-terminal region of $PLC\beta_1$ contained an unusually high number of charged residues (about 40%); $PLC\beta_2$ is also rich in charged residues in this region, approximately 40%, although the actual sequence identity is quite low. The *Drosophila* PLC gene, *NorpA*, isolated by Bloomquist et al (1988), most closely resembles the PLCβ family in that it also has a large, highly charged (40%) C-terminal domain. Another interesting feature of the PLCβ class is the presence of a string of nearly consecutive acidic residues in the region between domains I and II. The *NorpA* enzyme does not possess this acidic domain between domains I and II. It does, however, contain a string of alanine and proline residues at this location which may perform a similar function.

An additional member of this class, $PLC\beta_3$, has recently been cloned from a fibroblast cell line, WI-38. Preliminary sequence analysis of this gene has allowed us to classify it as a PLCβ family member, complete sequence determination is in progress.

PLCδ class

The last of the PLC-related cDNAs isolated in this study was also derived from the WI-38 cell library. It is most closely related to members of the PLCδ class and is referred to as $PLC\delta_3$. An alignment of $PLC\delta_3$ with $PLC\delta_1$ reveals sequence similarity over the N-terminal 300 amino acids as well as domains I and II. This class of isozymes is readily distinguishable by the lack of a C-terminal region after domains I and II. $PLC\delta_3$ also contains a predominantly acidic domain between domains I and II, as previously noted for the PLCβ enzymes.

An additional member of the PLCδ class of enzymes has recently been purified to homogeneity from bovine brain and cDNA clones for this enzyme have now been isolated (E. Meldrum, R. Kriz, and P. Parker). The preliminary sequence data confirm a close relationship with the PLCδ class and the enzyme is referred to as $PLC\delta_2$ (Fig. 2).

PLC isozyme characterization

Ryu et al 1987 have previously shown that $PLC\beta_1$, $PLC\gamma_1$ and $PLC\delta_1$ each have a particular Ca^{2+}-dependent hydrolysis profile for each of the three substrates PI, phosphatidylinositol phosphate (PI-P) and phosphatidylinositol 1,4-bisphosphate (PI-P_2). Therefore, we sought to determine if the newly described members of these individual families shared these same enzymological characteristics.

The full length cDNAs encoding $PLC\gamma_2$ and $PLC\beta_2$ were expressed in COS-1 cells and the lysates were fractionated by anion exchange chromatography. Both $PLC\gamma_2$ and $PLC\beta_2$ fractionated away from the endogenous $PLC\gamma_1$ activity. The peak fractions of each were then used to examine the Ca^{2+} dependence of the PLC activity towards PI, PI-P and PI-P_2 as indicated in Figs. 3 and 4. At optimal Ca^{2+} concentrations $PLC\gamma_2$ utilized PI and PI-P approximately eightfold better than PI-P_2. However, at 10^{-7} M Ca^{2+}, PI-P_2 was the preferred substrate.

At optimal Ca^{2+} concentrations $PLC\beta_2$ hydrolysed PI-P and PI-P_2 severalfold better than PI. At 10^{-7} M Ca^{2+}, however, PI-P_2 and PI-P were the preferred substrates with neligible activity towards PI. The relative activites of $PLC\gamma_2$ and $PLC\beta_2$ towards the three phosphoinositide substrates are similar to those

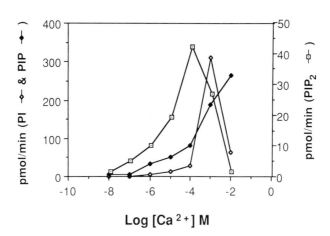

FIG. 3.   Effect of calcium concentration on the rate of PI, PI-P and PI-P_2 hydrolysis by $PLC\gamma_2$. Individual phospholipids were dried under nitrogen and resuspended in buffer containing 50 mM Hepes, pH 7.0, 200 mM NaCl, 2 mM EGTA, and 2 mg/ml deoxycholate. 50 μl of each lipid solution (containing 0.5 μCi of lipid), 40 μl of BSA (1.25 mg/ml) and 5 μl of enzyme were incubated at 37°C for 20 min. Five μl of $CaCl_2$ were added to give the free calculated Ca^{2+} concentration indicated. The final concentrations of PI, PI-P and PI-P_2 were 300, 275, and 264 μM, respectively.

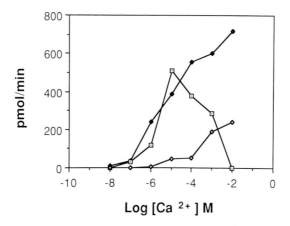

FIG. 4.   Effect of calcium concentration on the rate of PI, PI-P and PI-P$_2$ hydrolysis by PLCβ$_2$. This assay was performed as described in Fig. 3. ◇, PI; ◆, PI-P; ▫, PI-P$_2$.

observed for PLCγ$_1$ and PLCβ$_1$, respectively. Therefore the results suggest that within a family, the PLCs have a common substrate specificity.

Recently, Meisenhelder et al (1989) have shown that in 3T3 cells PLCγ$_1$ is phosphorylated on tyrosine residues in response to treatment with platelet-derived growth factor (PDGF). We have recently noted that PLCγ$_2$ expressed in COS-1 cells can also be phosphorylated on tyrosine in response to PDGF treatment (data not shown). However, no such phosphorylation is noted with PLCβ$_2$. This observation suggests that these two members of the PLCγ family may be regulated by similar mechanisms.

We and others have previously suggested that the SH2/SH3 region may serve a regulatory function in PLCγ$_1$ (Stahl et al 1988). Ralston & Bishop (1985) and Gould & Hunter (1989) have shown that PDGF treatment of 3T3 cells results in increased tyrosine phosphorylation of c-*src* in the SH2/SH3 region, as well as an increased *in vitro* kinase activity. The fact that PLCγ$_1$ and PLCγ$_2$ also have SH2/SH3 regions suggested the possibility that the SH2/SH3 regions of PLCγ may contain the PDGF-induced tyrosine phosphorylation sites.

We studied the PDGF-induced tyrosine phosphorylation of PLCγ$_1$ by coexpressing it with the PDGF receptor in COS-1 cells. We monitored the PDGF-induced tyrosine phosphorylation events by immunoprecipitating the [^{35}S]methionine-labelled proteins with anti-phosphotyrosine antibodies followed by SDS-polyacrylamide gel electrophoresis and autoradiography. Fig. 5 shows that in cells transfected with both PLCγ$_1$ and the PDGF receptor, two proteins of 170 kDa (the PDGF receptor) and 150 kDa (PLCγ$_1$) undergo increased tyrosine phosphorylation in response to PDGF treatment. Elution of these proteins from the anti-phosphotyrosine antibodies with phenylphosphate

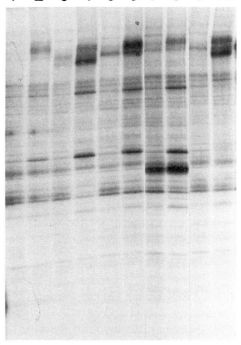

FIG. 5. PDGF-induced tyrosine phosphorylation of COS-1 cells transfected with PLCγ₁ mutants. Transfected cells were labelled with [³⁵S]methionine for six hours, treated with either PDGF (even-numbered lanes) or buffer (odd-numbered lanes), and lysed and immunoprecipitated with a monoclonal anti-phosphotyrosine antibody. The specifically bound products were eluted with phenylphosphate and analysed by SDS-polyacrylamide gel electrophoresis and autoradiography. All cells were transfected with the PDGF receptor and the following PLCγ₁ mutants, PLCγ₁ (lanes 3 & 4), PLCγ₁ΔSH2/SH3 (lanes 5 & 6), PLCγ₁-SH2/SH3 (lanes 7 & 8), PLCγ₁-Y-F802 (lanes 9 & 10). Details of these mutants are given in the text.

and re-immunoprecipitation with PLCγ₁ antisera reveals an increased tyrosine phosphorylation of PLCγ₁ (Fig. 6). In COS-1 cells transfected with a PLCγ mutant which has had the entire SH2/SH3 region deleted (PLCγ₁ΔSH2/SH3), no tyrosine phosphorylated product can be detected in either the single immunoprecipitation with anti-phosphotyrosine antibodies (Fig. 5) or in the double immunoprecipitation shown in Fig. 6. However, the [³⁵S]methionine-labelled product is present, as shown in Fig. 7. These results suggest that the SH2/SH3 domain may contain the tyrosine phosphorylation sites. Therefore we transfected cells with a PLCγ₁ mutant encoding only the SH2/SH3 domain (PLCγ₁-SH2/SH3) and found that this fragment of PLCγ₁ was phosphorylated on tyrosine in response to PDGF. Although the level of

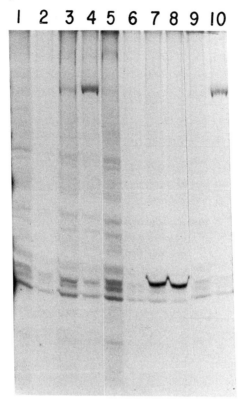

FIG. 6. PDGF-induced tyrosine phosphorylation of PLCγ_1. Cells were treated as described in Fig. 5 but the phenylphosphate eluates were immunoprecipitated with a PLCγ_1-specific antiserum before SDS-PAGE. All cells were transfected with the PDGF receptor plus PLCγ_1 (lanes 3 & 4), PLC$\gamma_1\Delta$SH2/SH3 (lanes 5 & 6), PLCγ_1-SH2/SH3 (lanes 7 & 8), and PLCγ_1-Y-F802 (lanes 9 and 10).

phosphorylation in the absence of PDGF was very high, a significant stimulation could be detected. The reason for the constitutively high levels of phosphorylation are unclear at present. We have made several point mutations in consensus tyrosine phosphorylation sites that are highly conserved in the SH2/SH3 domains. In one of these mutants, PLC-Y-F802, tyrosine 803 of PLCγ_1 was changed to a phenylalanine. We saw no effect of this mutation on tyrosine phosphorylation, indicating that tyrosine 803 is an unlikely PDGF-induced tyrosine phosphorylation site. These results, along with those from other deletion mutants (data not shown), indicate that the SH2/SH3 region contains the tyrosine phosphorylation sites. Our future studies will be directed towards identifying the tyrosine phosphorylation sites by a combination of deletion and point mutagenesis.

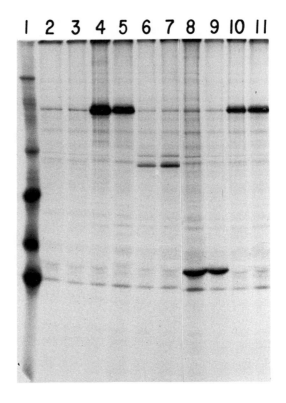

FIG. 7. Expression of PLCγ₁ mutants in COS-1 cells. The cell lysates used in the experiments described in Fig. 5 were immunoprecipitated with a PLCγ₁ antiserum and analysed by SDS-PAGE. Molecular weight markers are present in lane 1 and from the top of the gel are 200, 96, 77, 68 and 43 kDa, respectively. All cells were treated as described in Fig. 5 except that the odd-numbered lanes contain the PDGF-treated samples. All cells were transfected with the PDGF receptor and PLCγ₁ (lanes 4 & 5), PLCγ₁-ΔSH2/SH3 lanes (6 & 7), PLCγ₁-SH2/SH3 (lanes 8 & 9), and PLCγ₁-Y-F802 (lanes 10 & 11).

To date a number of investigators have correlated the increase in tyrosine phosphorylation of PLCγ₁ with an increased PI turnover in 3T3 cells (Wahl et al 1989, Meisenhelder et al 1989, Margolis et al 1989). However, 3T3 cells have a number of different PLCs, and it is possible that a PLC other than PLCγ₁ is responsible for the observed increase in PI turnover. We felt that if the level of PLCγ₁ was a limiting factor for the PDGF-induced PI breakdown, then increased levels of PLCγ₁ would result in a further increase in the PDGF-induced PI breakdown. Therefore, we overexpressed PLCγ₁ approximately 5–10-fold in Rat-2 cells and treated them with PDGF. We found increased PI breakdown as compared with the parental Rat-2 cells (Fig. 8). This strongly suggests that PLCγ₁ is responsible for the PDGF-induced PI turnover.

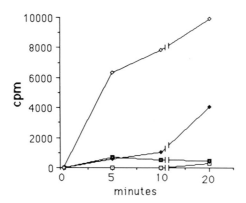

FIG. 8. Increased PI breakdown in Rat-2 cells overexpressing PLCγ₁. Stable transformants of Rat-2 cells overexpressing PLCγ₁ 5–10-fold were labelled with [³H]inositol for 48 hours, treated with recombinant PDGF for the indicated times and the amount of [³H]inositol 1,4,5-trisphosphate produced was determined. ▫, Rat-2 cells; ◆, Rat-2 cells plus PDGF; ■, Rat-2 cells overexpressing PLCγ₁; ◇, Rat-2 cells overexpressing PLCγ₁ treated with PDGF.

Several ligands which cause increased PI breakdown utilize G proteins to activate PLC. Aluminium fluoride has been reported to act as a universal activator of these G protein-dependent processes. We were interested in determining if PLCγ₁ could also be stimulated by aluminium fluoride. The results shown in Fig. 9 indicate that no increased responsiveness to aluminium fluoride can be detected in the overexpressing cell line, which argues that PLCγ₁ is not regulated by a G protein-dependent pathway. This observation both confirms and extends a number of reports which demonstrate that the PDGF-induced PI breakdown is not sensitive to various G protein-specific toxins.

In testing a number of ligands known to cause PI breakdown in fibroblasts, we saw little effect in Rat-2 cells with the notable exception of thrombin. Thrombin caused a marked increase in PI breakdown in the Rat-2 cells overexpressing PLCγ₁, suggesting that thrombin activates PLCγ₁. Thrombin has been reported to have both a pertussis toxin-sensitive and a pertussis toxin-insensitive component (Paris & Pouyssegur 1986). It seems likely from these studies that the insensitive component may be due to the activation of PLCγ₁, as there is little PLCγ₂ present in these cells. It will be of interest to examine the phosphorylation state of PLCγ₁ in response to thrombin. Such a multiplicity of effects in response to thrombin argue that thrombin activates many receptors, which are responsible for its diverse effects.

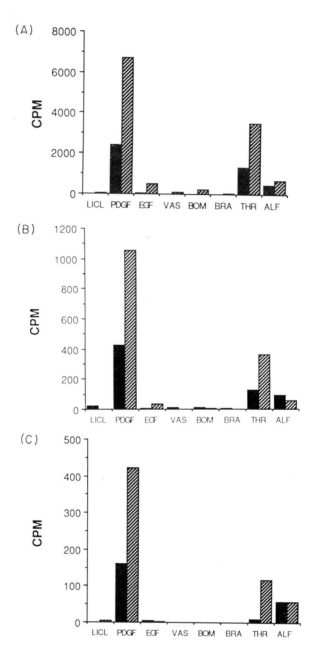

FIG. 9. Production of inositol breakdown products by known PLC agonists in Rat-2 cells overexpressing PLCγ$_1$. (A) Inositol phosphate. (B) Inositol 1,4-bisphosphate. (C) Inositol 1,4,5-trisphosphate. LICL, lithium chloride; EGF, epidermal growth factor; VAS, vasopressin; BOM, bombesin; BRA, bradykinin; THR, thrombin; ALF, aluminium fluoride. ■, Rat-2 cells; ▨, Rat-2 cells overexpressing PLCγ$_1$.

Acknowledgements

We thank Kalim Mir and Kevin Bean for assistance with the cDNA cloning and DNA sequencing.

References

Bennett CF, Crooke ST 1987 Purification and characterization of a phosphoinositide-specific phospholipase C from guinea pig uterus. J Biol Chem 262:13789–13797

Berridge MJ 1987 Inositol trisphosphate and diacylglycerol: two interacting second messengers. Annu Rev Biochem 56:159–194

Bloomquist BT, Shortridge RD, Schneuwly S et al 1988 Isolation of a putative phospholipase C gene of drosophila, norpA, and its role in phototransduction. Cell 54:723–733

Bristol A, Hall SM, Kriz RW et al 1988 Phospholipase C-148: chromosomal location and deletion mapping of functional domains. Cold Spring Harbor Symp Quant Biol vol 53:915–920

Gould KL, Hunter T 1989 Platelet-derived growth factor induces multisite phosphorylation of pp60c-src and increases its protein tyrosine kinase activity. Mol Cell Biol 8:3345–3356

Hofmann SL, Majerus PW 1982 Identification and properties of two distinct phosphatidylinositol-specific phospholipase C enzymes from sheep seminal vesicular glands. J Biol Chem 257:6461–6489

Jove R, Hanafusa H 1987 Cell transformation by the viral src oncogene. Annu Rev Cell Biol 3:31–56

Katan M, Parker PJ 1987 Purification of phosphoinositide-specific phospholipase C from a particulate fraction of bovine brain. Eur J Biochem 168:413–418

Margolis B, Rhee SG, Felder S et al 1989 EGF induces tyrosine phosphorylation of phospholipase C-11: a potential mechanism for EGF receptor signaling. Cell 57:1101–1107

Meisenhelder J, Suh P-G, Rhee SG, Hunter T 1989 Phospholipase C-γ is a substrate for the PDGF and EGF receptor protein tyrosine kinases in vivo and in vitro. Cell 57:1109–1122

Paris S, Pouyssegur J 1986 Pertussis toxin inhibits thrombin-induced activation of phosphoinositide hydrolysis and Na + /H + exchange in hamster fibroblasts. EMBO (Eur Mol Biol Organ) J 5:55–60

Ralston R, Bishop JM 1985 The product of the protooncogene c-src is modified during the cellular response to platelet-derived growth factor. Proc Natl Acad Sci USA 82:7845–7849

Rebecchi MJ, Rosen OM 1987 Purification of a phosphoinositide-specific phospholipase C from bovine brain. J Biol Chem 262:12526–12532

Ryu SH, Suh P, Cho KS, Lee K-Y, Rhee SG 1987 Bovine brain cytosol contains three immunologically distinct forms of inositolphospholipid-specific phospholipase C protein. Proc Natl Acad Sci USA 84:6649–6653

Stahl ML, Ferenz CR, Kelleher KL, Kriz RW, Knopf JL 1988 Sequence similarity of phospholipase C with the non-catalytic region of src. Nature (Lond) 332:269–272

Suh P-G, Ryu SH, Moon KH, Suh HW, Rhee SG 1988 Cloning and sequence of multiple forms of phospholipase C. Cell 54:161–169

Wahl MI, Nishibe S, Suh P-G, Rhee SG, Carpenter G 1989 Epidermal growth factor stimulates tyrosine phosphorylation of phospholipase C-11 independently of receptor internalization and extracellular calcium. Proc Natl Acad Sci USA 86:1568–1572

DISCUSSION

Hanley: Could you comment on the α isotypes of PLC?

Knopf: There is a putative PLCα of 68 kDa, which has been cloned (Bennett et al 1988). It has not yet been demonstrated to have biological activity, so I would have to reserve judgement on it. However, I do believe that there is a group of PLCs of about 68 kDa which will constitute the αs.

Noble: Could you say something about the effects of thrombin. Does it act as a mitogen? What concentration do you need to get these effects? Are they physiologically relevant?

Knopf: Thrombin has been used to stimulate PLC activity and mitogenesis by a number of investigators at a concentration of 1.0 units/ml. We included it to see whether or not it would turn on PLCγ_1 and it does. I have questioned the physiological relevance of the thrombin-induced mitogenic PI turnover response observed in fibroblasts. However, some argument can be made for thrombin being a physiological activator of PLC in the platelet system.

Westermark: In the Rat-2 cells transfected with PLC, with regard to PI turnover, did you check whether there is an increased responsiveness to stimulation of mitogenesis by PDGF, a shift in the dose–response relationship, for example?

Knopf: No, we haven't done that.

Brugge: John, in the experiments where you investigated the interaction of the PDGF receptor with the SH2/SH3 mutants of PLC, you immunoprecipitated [^{32}P]labelled cell lysates with antibody against phosphotyrosine and you didn't detect the mutant PLC. Is PLC bound to the PDGF receptor even though it's not phosphorylated? Is SH2 and/or SH3 necessary for binding?

Knopf: I would guess that the SH2/SH3 region is necessary for binding only because a mutant of PLCγ_1 expressing only the SH2/SH3 region is both bound to the PDGF receptor and phosphorylated by it. Presumably, the phosphorylation site and binding site are close to one another but that is not known. We haven't checked whether the PLCγ_1 mutant in which the SH2/SH3 region is deleted binds to the PDGF receptor.

Hanley: Do any of these PLC isoforms give stable transformation when transfected into cells?

Knopf: We have put a lot of these PLCγ_1 mutants into 3T3 cells and none of the mutants I described acts as a potent oncogene. The deletion of SH3 seems to activate c-*abl* but this does not seem to be the case with PLCγ_1. We have not tested the other isoforms.

Verma: Is anything known about the regulation of these different genes? When do the γ and δ isoforms appear? Are they co-ordinately regulated?

Knopf: There is not much known. These were isolated fairly recently.

Brugge: John, is there an explanation for the difference in the enhanced responsiveness of PDGF and EGF, given that both the EGF receptor and PDGF receptor are able to phosphorylate and to bind PLC?

Knopf: There has been some controversy as to whether or not EGF causes PI turnover in fibroblasts. It is generally agreed that the EGF receptor is present at lower levels in fibroblasts than is the PDGF receptor. That is thought to be a partial explanation for the absence or low level of EGF-induced PI breakdown as compared with the PDGF response.

Brugge: Can EGF stimulate PI turnover in Rat-2 cells?

Knopf: We have not looked very extensively, but we do see some stimulation. These experiments have to be repeated with the higher specific activity label and proper time courses in order to be verify the magnitude of the effect. This was just a 15 minute time point.

Hunter: In our experience the EGF receptor is simply a less good PLCγ kinase than the PDGF receptor. In A431 cells treated with EGF, where there are two million EGF receptors, we see a significantly lower level of phosphotyrosine in PLCγ_1 than in, for example, Swiss 3T3 cells treated with PDGF where there are 100 000 PDGF receptors (Meisenhelder et al 1989). However, Jossi Schlessinger has transfected the human EGF receptor into mouse 3T3 cells and sees good tyrosine phosphorylation of PLCγ (Margolis et al 1989). The fact that this is the human receptor in heterologous cells may make a difference, I don't know.

Sherr: Could someone clarify the specificity of Sue Goo Rhee's antisera for what is now called PLCγ (Suh et al 1988)? These antisera have been used by many investigators.

Knopf: We have not sent him any material, we probably should just to clarify things. We have made antisera to both PLCγ_1 and PLCγ_2 by expressing each gene in *E. coli* and raising antisera to the proteins in rabbits. Those antisera don't cross-react, the γ_1 antisera recognizes only PLCγ_1 and not PLCγ_2 and *vice versa.* Sue Goo Rhee has a monoclonal antiserum and it is possible that it recognizes a site conserved between isoforms of PLC.

Hunter: It's also a question of what was in his original preparation that was used to raise the antibodies. What are the relative levels of expression of PLCγ_1 and γ_2 in the brain?

Knopf: We don't see PLCγ_2 in brain.

Hunter: What about in fibroblasts?

Knopf: If there is PLCγ_2 there, it is at very low levels.

Hunter: Probably most people working on fibroblasts are looking at PLCγ_1.

Sherr: Have you looked in macrophages?

Knopf: We haven't looked in macrophages specifically. We have looked in the mononuclear fraction and we can detect PLCγ_1 and PLCγ_2. In peripheral blood lymphocytes we don't see any PLCγ_1. What proportion of those cells are macrophages, I don't know.

Westermark: You showed a picture of U937 cells, which are macrophage-derived cells; they had very low levels of PLCγ_2.

Knopf: It's very low but we have seen some differences in other enzymes in U937 cells as opposed to RAW cells, for example phospholipase A2 is present at 20-fold higher concentrations in RAW cells than in U937 cells. It may be that in real macrophages PLCγ_2 is expressed at much higher levels. We have not looked at the RAW2 cells yet for expression of γ_2.

Sherr: U937 cells are undifferentiated myeloid cells; they can be forced to differentiate along the macrophage lineage by a number of inducers, but they are analogous to HL-60 cells and would be expected to be quite different from mature macrophages. With an antiserum supplied by Joseph Schlessinger we were able to detect PLCγ in murine macrophages (Downing et al 1989). Your talk alerts me to the possibility that we could have detected PLCγ_1 or PLCγ_2. Obviously, the signal that we are looking at depends on the specificity of Jossi's antiserum.

Hunter: But this serum is against the C-terminal peptide of PLCγ_1 and it should be specific (Margolis et al 1989).

Knopf: I have looked at that—there's no homology between his peptide from PLCγ_1 and PLCγ_2.

Sherr: Then we could conclude that γ_1 is in macrophages?

Knopf: Yes.

Brugge: John, is it clear that the γ class of phospholipases are responsible for the PI-P$_2$ hydrolysis that takes place after receptor activation?

Knopf: It's not clear. The results from the cell lines in which PLCγ_1 is overexpressed provide the first good supportive evidence that that is the case. Overexpression of the enzyme leads to increased production of IP$_3$ (Fig. 8).

Hunter: With respect to how tyrosine phosphorylation might change activity, have you measured PLC activity *in vitro* in the COS cells after PDGF stimulation?

Knopf: No, but when we delete the whole SH2/SH3 region, which probably contains the tyrosine phosphorylation sites, there is not a great difference in activity, so I felt it was unlikely that tyrosine phosphorylation could have much effect on the *in vitro* activity.

Hunter: So how do you think tyrosine phosphorylation affects activity—by a direct increase in enzymatic turnover, by relief of negative inhibition or what?

Knopf: I don't know, but other cellular components may be necessary. For example, based on the levels of expression and the amount of tyrosine phosphorylation of PLCγ_1 that we see in COS cells, in spite of the relatively high background level of PI turnover, I would expect to see a further increase in PI turnover in response to PDGF, but we do not. If we add serum, there is a significant increase in PI turnover in these same cells. These results suggest that some additional factors may be involved in the activation of PLCγ_1.

Hunter: Is the level of expression in these COS cells greater than it is in the Rat-1 cells?

Knopf: Yes. It is greater by severalfold. In COS cells the enzyme is expressed in only 20% of the cells, so the actual expression in those cells is probably 10–20 fold higher. It's probably not the best system to look at, we may go back to it later to try and explain a few things, but at the moment we are using cell lines.

Tony Pawson made all the deletion mutants that I described. Currently, we know that the PLCγ_1 mutant with the SH3 deletion is still phosphorylated on tyrosine, and it will be interesting to see whether or not it is responsive to PDGF.

Hunter: Did you test a phenylalanine point mutation at the SH2/SH3 site of tyrosine phosphorylation identified by Sue Goo Rhee in your COS cells?

Knopf: No, we have not. Only one of these sites is a good candidate tyrosine phosphorylation site, being preceded by several aspartate residues; the other one is not.

Hunter: There's an equal increase in serine phosphorylation of PLCγ when cells are stimulated with PDGF. There is some question of whether that might be important in regulating the activity. Have you looked at serine phosphorylation?

Knopf: No, we thought we would try to understand the tyrosine phosphorylation, then take a look at serine phosphorylation.

References

Bennett, CF, Balcarek JM, Varrichio A, Crooke ST 1988 Molecular cloning and complete amino-acid sequence of form-I phosphoinositide-specific phospholipase C. Nature (Lond) 334:268–270

Downing JR, Margolis BL, Zilberstein A et al 1989 Phospholipase C-γ, a substrate for PDGF receptor kinase, is not phosphorylated on tyrosine during the mitogenic response to CSF-1. EMBO (Eur Mol Biol Organ) J 8:3345–3350

Margolis B, Rhee SG, Felder S et al 1989 EGF induces tyrosine phosphorylation of phospholipase C-11: a potential mechanism for EGF receptor signalling. Cell 57:1101–1107

Meisenhelder J, Suh P-G, Rhee SG, Hunter T 1989 Phospholipase C-γ is a substrate for the PDGF and EGF receptor protein tyrosine. Cell 57:1109–1122

Suh P-G, Ryn SH, Choi WC, Lee KY, Rhee SG 1988 Monoclonal antibodies to 3 phospholipase-C isozymes from bovine brain. J Biol Chem 263:4497–4504

fos-jun conspiracy: implications for the cell

Inder M. Verma, Lynn J. Ransone, Jane Visvader, Paolo Sassone-Corsi and William W. Lamph

Molecular Biology and Virology Laboratory, The Salk Institute, P.O. Box 85800, San Diego, CA 92138, USA

Abstract. Two nuclear oncoproteins, *fos* and *jun* (AP-1), cooperate in forming a very stable heterodimeric complex that binds to the AP-1 site on DNA with high affinity. The 'leucine zipper' domain of both *fos* and *jun* is necessary for the formation of this heterodimer. Mutations of single residues within the leucine zipper domain have no effect on protein complex formation. However, results from mutagenesis of the first leucine of the heptad repeat in either *fos* or *jun* basic regions and alteration of the spacing between the basic and leucine zipper domains indicate that the basic region of *fos* plays a crucial role in determining the DNA binding affinity of the transcriptional complex. Mutations of the basic amino acids in *fos* protein prevent binding to the tumour promoter response element (TRE) in the presence of wild-type *jun* protein. Thus *fos* protein appears to be dominant in *jun-fos* binding to DNA, even though *fos* alone cannot bind to TRE. Mutants in the basic region of *fos* and *jun* can be exploited as dominant-negative mutants to ablate the normal *fos* cellular function.

1990 Proto-oncogenes in cell development. Wiley, Chichester (Ciba Foundation Symposium 150) p 128–146

Proto-oncogenes have been postulated to play a pivotal role in cell growth, differentiation and development. They can be functionally divided into three broad classes, which also correspond to different cellular localizations: (1) growth factors and receptors, (2) mediators of intracellular signal transduction pathways, and (3) transregulators of transcription (Bishop 1987). In a scheme for the physiological interactions and cooperation between the products of the three different classes leading to cell proliferation or differentiation, growth factors presumably bind to their cognate receptors and activate a cascade of intracellular events culminating in modulation of gene expression. It is noteworthy that several oncogene products are critically placed in the pathway of transduction of external signals to the nucleus. Since the final destination of the impact of external stimuli is in the nucleus, our laboratory in the past several years has elected to study nuclear oncogenes. In particular, we have investigated the proto-oncogene *fos*,

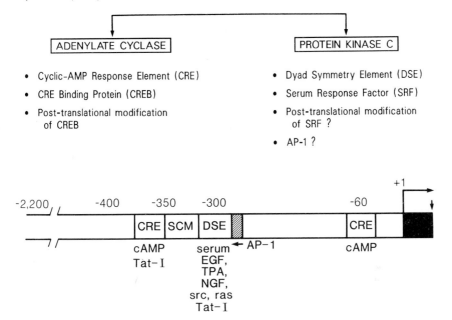

FIG. 1.   The two major intracellular pathways of *fos* induction in the cell in response to a variety of agents. The salient features of the upstream region of the c-*fos* promoter are also shown. EGF, epidermal growth factor; TPA, 12-*O*-tetradecanoylphorbol 13-acetate; NGF, nerve growth factor; SCM, *sis*-conditioned media; TAT-1, transcriptional activator of HTLV-1.

which can serve as a paradigm not only for other nuclear oncogenes but also for many early response genes.

fos oncogene

The *fos* oncogene was first identified as the resident transforming gene of FBJ murine osteosarcoma virus (FBJ-MSV) and was shown to cause bone tumours involving osteoblastic cells (Finkel et al 1966, Curran & Teich 1982, Verma & Graham 1987). The FBJ-MSV is a typical defective retrovirus, which has acquired normal cellular sequences encoding a protein of 381 amino acids. The normal cellular *fos* gene encodes a protein of 380 amino acids. During the biogenesis of the viral *fos* gene, 104 base pairs are removed, altering the reading frame. Thus, although viral and cellular *fos* proteins are similar in size, only the first 332 amino acids are the same, and the C-terminal 49 amino acids of viral *fos* are quite different from the 48 amino acids at the C-terminus of cellular *fos* (Van Beveren et al 1983). Despite the qualitative differences, both proteins have the ability to transform cells (Miller et al 1984).

If normal *fos* protein can transform cells, how is it regulated in the cell? The regulation can be at the level of transcription, which involves unique DNA motifs, or at a post-transcriptional level, which relates to the stability of mRNA. The third level of regulation is post-translational, by modification of the protein. The *fos* oncogene utilizes all levels of control for regulation in the normal cell. The cell has designed ways to control these proteins that have the ability to cause transformation, i.e. cancer, by a wide range of regulatory mechanisms.

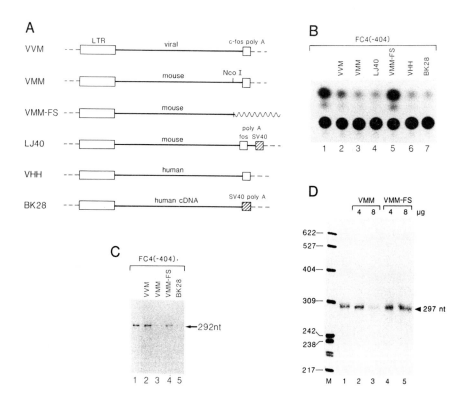

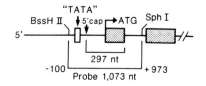

A variety of growth factors, including serum, differentiation agents like nerve growth factor (NGF) or 12-*O*-tetradecanoylphorbol 13-acetate (TPA), and forskolin, an agonist of adenylate cyclase, can induce expression of *fos*. There are two major signal transduction pathways, the protein kinase C/calcium-dependent pathway and the adenylate cyclase pathway. There are sequences immediately upstream of the *fos* promoter which are essential for the induction of the gene by agents that activate these pathways (Fig. 1).

Interestingly, the *fos* gene induction is extremely rapid and transient (Verma 1986). However, if cycloheximide is added to the cells, the synthesis of c-*fos* mRNA is sustained, suggesting that the cycloheximide prevents the synthesis of the protein which is necessary to shut off the transcription of the *fos* gene. In other words, there is a repressor protein synthesized during *fos* induction whose synthesis is blocked by cycloheximide.

Since one of the first proteins synthesized after induction of *fos* gene is *fos* itself, the *fos* protein is a good candidate to shut off its transcription as an autoregulatory phenomenon. Consequently, a *fos* promoter was linked to a reporter gene and transfected into starved cells (no serum was added to the cells, so that there was no induction of reporter CAT (chloramphenicol acetyl-transferase) gene product). When serum was added to these cells, the gene was induced. Co-transfection with a plasmid which has the ability to make *fos* protein

---

FIG. 2. Several *fos* expression vectors exert repression functions. (A) Structure of several *fos* chimaeric expression vectors used in this study. These recombinants have been described previously (Miller et al 1984). VVM expresses a viral *fos* protein, but contains the mouse c-*fos* 3' untranslated region. VMM expresses a viral–mouse *fos* fusion protein (p55); VMM-FS is a VMM derivative in which the *Nco*I site located upstream of the C-terminal domain has been filled with DNA polymerase, creating a frameshift mutation. LJ40 is equivalent to VMM, but contains a second poly(A) addition signal from SV40. VHH expresses a viral–human *fos* hybrid protein; BK28 contains the human c-*fos* full-length cDNA. All these coding sequences are under the transcriptional control of the FBJ-murine osteosarcoma virus long terminal repeat. When these recombinants were transfected into NIH/3T3 fibroblasts, *fos* protein was detectable in similar quantities by immunoprecipitation with anti-*fos* monoclonal antibodies and was localized to the nucleus by immunofluorescence (not shown). (B) Co-transfection of FC4 (Sassone-Corsi et al 1988a) with the *fos* expression vectors described in A. The extent of repression of c-*fos* promoter activity is eight- to tenfold for VMM, LJ40, VHH and BK28. The product of VVM represses with a much lower efficiency (about fourfold in this experiment, but in most cases no repression was observed), and VMM-FS shows no repression function. (C) RNAse protection mapping of *fos*-CAT RNA from cells transfected as in B. Same results were obtained as for the CAT assays; VMM and BK28 repress c-*fos* promoter activity, whereas both VMM-FS and VVM show no or little repression function. (D) Repression of endogenous *fos* gene transcription by *fos* protein; 4 or 8 µg of either VMM (lanes 2 and 3) or VMM-FS (lanes 4 and 5) were transfected into HeLa cells, and c-*fos* RNA levels were analysed. The bottom diagram shows the origin of the probe.

completely shuts off the synthesis of the reporter CAT protein. Transcription
of the endogenous *fos* gene was also shut off by c-*fos* protein. Frameshift
mutations in the C-terminus of the *fos* gene did not repress c-*fos* gene transcription. These results testify that the *fos* protein has the ability to suppress
the transcription from its own promoter (Sassone-Corsi et al 1988a; Fig. 2).

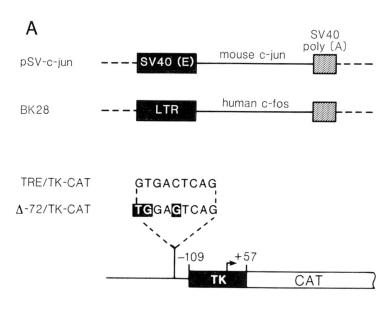

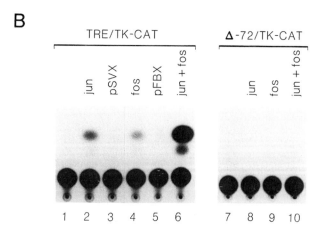

Does *fos* protein shut off only its own promoter or is it a general repressor of other promoters? The *fos* protein may be acting as a general transrepressor, and indeed, a number of promoters were shown to be affected by the *fos* protein. Interestingly, TPA-response element (TRE), also called the AP-1 site, showed the opposite effect of *fos*. When these regulatory sequences were linked to other reporter genes in the presence of *fos*, there was an extensive induction of transcription. Thus *fos* is a pleiotropic protein, acting as both a transrepressor and a transactivator (Sassone-Corsi et al 1988a).

jun oncogene

AP-1 is the site to which the *jun* product binds. The oncogene *jun* was isolated by P. Vogt and colleagues from retrovirus ASV17, which can cause tumours in chicken (Vogt et al 1987). The *jun* oncogene was shown to have sequence homology at the C-terminus with the yeast transcription factor GCN4. The GCN4 factor has a specific binding activity to DNA which is very similar to that of the factor AP-1 identified by R. Tjian and M. Karin's groups. The *jun* protein and AP-1 are essentially the same protein, p39 (Angel et al 1988, Bohmann et al 1988).

fos-jun interaction

If one immunoprecipitates *fos*, a number of proteins, called p39-related proteins, are brought down in addition to *fos* (Curran et al 1984). The antibody against *jun* also immunoprecipitates proteins very similar to p39. The *fos* and *jun* proteins actually collaborate with each other to augment transcription (Chiu

FIG. 3.   Cooperativity of *jun* and *fos* products in the transactivation of a TRE. (A) Structure of the recombinants used in the transfection experiments in mammalian cells. pSV-c-*jun* contains a mouse cDNA clone of c-*jun* under the regulatory control of the SV40 early promoter region. BK28 contains the c-*fos* human cDNA under the control of the FBJ-MSV long terminal repeat. Both pSV-c-*jun* and BK28 contain the SV40 early poly(A)-addition site. TRE/TK-CAT is a reporter plasmid containing the herpes thymidine kinase (TK) promoter (positions − 109 to + 57) linked to the CAT gene. An 18 bp oligodeoxynucleotide bearing the human metallothionein IIA TRE was inserted upstream of the TK promoter. ∆ -72/TK-CAT is equivalent to TRE/TK-CAT, but the TRE contains three point mutations that abolish TPA responsiveness. (B) Transfection experiments in mouse embryonal carcinoma F9 cells; 5 µg of either TRE/TK-CAT or ∆ -72/TK-CAT were co-transfected with 10 µg of pSV-c-*jun* (*jun*, lanes 2 and 8), 10 µg of BK28 (*fos*, lanes 4 and 9), or 5 µg of pSV-c-*jun* and 5 µg of BK28 (*jun + fos*, lanes 6 and 10). As control, co-transfection with recombinants equivalent to pSV-c-*jun* and BK28, but lacking the coding sequences (pSVX and pFBX, lanes 3 and 5) was performed. The total amount of DNA transfected was always 20 µg.

et al 1988, Sassone-Corsi et al 1988b). A TRE sequence was linked to a promoter and a reporter gene, and transfected into mouse embryonal carcinoma F9 cells, which have relatively low levels of *jun* or *fos*. There is very little activation when *jun* or *fos* is added independently; if *jun* and *fos* are mixed together, however, there is at least a tenfold increase in the induction of reporter gene expression. The *fos* and *jun* cooperate to increase transcription controlled by the TPA-response element by virtue of binding to each other (Fig. 3).

The next question is how do these nuclear oncoproteins interact with each other or with the DNA sequences. The leucine zipper motif was proposed by S. McKnight to account for the molecular mechanism of DNA binding by

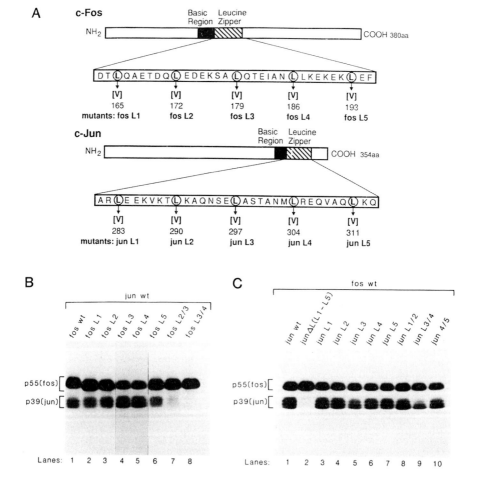

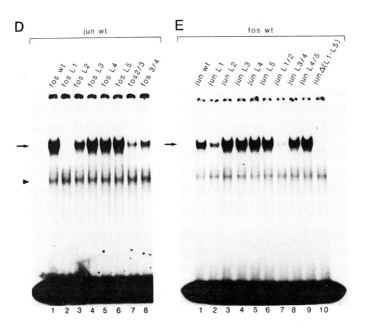

FIG. 4. *fos-jun* interactions. (A) Structure of the *fos* and *jun* proteins. Schematic representation of the *fos* and *jun* proteins showing the basic region and leucine repeat domains. Open boxes represent the primary amino acid sequences of *fos* and *jun*. The basic motifs, which contain a region of highly conserved basic residues proposed to be part of the DNA binding domain, are indicated by a solid box. The 'leucine zipper' structures indicated by the hatched boxes have been enlarged to show the sequence of the regions. The specific mutations from leucine to valine and the mutants generated with each change are indicated below each sequence. (B and C) Interaction of mutated *fos* and *jun* proteins. [³⁵S]methionine-labelled *fos* and *jun* wild-type and mutant proteins were synthesized *in vitro*, mixed and immunoprecipitated with anti-*fos* M2 antibody. (B) *jun* wild-type protein mixed with *fos* wild-type and mutant proteins. (C) *fos* wild-type protein mixed with *jun* wild-type and mutant proteins. The positions of both the *fos* and *jun* proteins are indicated. (D and E) DNA binding with *fos–jun* heterodimer: gel retardation analysis. [³²P]labelled TRE was incubated with unlabelled translation products of mixtures of the wild-type and mutant *fos* and *jun* proteins, then analysed by gel electrophoresis and autoradiography. (D) *jun* wild-type protein mixed with each *fos* mutant. (E) *fos* wild-type protein mixed with each *jun* mutant. The *fos–jun*–DNA complex is indicated by an arrow. The lower band (arrowhead) represents non-specific binding of proteins from the reticulocyte lysate.

proteins with this structure (Landschulz et al 1988). A number of proteins have this motif; the list includes *fos*, *fos*-related proteins, *jun*, *jun*-related proteins including GCN4 and AP-1, cAMP response element binding protein and c-*myc*. Another interesting structural feature is that the region immediately preceding the leucine domain is very rich in basic amino acids, and is thus known as the basic region.

To find which of the structures, the leucine zipper motifs or the basic regions, is responsible for the protein–protein association and which for protein–DNA association, mutants capable of making *fos* protein with the leucine zipper (*fos* 198) and without the leucine zipper (*fos* 160) were constructed. When the wild-type *fos* or *fos* 198 is mixed with *jun* and the TRE, there is binding; however, there is no binding with *fos* 160. Hence the zipper region is important for DNA binding (Sassone-Corsi et al 1988c, Kouzarides & Ziff 1988). In the transactivation assay, *fos* 198 has absolutely no effect on transcription regulated by the TRE, suggesting that the C-terminus region of the *fos* is important for transactivation (Sassone-Corsi et al 1988c). Furthermore, systematic mutation of all the leucine residues in the zipper region of *fos* indicates that the hydrophobic interactions are important in the binding, and the role of the leucine zippers is to bring the basic domains next to each other so that together they can bind to DNA (Ransone et al 1989) (Fig. 4).

The role of the basic domains was studied by mutational analysis. Mutations were made in the basic domain of the *fos* and *jun*. Again the protein–protein interaction was not affected in any of the mutants, but the DNA binding was completely eliminated. The *fos* protein has not been shown to bind directly to the TRE sequences, but must make a major contribution to the binding. In conclusion, the leucine zipper brings the *fos* and *jun* proteins together so that the basic regions are in opposition to each other and can interact with DNA. The orientation of the two proteins when bound is parallel.

fos and *jun* are part of a network of nuclear oncoproteins, much like that of lymphokines. The *jun* protein and AP-1 can bind not only to *fos*, but to *fos*-related antigens and *jun*-related proteins. An increase in *fos* or *jun* levels destabilizes the entire network. The oncogenes *fos* and *jun*, and perhaps other genes, are particularly interesting to study because they offer clues to the nature of genes whose activation or repression may, in turn, lead to stimulation or inhibition of transcription of other cellular genes, depending on the circumstances.

Acknowledgements

Supported by the NIH and ACS Grants to I.M.V.

References

Angel P, Allegretto EA, Okino ST et al 1988 Oncogene *jun* encodes a sequence-specific transactivator similar to AP-1. Nature (Lond) 332:166–171

Bishop JM 1987 The molecular genetics of cancer. Science (Wash DC) 235:306–311

Bohmann D, Bos TJ, Admon A, Nishimura T, Vogt PK, Tjian R 1988 Human proto-oncogene c-*jun* encodes a DNA binding protein with structural and functional properties of transcription factor AP-1. Science (Wash DC) 238:1386–1392

Chiu R, Boyle WJ, Meek J, Smeal T, Hunter T, Karin M 1988 The c-*fos* protein interacts with c-*jun*/AP-1 to stimulate transcription of AP-1 responsive genes. Cell 54:541–552

Curran T, Teich NM 1982 Product of the FBJ murine osteosarcoma virus oncogene: characterization of a 55,000-dalton phosphoprotein. J Virol 42:114–122

Curran T, Miller AD, Zokas L, Verma IM 1984 Viral and cellular *fos* proteins: a comparative analysis. Cell 36:259–268

Finkel MP, Biskis BO, Jinkins PB 1966 Virus induction of osteosarcomas in mice. Science (Wash DC) 151:698–701

Kouzarides T, Ziff E 1988 The role of the leucine zipper in the *fos-jun* interaction. Nature (Lond) 336:646–651

Landschulz WH, Johnson PF, McKnight SL 1988 The leucine zipper: a hypothetical structure common to a new class of DNA binding proteins. Science 240:1759–1764

Miller AD, Curran T, Verma IM 1984 c-*fos* protein can induce cellular transformation: a novel mechanism of activation of a cellular oncogene. Cell 36:51–60

Ransone LJ, Visvader J, Sassone-Corsi P, Verma IM 1989 *fos-jun* interaction: mutational analysis of the leucine zipper domain of both proteins. Genes & Dev 3:770–781

Sassone-Corsi P, Sisson JC, Verma IM 1988a Transcriptional autoregulation of the proto-oncogene *fos*. Nature (Lond) 334:314–319

Sassone-Corsi P, Lamph WW, Kamps M, Verma IM 1988b *Fos*-associated cellular p39 is related to nuclear transcription factor AP-1. Cell 54:553–560

Sassone-Corsi P, Ransone LJ, Lamph WW, Verma IM 1988c Direct interaction between *fos* and *jun* nuclear oncoproteins: role of the leucine zipper domain. Nature (Lond) 336:692–695

Van Beveren C, van Straaten F, Galleshaw JA, Verma IM 1983 Analysis of FBJ-MuSV provirus and c-*fos* (mouse) gene reveals that viral and cellular *fos* gene products have different carboxy termini. Cell 32:1241–1255

Verma IM 1986 Proto-oncogene *fos*: a multifaceted gene. Trends Genet 2:93–96

Verma IM, Graham WR 1987 The *fos* oncogene. Adv Cancer Res 49:29–52

Vogt PK, Bos TJ, Doolittle RF 1987 Homology between the DNA-binding domain of the GCN4 regulatory protein yeast and the carboxy terminal region of a protein coded for by the oncogene *jun*. Proc Natl Acad Sci USA 84:3316–3319

DISCUSSION

Hunter: Why doesn't the *fos* basic mutation work as a dominant-negative mutation?

Verma: I don't know. It is important to discover the exact nature of the protein–DNA interactions. Is it only the *jun* that binds to the DNA? I would have predicted that the *fos* mutant was dominant, because this mutation abolished binding to the DNA. From other experiments we have done, it would appear that this *fos* mutant is not a dominant-negative. It might be in other situations, but not in this kind of experiment. We think that *jun* basic domain mutants act dominantly because they bind more tightly to wild-type *fos* protein than do *jun* wild-type proteins.

Wagner: If you make the homodimer of *fos*, does it bind to DNA?

Verma: I don't think you can make a homodimer of *fos*. One can take the basic region of *fos* and link it to GCN4 leucine zippers. This chimaeric protein can bind to the TRE. That is not a true homodimer of *fos*, but it does show that the *fos* basic region can bind to DNA. I suspect that *fos* does not make a homodimer because its N-terminal region prevents formation of the homodimer. I think that *fos* is one of the few proteins for which it is an advantage not to make homodimers and to form only heterodimers. Because it must interact with many other proteins, it has no reason to be a homodimer. People who do the experiments of putting parts of different proteins together run the risk of ignoring the whole protein.

If you make a mutant in the *fos* at residues 144 or 146, and mix it with wild-type *jun* protein and add it to the TRE, you see very poor binding to the DNA. That suggests that *fos* protein must have a formal direct interaction with the DNA. It doesn't say anything more than that. Whether that is due to the *fos* protein making direct contact with the DNA, one cannot say from that experiment.

If you make a mutation of *jun* in the basic domain (ΔRK*jun*), and mix it with wild-type *fos*, there is no DNA binding. If you add increasing amounts of wild-type *jun*, you still don't see any DNA binding, suggesting that once the *fos*–ΔRK*jun* complex is made, *jun* cannot substitute into the complex or affect the binding in a dominant way. If you do the reverse experiment, add mutant *fos* to wild-type *jun*, it competes and abolishes DNA binding.

Harlow: In the mixing experiment, if mutant *jun* works like a normal transdominant, you should be able to recover wild-type activity by adding large amounts of wild-type *jun*.

Verma: The question is what is the binding affinity of each of these mutants to their cognate TRE. Could it be that the mutation of the basic domain does not abolish DNA binding but increases the binding affinity? Could it be that the mutant binds so much more strongly than the wild-type that once the mutant *jun* has bound to *fos* the wild-type *jun* cannot displace it. That's why we did the co-translation experiment, in which the mutant and the wild-type *jun* have equal opportunity to bind to *fos*, and even then the mutant is dominant.

Harlow: It is possible but it doesn't seem very likely.

Verma: We have done some preliminary experiments on the binding affinity of *fos* and *jun* mutants to their cognate TRE. I think there is going to be a difference. I think the idea that the basic domain is absolutely limited to binding to DNA and has nothing to do with formation of the leucine zipper is probably not correct.

Harlow: How extensively have you mutated the DNA-binding domain of *fos* protein to try to generate dominant-negative mutants?

Verma: We have mutated everything in the basic domain. It could be that the DNA binding region is much larger than anticipated in the basic region. Within the basic region we have mutated seven residues of both *jun* and *fos*. All of them have an effect, but the effect is different, which is interesting because you can use them to make cell lines with different transdominant-negative mutants. We have also deleted the entire basic domain and *fos* is still not dominant.

Harlow: You have been transfecting all of these dominant-negatives into cells, do you see any phenotypic effects?

Verma: We made fibroblast cell lines containing *fos* mutants and nothing happened to them. In v-*fos*-transformed cells transfected with the dominant-negative c-*fos*, nothing happened. Then we realized that it is the *jun* we should be looking at and that is being done now.

Brugge: In the experiment where you co-express a *fos*-containing plasmid and a CAT reporter plasmid, do you think there is excess *jun* in a cell that binds to this exogenous *fos*?

Verma: The *fos–jun* complex is needed only for induction via the TRE. You can take the *fos* dyad symmetry element (DSE) alone and see transrepression without the AP-1 element. We removed the AP-1 element and the suppression was maintained, which suggests that the *fos* suppression is not linked directly to the AP-1 element, the DSE is enough. The DSE is immediately followed by a sequence, TGACTCA, which is very like the AP-1 element and the CRE (the only difference between the AP-1 element and the CRE is an extra nucleotide; they bind similar proteins). If you put c-*fos* protein into cells, the c-*fos* protein and AP-1 element present in the *fos* promoter can form a complex to shut off transcription.

Brugge: But is *fos* alone sufficient or do you need *fos* and *jun*?

Verma: fos alone is sufficient. But there are always basal constitutive levels of *jun* in every cell.

Brugge: But if DNA binding is required, the *fos* protein alone can't bind DNA.

Verma: If you take a DNA binding mutant of *fos*, there is no suppression. Another protein may be needed, protein X.

Hunter: Which may be *junB* or *junD*?

Verma: There's no distinction between *junB*, *junD* and c-*jun* in terms of binding to *fos* protein.

Hunter: So it could be anyone of those.

Verma: Yes, or it could be *fos*-related. When I say *fos* I mean generic *fos* and similarly generic *jun*. There are three *fos* proteins known at the moment.

Hunter: Michael Karin has some evidence that *junB* acts as a repressor rather than as an activator (Chui et al 1989).

Land: With regard to transformation, it is formally possible that the inability of v-*fos* to repress transcription could lead to transformation. Do you know if v-*fos* could be a dominant repressor of the ability of c-*fos* to repress genes?

Verma: No. It is possible that it could act as a competitor by removing *jun* from the system, if *jun* was the only factor required.

Hunter: It is a good question, whether the loss of the transrepression activity by the altered C-terminus of v-*fos* is important.

Verma: c-*fos* transforms fibroblasts as efficiently as does v-*fos*, if you can make equivalent amounts of it.

Wagner: But c-*fos* does not transform many other cell types. In transgenic mice whose cells produce high levels of *fos*, you don't see transformation of many cell types.

Verma: fos causes only bone tumours. Why doesn't *fos* cause other kinds of tumour if it is expressed in different cell types? I presume there are other genes involved, not just *jun*, which may not be present in all cell types. *fos* doesn't have the ability to transform every cell, it depends on what other genes are expressed in the cells.

Hunter: Has anyone made a c-*fos* FBJ virus and asked whether it can cause osteosarcomas?

Verma: Yes, we made a c-*fos* retrovirus and it causes bone tumours.

Wagner: The FBJ-LTR at the 3′ end of c-*fos* in transgenic mice is necessary and sufficient to cause specific bone tumours. If you remove the LTR, at present we have never seen bone tumours.

Vande Woude: If the FBJ-LTR constructs are expressed early in embryogenesis, they might be lethal. When are they expressed during development?

Wagner: This transgene is expressed at embryonic day 17 (E17). To get around this problem, we have done the same experiments in embryonic stem cells, where we characterized the expression of c-*fos* in various embryonic stem cell clones. We expected to see a dominant 'embryonic' effect, as seen with other constructs. However, at three or four weeks of age the chimaeric mice develop these bone tumours. We are therefore convinced that the *fos* gene is expressed during development.

Hunter: It's interesting that both FBJ and FBR MSVs have altered C-termini, so perhaps there is some selective advantage in the loss of the C-terminus.

Verma: The c-*fos* protein is 380 amino acids long; the v-*fos* protein is about the same size, it is 381 amino acids long but the C-terminal 49 amino acids are from a different reading frame. There is another *fos*, called FBR, which is missing 98 amino acids from the C-terminus compared to the c-*fos*, has two amino acid substitutions, and is missing 28 amino acids from the N-terminus. That FBR causes transformation and bone tumours. The changes that are present in v-*fos* with respect to c-*fos* are not the same in FBR, suggesting that those changes are not important.

Vande Woude: Do you have any explanation for why *fos* transforms fibroblasts so poorly?

Verma: fos is not unique. I think *myc* has a hard time transforming fibroblasts. If you add *fos* to rat fibroblasts, you see foci; in mouse fibroblasts FBJ makes very poor foci, FBR does much better. In that assay using *jun* we do not see the same kind of changes in the cell. *myc* in our hands also doesn't do that.

Westermark: There is one point I don't understand in your autorepression system in which *fos* regulates its own expression. If the mRNA and the protein have a short half-life, why do you still see suppression after several hours in the continuous presence of growth factors?

Verma: The repression can't be due to the *fos* protein, because that can no longer be detected. We know nothing about what happens from when the growth factor is added to cells to when the nuclear proteins are turned on. There could be intermediary proteins which are equally important for the induction, whose synthesis is blocked after the first expression of *fos*.

The only other system in which continuous expression of *fos* is seen is in the placenta.

Wagner: It is interesting that in several different cell systems where we have introduced exogenous c-*fos* genes, we have hardly ever down-regulated expression of the endogenous *fos*. However, if you establish cell lines from bone tumours of transgenic mice, only the transgenic *fos* is expressed, the endogenous *fos* gene is repressed.

Verma: Do you think that repression is due to the exogenous gene?

Wagner: Yes, but we have no easy way to prove it.

Sherr: Rodrigo Bravo and Rolf Muller reported an interesting series of experiments with peripheral blood monocytes and some macrophage cell lines. After induction with cAMP, there was a persistent appearance of *fos* transcripts lasting over 6–8 hours. Have you analysed that particular situation to ask why you don't see transformation?

Verma: We haven't looked at the details of that. We induced U937 cells or HL-60 cells to differentiate into macrophages, then looked at the induction of the *fos*. The level of *fos* mRNA normally falls within two hours. In these cells there was a little decline after two hours but then a second increase in *fos* expression after about 7–9 hours, which is unusual.

We did immunoprecipitations, but we could never see any *fos* protein in those cells after the two hours, even though the message was there. We thought that maybe the *fos* mRNA was unable to be translated. Bravo & Muller said that the second burst of *fos* expression is not due to the induction by TPA but due to the cAMP. It's not clear to me whether that's regulated at the transcriptional level or via the stability of the mRNA or the protein. We have not formally taken macrophages and asked whether *fos* is regulated differently in them. That's a good experiment to pursue.

Feramisco: In terms of the ability of the transacting CRE binding protein (CREB) mutants to be dominant-negative mutants, can you get those to go to zero in the CAT assays?

Verma: They go to about 4%; there is always some background activity. The difficulty is the way we do the experiments. We need 10 µg of CREB in order to see the CREB activity. The maximum we can handle in these transfection assays is 20–30 µg. If we could use a tenfold excess of the mutant, the experiment should work well, but that is much more than the transfection assay system can handle. We are trying to use 3–4 µg in an RNA protection assay where we can use a fivefold excess of the mutant and I bet the background will be absolute zero.

Hafen: How great is the specificity of the interaction in the leucine zipper region? Can the dominant repressors bind to other leucine zipper-containing proteins and repress their function?

Verma: This question is asked frequently: if all these proteins contain the leucine zipper, why does the *fos* protein not make heterodimers with all of them. *fos* protein is known to make a dimer only with *jun*. CREB doesn't make a dimer with *jun*; CREB doesn't bind to *fos* even though it has sequences similar to *jun*. Another CREB protein, CREB P1, binds to *jun*, and it has the same leucine zipper residues as the wild-type CREB which doesn't bind to *jun*.

We asked, why does *fos* not make homodimers but *jun* does? Aaron Klug, R. Doolittle and L. Orgel looked at the sequence and suggested what may be happening. Looking at the structure of the leucine zipper, if it forms coiled coils, then leucines 4 and 5 will be next to each other. In all cases where a homodimer was made, the position 5 was hydrophobic. For *fos*, this position is basic. We thought that if we switched the position 5 residues of *fos* and *jun*, we might be able to make *fos* homodimers. It turns out that a *jun* protein containing *fos* position 5 residues makes homodimers, but the reciprocal *fos* protein does not.

Hunter: Has anyone modelled these as coiled coils?

Verma: They are all coiled coils. There is a question about the polarity of the proteins. In the homodimer or heterodimer, do the subunits have the same polarity or antipolarity? The leucine point mutations in the leucine zipper suggested that they probably have the same polarity.

Peter Kim synthesized the GCN4 leucine zipper peptide and put cysteine residues either at the N-terminus or the C-terminus, and then mixed them together. Only when the cysteines were at the same end of the peptide did they make a disulphide bridge, suggesting that their polarities must be parallel. He then looked at these by NMR and showed that the structure is a coiled coil and in the parallel orientation. So you can imagine two homodimers making a coiled coil and then the basic residues sticking out.

Hunter: But in that model, how do the basic VNVV residues interact in the dimer?

Verma: They can interact only if the subunits do form a coiled coil.

Hunter: With regard to the N-terminus of *fos*, what sort of mutations have been made?

Verma: The first 28 amino acids are probably not important for transformation, because they are missing from FBR. I don't know of any mutation of the region from 28 to 166.

Hunter: But are you presuming this region is involved in transactivation?

Verma: No, I assume the N-terminus is involved in prevention of formation of the homodimers. The transactivation domain, I think, is between position 198 and something around position 310. To remind you: 135–160 is the basic region; 162–198 is the leucine zipper; from 198 to 340 may be the transactivation domain and the remaining 40 amino acid residues form the transrepression region. The transactivation region should be 198–310, but since FBR doesn't have 98 amino acids, we can assume that they are not important. The first 28 are missing from FBR, so I don't know what they do. If you put a leucine zipper from *jun* in place of the *fos* leucine zipper, there is no homodimer formation. I think there's something in the protein that prevents the formation of homodimers.

Hunter: Have you deleted that region?

Verma: No.

Feramisco: Does the transformation activity map to the same regions as the transactivation ability?

Verma: Grossly speaking, yes. Rolf Muller reported that a protein with a mutation in the leucine zipper didn't transform but that's because presumably it cannot form dimers.

Vande Woude: Can you tell us something about the transforming activity of *jun*?

Verma: jun was isolated as the transforming gene of avian sarcoma virus 17. v-*jun* transforms chicken cells; c-*jun* does so only poorly. Peter Vogt has done the most work on it. In our hands and in Peter's neither c-*jun* nor v-*jun* transforms mammalian cells. John Minna reported that they could see transformation of mammalian cells (Schütte et al 1989).

Hunter: The c-*jun* transformation is pretty weak. The chick cells are marginally transformed. Tim Bos and Peter Vogt have been making c-*jun*/v-*jun* chimaeras to determine which changes in v-*jun* are necessary for transformation. It looks like the deletion at the N-terminus of v-*jun* is the major change necessary, but one or more of the point mutations in v-*jun* are also important (personal communication).

Wagner: Is there any convincing experiment which shows that a *fos*-responsive gene has been activated by binding of the *fos–jun* complex to its promoter?

Verma: We were hoping to have an inducible promoter but every promoter that we have investigated is constitutively active at a low level and that's enough to cause problems for us.

Wagner: Are all TRE-dependent genes AP-1 responsive?

Verma: I don't know how many of them will be turned on but there could be a very large number.

Wagner: Has anyone shown convincingly that induction of *fos* has turned on an endogenous gene, not a co-transfected one?

Verma: The only one I know about is strictly repressed and that's the *ap2* gene during adipocyte differentiation shown by Bruce Spiegelman.

McMahon: Tom Curran and Jim Morgan (personal communication) have suggested that proenkephalin might be a target for *fos*.

Verma: That's because its promoter has a CRE element and *fos–jun* may bind to the CRE element.

Hunter: The key question is, given that you can clearly demonstrate binding of the heterodimer, how does this activate transcription? Why is *fos–jun* better than the *jun–jun* complex?

Verma: The DNA binding affinity is much higher. I think *jun–jun* probably just doesn't bind to DNA very well.

Hunter: You assay that in an artificial system where the levels of protein are rather low.

Verma: You can also do it *in vivo* with the protein synthesized in a bacterial system and the binding is still very poor. Tom Curran and Paulo Sassone-Corsi believe there is something in the *in vitro* reticulocyte extracts that has the potential to inactivate DNA binding.

Vande Woude: fos will displace *jun* in the dimer, so bonding in the heterodimer is stronger.

Verma: The *jun–jun* complex must be a dynamic homodimer.

Nusse: What about the evidence that *rel* is a transcriptional activator?

Verma: Paulina Bull in Tony Hunter's and my laboratory has been studying the *rel* protein. The proto-oncogene c-*rel* is a homologue of the *Drosophila* gene, *dorsal*. c-*rel* is a serum-inducible gene, its mRNA is made within 30 minutes of serum activation; it is also inducible by TPA. It is super-induced by cycloheximide, so it has all the characteristics of an early response gene.

Howard Temin and colleagues have shown that *rel* protein is both nuclear and cytoplasmic. *rel* is also lymphoid specific. Aficionados of NF$\varkappa$B immediately thought that c-*rel* might be like NF$\varkappa$B. So we decided to look at its trans-activation properties. Paulina Bull linked the c-*rel* gene to a GAL4 DNA-binding domain (1–147 amino acids) and β-galactosidase and transformed yeast cells and tested whether this construct can transactivate a *lacZ* gene regulated by a GAL4 response element. c-*rel* protein is about 580 amino acids long. If she took the coding sequence for the total c-*rel* protein, there was some expression of β-galactosidase, but the 5′ half of c-*rel* gave no expression of β-galactosidase. If Paulina took the 3′ half, it had activity similar to that of the wild-type GAL4 protein.

Tom Gilmore at Boston University has done the same experiment using the LexA fusion protein promoter and found that chicken c-*rel* can transactivate; however, his results differ in that Tom can show activity with full length c-*rel* as well as with just the 5′ end.

Sherr: There's one publication by Gelinas & Temin (1988) where they claim that *rel* is a transactivator (see also Hannink & Temin 1989).

Verma: They used the polyoma enhancer region and cotransfected it with v-*rel* and found fourfold induction of the polyoma enhancer.

Vande Woude: What is the correlation between the cellular gene and the viral one? Is the C-terminus conserved between v-*rel* and c-*rel*?

Verma: Yes, but I'm not sure to what extent.

Hunter: Tom Gilmore's experiments showed that the N-terminal two thirds of c-*rel* had transactivating activity, so the transactivating region may lie in the middle of the protein.

Vande Woude: The question is whether v-*rel* contains the same enhancer sequences as c-*rel*.

Verma: Yes, we need to test v-*rel*.

Nusse: In *Drosophila*, there is evidence for a protease activity that regulates the ability of the *dorsal* protein to translocate to the nucleus, so perhaps there is a correlation with these different domains—part of the protein may have to be cleaved to make it active.

Verma: What part of *dorsal* is homologous to c-*rel*?

Nusse: I don't know, it's only one domain, not the whole protein.

Verma: But if you label *Drosophila* embryos with anti-*dorsal* antibody, all the nuclei are labelled on the ventral side of the embryo. It is called *dorsal* because in the mutants the dorsal side doesn't form.

Nusse: The protein is also made in the upper part of the embryo but not in the nuclei. A protease regulates its translocation to the nucleus.

Hunter: That's from genetic evidence?

Nusse: Yes. There are some mutants in which the protein does not localize to the nucleus and those are presumably mutants in genes regulating the protease activity.

Verma: The surprising part is from Temin's data that the transforming activity is due to cytoplasmic *rel* not nuclear *rel*.

Hunter: They put a nuclear transfer sequence on v-*rel* and it still transforms lymphoid cells, so I think that the *rel* protein functions in the nucleus.

Verma: They couldn't show c-*rel* was in the nucleus, they were very clear about that. But there could still be some there.

Sherr: Temin's lab has replaced *rel* sequences with *dorsal* and tested these constructs in a transactivation assay. I believe that those constructs work. There's apparently a very high degree of homology between *dorsal* and v-*rel*, something like 50%.

Hunter: Only in the N-terminal half.

Verma: But it's all patches, there is not stretch of homology.

References

Chiu R, Angel P, Karin M 1989 Jun-B differs in its biological properties from, and is a negative regulator of, c-jun. Cell 59:979–986

Gelinas C, Temin HM 1988 The v-*rel* oncogene encodes a cell-specific transcriptional activator of certain promoters. Oncogene 3:349–355

Hannink M, Temin HM 1989 Transactivation of gene expression by nuclear and cytoplasmic *rel* proteins. Mol Cell Biol 9:4323–4336

Schütte J, Minna JD, Birrer MJ 1989 Deregulated expression of human c-jun transforms primary rat embryo cells in cooperation with an activated c-ha-ras gene and transforms Rat 1A cells as a single gene. Proc Natl Acad Sci USA 86:2257–2261

mos proto-oncogene function

George F. Vande Woude*, Roberto Buccione°, Ira Daar*, John J. Eppig°, Marianne Oskarsson*, Richard Paules*, Noriyuki Sagata°† and Nelson Yew*

*BRI-Basic Research Program, NCI-Frederick Cancer Research Facility, Frederick, MD 21701, °The Jackson Laboratory, Bar Harbor, ME 04609, USA, †Tsukuba Life Science Center, Tsukuba, Ibaraki 305, Japan

Abstract. Maturation promoting factor (MPF) is a cytoplasmic activity that causes oocytes arrested in prophase to resume meiosis. An inactive form of MPF termed pre-MPF exists in fully grown oocytes. In *Xenopus* oocytes, progesterone induces maturation and pre-MPF activation. These early maturation events require protein synthesis. We have shown that p39^{mos} synthesis is rapidly induced in progesterone-treated *Xenopus* oocytes during the protein synthesis sensitive period and prior to activation of pre-MPF or germinal vesicle breakdown (GVBD). p39^{mos} may qualify, therefore, as an 'initiator' of maturation. Mouse oocytes undergoing meiotic maturation also express p39^{mos}. Microinjection of antisense *mos* oligodeoxynucleotides into fully grown mouse and *Xenopus* oocytes results in the block of meiotic maturation. In *Xenopus*, antisense-injected oocytes not only lack p39^{mos}, but also lack MPF and fail to undergo GVBD. In the mouse, the microinjected oocytes progress through GVBD, but fail to produce the first polar body; cytogenetic analysis shows they are arrested at the bivalent chromosome stage of metaphase I. This and additional studies with *Xenopus* oocytes indicate that p39^{mos} is also required throughout maturation. We have shown that p39^{mos} is indistinguishable from the protein product constitutively expressed in NIH/3T3 cells transformed with activated c-*mos*. It is likely that its activity as a transforming gene may be due to activation of pre-MPF activities in interphase in the somatic cell cycle.

1990 Proto-oncogenes in cell development. Wiley, Chichester (Ciba Foundation Symposium 150) p 147–162

Our interests over the past 10 years have been concerned primarily with mechanisms of oncogene activation and the role of oncogenes in neoplastic transformation. It was obvious when we began these investigations that to understand how oncogenes participate in neoplastic transformation, it was essential first to determine the function of the normal cellular proto-oncogene. Thus, our major goal has been to identify how proto-oncogenes work, with the expectation that this knowledge will lead to an explanation of the transformed phenotype and ultimately to the development of protocols for specifically targeting cancer cells. We assumed that the oncogene product would have the same targets or substrates as the normal product, but would be active at an inappropriate time during the cell cycle. Determining the normal function of a proto-oncogene, however, has not been trivial. Aside from those proto-oncogenes which encode growth factors, progress in understanding signal transduction has been slow. During the past several years, there have been significant breakthroughs, however, in understanding cell cycle regulation, and now it appears that some proto-oncogenes control these processes.

mos was the first cellular proto-oncogene to be characterized, as well as the first shown to have transforming potential (Jones et al 1980, Oskarsson et al 1980). In spite of this early start, many laboratories, including our own, failed to detect its normal expression and interest waned. It wasn't until 1985 that we discovered it was expressed in certain normal tissues at low levels and at much higher levels in gonadal tissue (Propst & Vande Woude 1985). We and others subsequently showed that expression was specific to germ cells (Propst et al 1987, Goldman et al 1987, Mutter & Wolgemuth 1987, Keshet et al 1988, Propst et al 1988). Recently, we provided the first evidence for the requirement

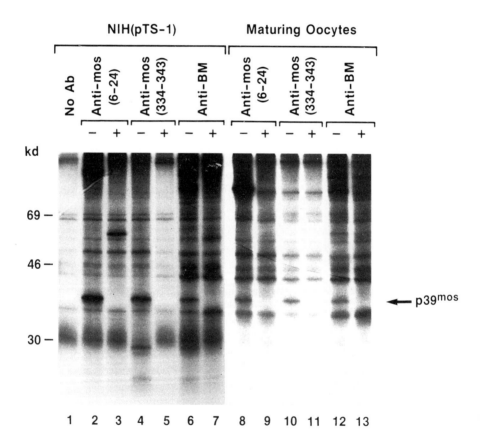

of the *mos* proto-oncogene product during meiotic maturation in *Xenopus* and mouse oocytes (Sagata et al 1988, Paules et al 1989) and showed that it is a candidate initiator of maturation or M phase promoting factor (MPF) in amphibian oocytes (Sagata et al 1989). MPF is a regulator of both meiosis and mitosis. These studies indicate that the *mos* product functions during meiotic maturation in vertebrates.

mos protein product in transformed cells and mouse oocytes

We characterized the *mos* product expressed in NIH/3T3 cells and compared it to the product expressed in maturing mouse oocytes (Paules et al 1989). Using three different *mos*-specific antisera, we immunoprecipitated a protein with an apparent molecular weight of 39 kDa from metabolically [^{35}S] labelled NIH/3T3 cells transformed with a mouse *mos* construct activated by a viral long terminal repeat, pTS-1 (Blair et al 1982) (Fig. 1, lanes 2, 4 and 6). Pulse-chase studies revealed that p39^{mos} has a short half-life ($t_{1/2} = 30$ minutes) and *in vivo* phosphorylation studies show that p39^{mos} is a phosphoprotein (data not shown). Proteins isolated from mouse oocytes undergoing meiotic maturation were metabolically [^{35}S] labelled and analysed for the presence of *mos* proto-oncogene product. Using the same three mouse *mos*-specific antisera, we detected a 39 kDa protein by immunoprecipitation analysis (Fig. 1, lanes 8, 10 and 12). As with protein extracts from NIH(pTS-1) cells, immunoprecipitation of p39^{mos} from mouse oocyte extracts is blocked by competing antigen (Fig. 1, lanes 3, 5, 7, 9, 11 and 13). In addition, p39^{mos} in meiotic mouse oocytes also has a short half-life ($t_{1/2} < 2$ hours) (data not shown). Thus, the *mos* protein normally expressed in mouse oocytes during meiosis appears indistinguishable, in size, antigenically and in stability, from the transforming *mos* protein detected in NIH(pTS-1) cells.

FIG. 1.   Immunoprecipitation analyses of *mos* protein from *mos*-transformed NIH/3T3 cells and meiotic mouse oocytes. Cells were metabolically labelled with [^{35}S] methionine and [^{35}S] cysteine for one hour [NIH(pTS-1) cells] or three hours (maturing oocytes). Aliquots of clarified lysates containing the equivalent of approximately 26×10^6 TCA-precipitable c.p.m. from NIH(pTS-1) cells or approximately 5.5×10^6 TCA-precipitable c.p.m. from about 560 oocytes, either in the absence ($-$) or presence ($+$) of competing antigen, were subjected to immunoprecipitation analyses using antiserum against the N-terminal portion of *mos* [anti-*mos*(6–24)], against the C-terminal portion of *mos* [anti-*mos*(334–343)], or against bacterially expressed v-*mos*$^{HT-1}$. Fully grown mouse oocytes were released from IBMX (1-isobutyl-3-methylxanthine)-maintained arrest and allowed to progress through meiotic maturation *in vitro* for eight hours prior to labelling. One aliquot of NIH(pTS-1) cell lysate received only protein A-Sepharose treatment (lane 1). Immunoprecipitated proteins were analysed by 12.5% SDS-polyacrylamide gel electrophoresis followed by autoradiography for either three days (lanes 1 to 7) or three weeks (lanes 8 to 13). The arrow points to the protein specifically precipitated by all three *mos* antisera from c-*mos*^{mu}-transformed NIH/3T3 cells [NIH(pTS-1)] and meiotically maturing mouse oocytes. (From Paules et al 1989.)

p39^{mos} expression in oogenesis

We examined the pattern of p39^{mos} expression during oogenesis (Paules et al 1989). Oocytes were collected and proteins were metabolically [^{35}S]labelled for three hours during the following stages of oogenesis: growing oocytes; fully grown prophase It-arrested oocytes; maturing oocytes released from prophase I-arrest from 2–5 hours, from 5–8 hours, and from 8–11 hours; metaphase II oocytes (ovulated eggs); and pronuclear-stage embryos (zygotes). We have not detected p39^{mos} in growing oocytes, but we do detect a low level in prophase I-arrested, fully grown oocytes (Fig. 2, lanes 1 and 3). Higher levels of p39^{mos} are detectable in oocytes metabolically labelled during meiotic maturation from 2–5 hours after release from prophase I arrest (data not shown), from 5–8 hours (Fig. 2, lanes 5 and 7), from 8–11 hours (Fig. 1, lanes 8, 10 and 12), and in ovulated eggs, but not in zygotes (Paules et al 1989). Moreover, there is no apparent difference between the levels of p39^{mos} synthesized in oocytes matured *in vivo* and *in vitro* (data not shown).

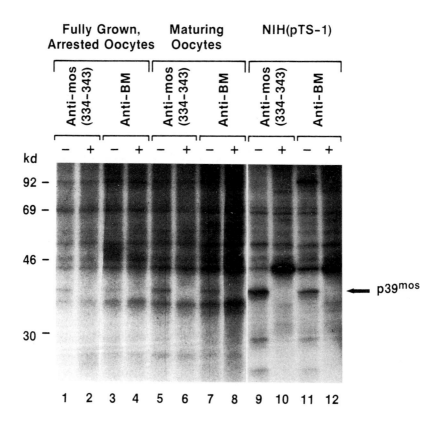

mos antisense oligodeoxyribonucleotides interrupt meiotic maturation

We tested whether injection of *mos* antisense oligodeoxyribonucleotides (oligonucleotides) would have any effect on maturing mouse oocytes (Paules et al 1989), since injection of *mos* antisense nucleic acids into amphibian oocytes has been shown to interrupt maturation (Sagata et al 1988). A mixture of three different *mos* antisense or sense oligonucleotides corresponding to unique coding regions within *mos* RNA transcripts was injected into cumulus-enclosed, fully grown, prophase I-arrested oocytes. In addition, we tested similar mixtures of oligodeoxyribonucleotide phosphorothioates (S-oligonucleotides) that are resistant to nuclease digestion within the cellular compartment (Paules et al 1989). Microinjection of these antisense oligonucleotide mixtures into fully grown oocytes inhibited completion of the first meiotic division and production of the first polar body (PB). While microinjection of sense normal or S-oligonucleotides did not significantly alter the percentage of oocytes producing the first PB (58% and 60%, respectively, compared to 72% in uninjected control oocytes), the microinjection of antisense normal or S-oligonucleotides markedly inhibited PB emission (15% and 10% of the injected oocytes, respectively) (Fig. 3A).

To confirm that this inhibition of meiotic maturation was *mos* specific, we tested the three sets of sense and antisense oligonucleotides individually (Fig. 3B). Again, oocytes injected with any of the three different *mos* antisense oligonucleotides were markedly inhibited in the completion of meiotic maturation (10–27% PB emission), while oocytes injected with sense oligonucleotides showed only a slight reduction in PB emission (58–62%, compared to 75% in control oocytes).

Chromosomal preparations from antisense oligonucleotide-injected oocytes revealed bivalent-stage chromosomes corresponding to chromosomes at or around metaphase I, while sense-injected or uninjected oocytes were at metaphase II. During the normal process of meiotic maturation, cytoplasmic organelles and cytoskeletal components cluster in the perinuclear region at

---

FIG. 2.   p39^{mos} expression during oogenesis. Metabolically [^{35}S] labelled proteins were precipitated from extracts of oocytes in various stages of oogenesis. Fully grown oocytes were either held in the presence of IBMX during labelling for 2.5 hours or allowed to progress through meiosis *in vitro* for five hours prior to labelling for three hours. Aliquots of clarified lysates containing the equivalent of 4.1×10^6 TCA-precipitable c.p.m. from about 350 fully grown, GV-arrested oocytes, 6.3×10^6 TCA-precipitable c.p.m. from about 570 meiotically maturing oocytes, or 24×10^6 TCA-precipitable c.p.m. from NIH(pTS-1) cells were subjected to immunoprecipitation analyses using antiserum against the C-terminal portion of *mos* [anti-*mos* (334–343)] or against bacterially expressed v-*mos*$^{HT-1}$, either in the absence ($-$) or presence ($+$) of competing antigen. Immunoprecipitated proteins were analysed by 12.5% SDS-polyacrylamide gel electrophoresis followed by autoradiography for either five weeks (lanes 1 to 8) or one week (lanes 9 to 12). (From Paules et al 1989.)

metaphase I, just before germinal vesicle breakdown (GVBD), and disperse shortly thereafter. Strikingly, cytoplasmic organelles remained clustered in the pericentral region following GVBD and did not disperse in antisense oligonucleotide-injected oocytes (Fig. 4). Examination of electron micrographs revealed that oocyte cytoplasmic organelles, such as mitochondria, appear morphologically normal.

The p39^{mos} proto-oncogene product is a candidate 'initiator' for oocyte maturation in *Xenopus*

We have reported that p39^{mos} is expressed in maturing *Xenopus* oocytes and demonstrated that *mos*-specific antisense oligonucleotides microinjected into *Xenopus* oocytes inhibit p39^{mos} expression and progesterone-induced maturation (Sagata et al 1988). This result, as well as the evidence that synthesis of p39^{mos} occurs during steroid-induced oocyte maturation, strongly suggests

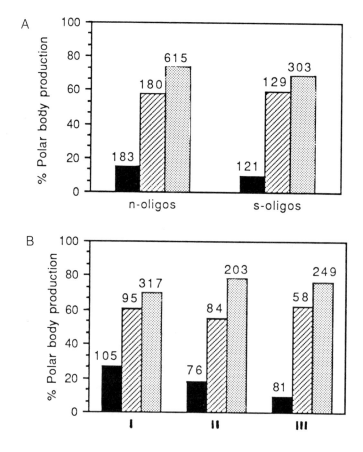

that the *mos* proto-oncogene product is required for MPF activity. This activity appears after fully grown oocytes are released from their naturally arrested state at meiotic prophase (Masui & Markert 1971). MPF exists in immature *Xenopus* oocytes in an inactive form (pre-MPF) (Cyert & Kirschner 1988), and during the first few hours after progesterone treatment, *de novo* protein synthesis is required for MPF activation. This protein synthesis-dependent step led Wasserman & Masui (1975) to propose that an 'initiator' of maturation was required for MPF activation. Such an initiator must be synthesized early and in low abundance, because a detectable increase in protein synthesis occurs only after MPF is activated, at a time when cycloheximide no longer inhibits GVBD. We were interested in determining when p39^{mos} was synthesized after steroid treatment and whether it had the properties of an initiator.

Low levels of p39^{mos} were detected in fully grown oocytes during a one-hour labelling period in oocytes defolliculated by collagenase, but not in manually defolliculated oocytes (N. Sagata, unpublished data) (Fig. 5A). The level of p39^{mos} increases slightly during the first hour after addition of progesterone, but by the second hour the levels increase dramatically and approach the high levels observed during the third and fourth hours (Fig. 5A). In the same group of oocytes, we determined the time after progesterone treatment when GVBD occurs in the presence of cycloheximide, a time that corresponds to pre-MPF activation. GVBD did not become protein synthesis independent until more than two hours after steroid treatment. Thus, the greatest increase in p39^{mos} precedes both GVBD and pre-MPF activation (Figure 5B). These properties of p39^{mos} fulfill those required by a candidate initiator of MPF.

To test whether p39^{mos} can activate maturation in the absence of progesterone, the *Xenopus* proto-oncogene (c-*mos*^{xe}) RNA was transcribed from an SP6 vector with or without a cap structure and injected into fully grown

FIG. 3. (A) Effect of injections of mixtures of *mos* oligodeoxyribonucleotides in mouse oocytes on the production of the first polar body. Cumulus cell-enclosed oocytes blocked at the GV stage in IBMX-containing medium were microinjected with a mixture of three different *mos* antisense (solid bars) or sense (hatched bars) normal oligodeoxyribonucleotides (n-oligos) or oligodeoxyribonucleotide phosphorothioates (s-oligos). These were then allowed to mature overnight. A group of uninjected control oocytes (stippled bars), cultured in the same manner, was examined in each experiment. χ^2 analysis showed significant differences ($P < 0.01$) for sense-injected *versus* control oocytes and for sense-injected *versus* antisense-injected oocytes. (B) Effect of injections of individual *mos* oligodeoxyribonucleotides in mouse oocytes on the production of the first polar body. Cumulus cell-enclosed oocytes blocked at the GV stage in IBMX-containing medium were microinjected with one of three different antisense (solid bars) or sense (hatched bars), normal oligodeoxyribonucleotides and then allowed to mature overnight. A group of uninjected control oocytes (stippled bars), cultured in the same manner, was examined in each experiment. χ^2 analysis revealed significant differences with $P < 0.01$ for sense-injected *versus* uninjected oocytes and for sense-injected *versus* antisense-injected oocytes in all but two cases (sense-injected *versus* uninjected oocytes with oligonucleotides II and III, for which P was < 0.05). (Fig. 3B taken from Paules et al 1989).

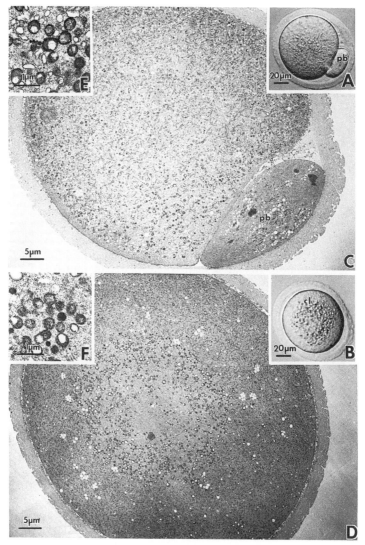

FIG. 4. Mouse oocytes injected with *mos* oligodeoxyribonucleotides. Cumulus cell-enclosed oocytes blocked at the GV stage in IBMX-containing medium were microinjected with a mixture of normal sense or antisense oligonucleotides and then allowed to mature overnight. Nomarski optics and low-magnification electron micrographs of sense-injected oocytes (panels A and C, respectively) show polar body (pb) emission and organelles distributed throughout the cytoplasm. Antisense-injected oocytes (panels B and D) show organelles clustered (cl) in the paracentral region. Such clustering occurs in the normal progress of maturation, just before GVBD, with dispersal occurring shortly thereafter. Otherwise, the cytoplasm of the antisense-injected oocytes appears normal, with no apparent degradation of organelles, such as mitochondria (panels E and F) which are particularly sensitive to toxic conditions.

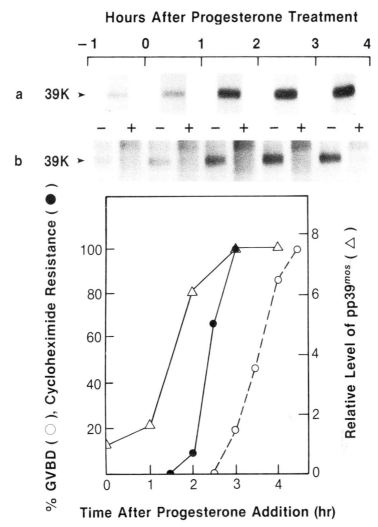

FIG. 5. Fully grown Stage VI *Xenopus* oocytes were defolliculated by treatment with collagenase and cultured in modified Barth solution (MBS). (A) Groups of 10 oocytes were metabolically labelled with [^{35}S]methionine at 2 mCi/ml in MBS for one hour at one hour intervals before and after progesterone treatment. Oocyte extracts from the equivalent of four oocytes were subjected to immunoprecipitation analysis with either (a) a monoclonal antibody raised against the *Xenopus* c-*mos* product expressed in *Escherichia coli* or (b) a peptide antiserum in the presence (+) or absence (−) of the peptide antigen. (B) Thirty oocytes were scored for GVBD every 30 minutes after progesterone addition (○---○). Groups of 20 oocytes were also treated with cycloheximide (100 μg/ml) at 30-minute intervals after progesterone addition. After 12 hours, oocytes were examined for % GVBD, which is expressed as cycloheximide resistance of GVBD (●---●). Densitometric measurements of p39^{mos} on autoradiographs performed in parallel are shown (△---△). (From Sagata et al 1989).

oocytes. Oocytes injected with capped c-mos^{xe} RNA made two products, p39^{mos} and p38^{mos} (data not shown). Levels of these products were tenfold greater than that of the endogenous p39^{mos} synthesized in response to progesterone treatment. The smaller *mos* product co-migrates with the p38^{mos} primary translation product detected in reticulocyte lysates (data not shown), while the larger product most likely corresponds to the phosphorylated form of p39^{mos}. In this system, only capped RNA is efficiently translated and oocytes injected with uncapped c-mos^{xe} RNA show no increase above the level of p39^{mos} detected in fully grown oocytes. Capped c-mos^{xe} RNA induced GVBD in a dose-dependent manner (Fig. 6), whereas uncapped c-mos^{xe} RNA or capped RNA annealed with antisense oligonucleotide and then injected showed no significant increase in GVBD (Fig. 6). These results demonstrate that overexpression of p39^{mos} triggers GVBD in the absence of progesterone.

We observed that the induction kinetics of GVBD by c-mos^{xe} RNA was slower than that of progesterone (data not shown). Oocytes microinjected with c-mos^{xe} RNA and simultaneously treated with progesterone showed no significant acceleration and had similar GVBD kinetics as oocytes treated with progesterone alone. However, if fully grown oocytes are preincubated with c-mos^{xe} RNA for one hour and then stimulated with progesterone, GVBD is accelerated and the time required for GVBD is reduced by approximately 25%. Thus, preincubation with p39^{mos} diminishes the requirement for one or more progesterone-dependent steps.

We also tested whether extracts from oocytes injected with c-mos^{xe} RNA possess MPF activity. For this purpose, we prepared crude MPF extracts from test oocytes, injected them into cycloheximide-treated oocytes, and examined the oocytes for GVBD after eight hours (Table 1). Extracts from oocytes injected with sense *mos*-specific oligonucleotides induced 100% GVBD in recipient cycloheximide-treated oocytes, whereas those injected with *mos* antisense oligonucleotides did not, nor did extracts from control uninjected oocytes (Table 1). These results show that p39^{mos} is required for the activation of pre-MPF and that overexpression of the product can induce activation. In response to progesterone treatment, p39^{mos} levels increase before pre-MPF activation; thus p39^{mos} has the properties of a putative initiator protein. This is the first association of the *mos* proto-oncogene product with a cellular function. The known activities of MPF provide us with the first reasonable picture of downstream events of a proto-oncogene function. Also, while all the studies show that the *mos* gene product functions during meiosis, we have shown earlier that *mos* RNA is expressed in somatic cells and we cannot exclude that it may also function during mitosis.

We have shown in the *Xenopus* oocyte maturation system that p39^{mos} acts downstream from p21^{ras}. Studies are under way to identify the substrates for p39^{mos}. Since microinjection of synthetic *mos* RNA into *Xenopus* oocytes can activate MPF, the *mos* protein may be interacting with one or more of the

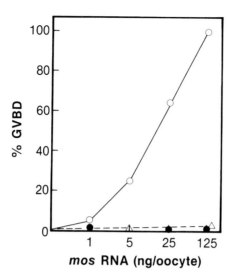

FIG. 6. Groups of 20 fully grown oocytes were microinjected with either capped (○---○) or uncapped (△ --- △) c-*mos*^xe RNA at the indicated doses (Sagata et al 1989). Capped c-*mos*^xe RNA pre-annealed with an excess (30 ng) of *mos*-specific antisense oligodeoxyribonucleotide was also microinjected (●---●). After 12 hours, each oocyte was examined for GVBD both externally and internally. (From Sagata et al 1989.)

TABLE 1 MPF activation in c-*mos*^xe *Xenopus* oocytes injected with RNA or oligodeoxyribonucleotides

Treatment[a]	%GVBD	No. Experiments
Progesterone (100)	100	2
C-SP6 (20 ± 5%)	35 ± 10	3
S/P (90 ± 8%)	100	3
AS/P (0)	0	3

[a]Groups of ten donor oocytes were either tested with progesterone, or microinjected with 125 ng of capped c-*mos*^xe RNA (C-SP6), or microinjected with mixtures of *mos* sense (S/P) or antisense (AS/P) oligonucleotides. The mixtures contained 130 ng of each oligonucleotide. After four hours preincubation, the oligonucleotide-containing oocytes were treated with progesterone. All groups of oocytes were cultured for six hours after progesterone treatment or c-*mos*^xe RNA microinjection, and crude MPF extracts were prepared. The % GVBD in donor oocytes at the time of MPF preparation is indicated in parentheses. The variation where observed is given.

components of MPF. *In vitro* immune complex kinase assays are being performed with p39^{mos} precipitated from either progesterone-induced oocytes or transformed NIH/3T3 cells that overexpress the *Xenopus* c-*mos* product. In collaboration with Noriyuki Sagata, we have shown that the *Xenopus* c-*mos* protein has autophosphorylation activity *in vitro* (Fig. 7).

Discussion

In most species, fully grown oocytes arrest in the first prophase of meiosis and resume meiotic maturation by progressing through a series of well-defined events that includes dissolution of the nuclear membrane (or GVBD), chromosome condensation, spindle formation, cytokinesis, and arrest again in the second meiotic metaphase. We have shown that *mos* protein is necessary for the resumption of meiotic maturation and in mouse oocytes we show that *mos* is needed for oocytes to continue to progress through meiotic maturation beyond metaphase I. These results suggest a continual need for *mos* protein for normal meiotic maturation to progress to completion. Shortly before GVBD, the

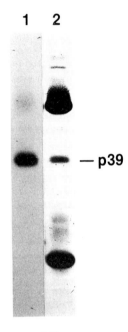

FIG. 7.   Immune complex kinase assay of *Xenopus mos*-transformed NIH/3T3 cells. The *Xenopus mos* protein was immunoprecipitated from a clarified cell extract using a *mos*-specific monoclonal antibody (5S). The kinase reaction was performed with 20 µCi of [γ ^{32}P]ATP for 15 minutes at room temperature, then electrophoresed on a 12% SDS-polyacrylamide gel. Lane 1, autoradiograph after one hour exposure; lane 2, Coomassie blue stain.

cytoplasmic activity MPF appears, which, when transferred to fully grown, prophase I-arrested *Xenopus* oocytes, induces resumption of meiotic maturation and drives maturation to completion. It appears from the results of our studies with *Xenopus* and mouse oocytes that *mos* protein is necessary to maintain active MPF throughout meiotic maturation. Furthermore, since MPF is a universal regulator of the transition from late G2 into mitosis (G2/M), constitutive expression of *mos* in somatic cells, such as NIH(pTS-1) cells, may induce the transformed phenotype through activation of certain elements of pre-MPF at an inappropriate time in the cell cycle. One other consequence of this model would be that G2/M signals could aberrantly regulate the types of genes that are expressed during interphase, affecting both phenotype and the ability of the cell to differentiate normally.

Acknowledgements

Research sponsored in part by the National Cancer Institute, DHHS, under contract No. N01-C0-74101 with Bionetics Research, Inc (GVW, ID, MO, RP, NS, NY) and Grant HD20575 from the National Institute of Child Health and Human Development (JE, RB). The contents of this publication do not necessarily reflect the views or policies of the Department of Health and Human Services, nor does mention of trade names, commercial products or organizations imply endorsement by the US Government.

References

Blair DG, Cooper CS, Oskarsson MK, Eader LA, Vande Woude GF 1982 New method for detecting cellular transforming genes. Science (Wash DC) 218:1122–1125

Cyert MS, Kirschner MW 1988 Regulation of MPF activity *in vivo*. Cell 53:185–195

Goldman DS, Kiessling AA, Millette CF, Cooper GM 1987 Expression of c-*mos* RNA in germ cells of male and female mice. Proc Natl Acad Sci USA 84:4509–4513

Jones M, Bosselman RA, van der Hoorn FA, Berns A, Fan H, Verma IM 1980 Identification and molecular cloning of Moloney mouse sarcoma virus-specific sequences from uninfected cells. Proc Natl Acad Sci USA 77:2651–2655

Keshet E, Rosenberg M, Mercer JA et al 1988 Developmental regulation of ovarian-specific *mos* expression. Oncogene 2:235–240

Masui Y, Markert CL 1971 Cytoplasmic control of nuclear behavior during meiotic maturation of frog oocytes. J Exp Zool 177:129–145

Mutter GL, Wolgemuth DL 1987 Distinct developmental patterns of c-*mos* proto-oncogene expression in female and male mouse germ cells. Proc Natl Acad Sci USA 84:5301–5305

Oskarsson M, McClements W, Blair DG, Maizel JV, Vande Woude GF 1980 Properties of a normal mouse cell DNA sequence (sarc) homologous to the src sequence of Moloney sarcoma virus. Science (Wash DC) 207:1222–1224

Paules RS, Buccione R, Moschel R, Vande Woude GF, Eppig JJ 1989 The mouse c-*mos* proto-oncogene product is present and functions during oogenesis. Proc Natl Acad Sci USA 86:5395–5399

Propst F, Vande Woude GF 1985 c-*mos* proto-oncogene transcripts are expressed in mouse tissues. Nature (Lond) 315:516–518

Propst F, Rosenberg MP, Iyer A, Kaul K, Vande Woude GF 1987 c-*mos* proto-oncogene RNA transcripts in mouse tissues: structural features, developmental regulation and localization in specific cell types. Mol Cell Biol 7:1629–1637

Propst F, Rosenberg MP, Oskarsson MK et al 1988 Genetic analysis and developmental regulation of testis-specific RNA expression of *Mos*, *Abl*, actin and *Hox-1.4*. Oncogene 2:227–233

Sagata N, Oskarsson M, Copeland T, Brumbaugh J, Vande Woude GF 1988 The c-*mos* proto-oncogene product functions during meiotic maturation in *Xenopus* oocytes. Nature (Lond) 335:519–525

Sagata N, Daar I, Oskarsson M, Showalter S, Vande Woude GF 1989 c-*mos* proto-oncogene product is a candidate "initiator" for oocyte maturation. Science (Wash DC) 245:643–646

Wasserman WJ, Masui Y 1975 Effects of cycloheximide on a cytoplasmic factor initiating meiotic maturation in *Xenopus* oocytes. Exp Cell Res 91:381–388

DISCUSSION

Méchali: When you inject c-*mos* RNA into *Xenopus* oocytes, it is translated and there is oocyte maturation. But oocytes contain a large amount of c-*mos* RNA that is not usually translated; do you think there is some factor that facilitates translation of endogenous *mos* mRNA which is induced by progesterone?

Vande Woude: We tried to find whether the *mos* RNA in fully grown oocytes was associated with polysomes, but all we know is that it is associated with protein complexes. A large percentage of the maternal mRNA pool is associated with protein complexes. It is very interesting how the expression of specific maternal mRNAs is induced in response to progesterone or other reagents that activate maturation.

Verma: George, is there only one size of *mos* mRNA in oocytes, as opposed to the different sized transcripts in the testis?

Vande Woude: In higher vertebrates, birds and mammals, we find that *mos* transcripts vary in size in a tissue-specific manner. In *Xenopus* the transcripts in brain, testis and ovary are all the same size, and the testis and ovary transcripts have the same 5' ends. Upstream genome sequences are quite interesting. We showed a number of years ago that the human *mos* locus has an N-terminal overlapping open reading frame. *Xenopus* also has one.

Verma: So this is like the one where you have inhibition by the upstream sequences.

Vande Woude: Right. We have shown that it is necessary to interrupt the overlapping open reading frame in order to activate the transforming potential of both the *Xenopus* and human *mos* genes.

Méchali: Is it possible to induce premature maturation by injection of *mos* mRNA into immature oocytes?

Vande Woude: We have tried experiments similar to those you suggest but the results were negative. Other things are required, probably pre-MPF, which may not be present.

Brugge: George, have you been able to express enough *mos* to inject the purified protein into oocytes to see whether it can overcome the cycloheximide block?

Vande Woude: The *mos* made in baculovirus is insoluble, therefore we have been unable to use that approach.

Maness: Geoffrey Cooper has results using *mos* antisense oligonucleotides that indicate blockage of germinal vesicle breakdown at metaphase II in mouse oocytes. Your results show normal germinal vesicle breakdown. How do you reconcile these findings?

Vande Woude: We seem to be doing it the same way, but he is seeing effects later in maturation. John Eppig has been doing our injections and he developed that system; I am not sure what the differences are due to. It would be consistent with our belief that *mos* is required throughout the maturation process, and not just at the beginning. In *Xenopus*, Ira Daar in my lab has demonstrated that *mos* has to be present for maturation.

Hunter: George, what about regulation of protein kinase activity during maturation? Have you shown that you need the kinase activity by using a kinase-deficient mutant of *mos*?

Vande Woude: No, we haven't, but that has been shown in Dan Donoghue's lab using a *mos* gene with a mutation in the ATP-binding domain. Nori Sagata has shown that there is pp39^{mos} autophosphorylation activity in product precipitated from mature oocytes. That is the first demonstration that the *mos* proto-oncogene product has kinase autophosphorylation activity.

Hunter: But you can't say whether its activity changes during maturation? Is there an additional level of regulation of kinase activity?

Vande Woude: That hasn't really been done.

Brugge: George, have you incubated *mos* protein with MPF to see whether it's a substrate?

Vande Woude: We haven't even excluded the possibility that *mos* may be part of MPF.

Brugge: Have you assayed the phosphorylation of MPF in an immune complex assay with *mos*?

Vande Woude: Those kinds of experiments are in progress.

Hanley: Can you comment on *mos* expression in mammalian sperm maturation? It is the only other situation I know where there is some evidence that there may be a cell surface progesterone receptor.

Vande Woude: I don't know.

Hunter: What about the possibility that *mos* functions in later cell cycles?

Vande Woude: Fritz Propst first found *mos* transcripts in adult tissues several years ago. We also found expression of *mos* in NIH/3T3 cells by S1 nuclease analysis. We have some preliminary data which suggest that *mos* transcripts may be specifically expressed during G2/M and M stages of the cell cycle. We are not completely happy with these experiments because the level of transcription is so low. I think it will be necessary to show that *mos* functions during this period. We have tried using antisense oligonucleotides in fibroblasts but without any demonstrable effect on growth. We have used *mos*-transformed cells as controls and not observed reduction of *mos* product in these cells. We believe that the oligonucleotides are not entering the cells.

Maness: In the oocytes injected with antisense oligonucleotides to what extent is the accumulation of cytoplasmic particulates over the nucleus associated with an alteration of the underlying cytoskeleton architecture?

Vande Woude: A good question; we are only just beginning to address that issue.

Hunter: In *Xenopus*, is the *mos* protein degraded?

Vande Woude: Nori Sagata has shown that 15 minutes after fertilization all the maternal pp39^{mos} product has been specifically degraded by a calpain protease. However, *mos* RNA persists through late blastula and it is possible that it is translated at undetectable levels during the rapid mitotic cleavage.

Hunter: Is that RNA on polysomes?

Vande Woude: I don't know.

Verma: Have you looked at *mos* in other species?

Vande Woude: We have tried but it is difficult because it is not well conserved. For example, we have tried to identify the *Drosophila* homologue using the polymerase chain reaction, but so far we have not succeeded.

Hunter: What about the question of a family of *mos* kinases?

Vande Woude: That is a very important issue. If the *mos* gene is not highly conserved, I would expect that the homology with other potential members of the family is low and they will be difficult to detect by conventional methods.

Norbury: It would be interesting to know if the toxicity of *mos* overexpression could be relieved in a transformed cell where other aspects of growth control are already deregulated. One interpretation of the toxicity would be that *mos* activates some aspect of cell cycle control without simultaneous activation of growth control.

Vande Woude: Yes, there are two possible explanations. Either overexpression is toxic and selection is for cells that express lower levels of *mos* that do not interfere with proliferation, or we could be selecting for cells that suppress the high level of *mos* expression. We suspect that toxicity, however, is due to activation of M phase events.

Feramisco: Does *mos* or MSV injected into quiescent somatic cells induce a transition through the cell cycle?

Vande Woude: I don't think that has been done.

Brugge: George, have you incubated cells expressing a temperature-sensitive *mos* gene at the non-permissive temperature in the absence of serum and then shifted the temperature to see if *mos* can induce DNA synthesis in the absence of serum?

Vande Woude: That hasn't been done for *mos*.

General discussion I

The role of proto-oncogenes in amphibian development I

Méchali: The three main characteristics of the proto-oncogene c-*myc* are that the mRNA and protein it encodes have a short half-life, the protein is localized in the nucleus, and there is a strong correlation between expression of c-*myc* and cell proliferation. However, during early embryonic development in *Xenopus*, all of these characteristics are reversed.

First, c-*myc* is expressed in *Xenopus* oocytes, which remain in the ovary for up to two years without dividing, and thus expression of c-*myc* is uncoupled from cell proliferation (Taylor et al 1986, King et al 1986, Hourdry et al 1988, Gusse et al 1989). A single oocyte contains 5×10^5 times the amount of c-*myc* RNA and protein found in a somatic proliferative cell (Taylor et al 1986, Gusse et al 1989).

Second, both the mRNA and protein are stable in the oocyte. This large stockpile of c-*myc* protein is preserved only during the cleavage stage of development; most of the maternal c-*myc* protein is degraded within two hours of the start of gastrula stage.

The third characteristic that is reversed is that the c-*myc* protein is not found in the nucleus as in somatic cells, but in the cytoplasm of the oocyte. Oocytes are cells that grow without division. After maturation and fertilization, the eggs undergo the series of rapid cell divisions characteristic of early vertebrate development. We thought that the stockpiling of c-*myc* in the cytoplasm of oocytes might be necessary to avoid activation of its function in oocyte nuclei. Thus we expected the protein to translocate into the nuclei after fertilization. Indeed we observed that fertilization triggers the rapid migration of the maternal c-*myc* protein into the nuclei of the cleavage embryo (Gusse et al 1989). This transfer is not proportional to cell number; there is a rapid entry of c-*myc* protein into the nucleus during the very early cleavages, then the amount of c-*myc* protein per nucleus decreases exponentially until an equilibrium is reached when the embryo is between 4000 and 8000 cells. This is at the 12th cell cycle, a time precisely corresponding to the activation of zygotic transcription. So there is a strong correlation between the titration of cytoplasmic c-*myc* protein by the embryonic nuclei and the onset of new transcription in the embryo at the mid-blastula transition. Some time ago, Newport & Kirschner (1982a,b) proposed that the mid-blastula transition was controlled by a maternal cytoplasmic factor which might be titrated by the nuclei at that stage. The properties of c-*myc* protein during *Xenopus* development fulfil these requirements.

These results suggest that c-*myc* may play a role in nuclear replication. Gerard Evan has raised an antibody against a *Xenopus* c-*myc* peptide that is fully conserved in all species. We have injected the purified antibody into one of the two blastomeres of an embryo at the 2-cell stage. The injected blastomere stops dividing, whereas the other half of the embryo divides normally (M. Méchali, M. Gusse, unpublished). As one of the controls we injected the antibody preincubated with the corresponding antigenic peptide and we observed that development proceeded normally. We also confirmed that the antibody reacted specifically with the c-*myc* protein inside the egg.

Control experiments show that injection of rabbit IgG or an antibody against dinitrophenol (a neutral component) leads to 75–80% of normal development, whereas in 65–70% of the anti-c-*myc* antibody-injected embryos cleavage is blocked or strongly inhibited.

In conclusion, we have observed a specific developmental regulation of c-*myc* characterized by an unusual uncoupling of c-*myc* expression and cell proliferation. This uncoupling leads to overexpression of c-*myc* protein in the oocyte, which makes this cell competent for the rapid cleavages after fertilization. We are now working on the idea that c-*myc* drives the onset of zygotic expression by controlling the frequency of DNA replication during early development.

Vande Woude: Do you know where the block is in the blastomere whose division has been blocked by the anti-c-*myc* antibody?

Méchali: In sections of the blocked embryo observed by light microscopy, we do not see a nuclear syncytium; there is no nuclear replication. It is not a block of cell division in which nuclear division continues.

Vande Woude: Have you tried labelling with [³H]thymidine?

Méchali: There are *in vitro* cell systems for DNA replication derived from *Xenopus* eggs that are used in many laboratories. A high speed extract mimics all the events at the replicative fork in eukaryotes: RNA priming, elongation of DNA chains and chromatin assembly (Méchali & Harland 1982, Almouzni & Méchali 1988). All these reactions proceed normally after addition of anti-c-*myc* antibody. In low speed egg extracts, initiation of double-stranded DNA replication occurs in the presence of sperm nuclei (Blow & Laskey 1986). This process is inhibited 60–80% by anti-c-*myc* antibody. As elongation of the DNA is not inhibited, the event affected must be a step that is important to trigger initiation.

Hunter: How do you think that c-*myc* protein is retained in the cytoplasm?

Méchali: We don't know. We looked at the phosphorylation and found if you inject ATPγ³²P into an oocyte, the protein is not phosphorylated during a 24 hour labelling period. If you inject ATPγ³²P into a fertilized egg, c-*myc* is phosphorylated within six hours. I don't know if this is *de novo* phosphorylation or removal of phosphate from the oocyte protein and turnover. However, there is clearly a change in the phosphorylation state of the protein which correlates with the translocation of c-*myc* protein from the cytoplasm into the nucleus.

Hunter: Is it complexed with another protein for instance?

Méchali: I don't know. It could be active repression of the translocation or simply that the nuclear membrane of the oocyte excludes c-*myc* protein but that of the egg doesn't, or that a positive factor required for translocation is induced at fertilization. All these things are possible.

Harlow: Are other proteins getting transported to the nucleus normally?

Méchali: Yes, the proliferating nuclear antigen (PCNA), a protein that is required for DNA replication, is almost entirely localized in the nucleus, so the oocyte does not exclude all proteins required for DNA replication from the nucleus.

McMahon: Have you tried injecting labelled c-*myc* made in reticulocyte lysate to see if it enters the nucleus?

Méchali: No, we have not yet. We also want to inject c-*myc* protein directly into the nucleus of the oocyte. I don't think it will stimulate DNA replication in the oocyte, but we can try.

Although the oocyte contains a huge amount of c-*myc* protein, the concentration is not greater than that in somatic cells, because the egg is very large. However, the amount is large relative to the genome. There is also this massive entry into the nucleus after fertilization. Our working model is that in the fertilized egg, the many DNA replication sites, which are at close intervals in the genome, are controlled by c-*myc*. This might explain the very rapid cell cycle during early development. Such a short interval between DNA replication sites might not be compatible with transcription. At each round of egg cleavage that proceeds without transcription, the amount of c-*myc* protein relative to the genome is decreased, and this might lead to a reduction in the number of DNA replication sites. When the interval between replication sites exceeds a critical distance, transcription might be switched on.

Nurse: If the level of c-*myc* is the basis of the timing of the mid-blastula transition, then injecting c-*myc* protein into the nucleus of the oocyte should delay that transition.

Méchali: We are going to do that experiment. It should also be possible to do the reverse, the antibody should sequestrate the protein and accelerate the initiation of zygotic gene transcription.

Vande Woude: In the experiments where you block cleavage by injecting the antibody, if you inject lower amounts of antibody, can you see the effect at later cleavage stages?

Méchali: If we inject half the usual amount of the antibody, we do not get anything at the beginning of the cleavage stage and then there is a delaying effect. For example, the control half of the embryo is at stage 8, the injected half is at stage 6. In most of the totally arrested blastomeres, you can observe abortive cleavages at the morula stage, which suggest that the cytoplasmic clock is still working.

Role of proto-oncogenes in amphibian development II

Verma: This is the work of Mark Kindy. We cloned the *Xenopus fos* gene to see how conserved the gene is between species. We then thought it would be interesting to look at its pattern of expression.

The mRNA is made for only a very short period of time. It is present in oocytes and unfertilized eggs. It disappears immediately after fertilization. At stage 23 of development *fos* mRNA comes back. There is also some at stage 45.

We tried *in situ* hybridization in the oocytes; the best we could do was to show that it is found at both the animal and vegetal poles. During oogenesis there is a lot of *fos* mRNA and you can see some protein too, but very little. The problem with looking for the protein is that the antibodies we have at the moment are not very good. At Stage 2 there is some *fos* mRNA and at Stage 3 there is practically none left. By the first cleavage the *fos* mRNA is practically all gone. So the RNA is present for a very short time; the protein is also present for a very short time, although inability to detect the protein doesn't necessarily mean much here, it just means we can't detect it with our assay system.

We then asked whether this *Xenopus fos* protein behaved like the mammalian one. Is there a *jun*-like protein? Is the *fos* involved in the general suppression of transcription during oogenesis? The work was done on the *Xenopus* A6 cell line. If you add serum or TPA to this cell line, the *fos* gene behaves exactly as it does it mammalian cells, all the induction patterns are maintained. The *Xenopus fos* makes a complex with mouse *jun* protein that will recognize DNA, specifically the TRE. If you immunoprecipitate proteins from A6 cells with anti-*fos* antibody, you see a series of 39 kDa proteins. Mark Kindy thinks that there is a *Xenopus jun*-like protein that may interact with the *Xenopus fos* protein.

Mark has tried injecting *fos* antisense oligonucleotides into oocytes, but has no clear results yet on the effects on development. We need to know whether the protein is present throughout the times when we can detect the RNA or are there times when the RNA is present but not translated.·

Hunter: So you can't say whether the protein is in the cytoplasm?

Verma: The protein is nuclear, as shown by immunofluorescence in oocytes and in tissue culture cells. If there were low concentrations present in the cytoplasm, we wouldn't be able to detect it with our antibodies.

Vande Woude: Is it hyperphosphorylated?

Verma: I don't know about in the oocyte. In the cell line it is exactly the same as in mammalian cells. It has the same C-terminus as the mouse *fos* protein, which is needed for transrepression. So an attractive hypothesis is that *fos* acts as a general repressor, which explains why there is so little transcription in the oocyte.

Méchali: The protein is lost by oocyte Stage 3 which is when ribosomal RNA synthesis begins, which might be significant.

References

Almouzni G, Méchali M 1988 Chromatin assembly during DNA synthesis in a cell free system from Xenopus eggs. EMBO (Eur Mol Biol Organ) J 7:665–672

Blow J, Laskey RA 1986 Initiation of DNA replication in nuclei and purified DNA by a cell-free extract of *Xenopus* eggs. Cell 47:577–587

Gusse M, Ghysdael J, Evan G, Soussi T, Mechali M 1989 Translocation of a store of maternal cytoplasmic c-myc protein into nuclei during early development. Mol Cell Biol 9:5395–5403

Hourdry J, Brulfert A, Gusse M, Schovaert D, Taylor MV, Méchali M 1988 Localization of c-myc expression during oogenesis and embryonic development in Xenopus laevis. Development 104:631–641

King MW, Roberts JM, Eisenman RN 1986 Expression of the c-myc proto-oncogene during development of Xenopus laevis. Mol Cell Biol 6:4499–4508

Méchali M, Harland R 1982 DNA synthesis in a cell-free system from *Xenopus* eggs. Priming and elongation on single-stranded DNA *in vitro*. Cell 30:93–101

Newport JW, Kirschner M 1982a A major developmental transition in early *Xenopus* embryos: 1. Characterization and timing of cellular changes at the midblastula stage. Cell 30:675–686

Newport JW, Kirschner M 1982b A major developmental transition in early *Xenopus* embyros: 2. Control of the onset of transition. Cell 30:687–696

Taylor MV, Gusse M, Evan GI, Dathan N, Méchali M 1986 *Xenopus* myc proto-oncogene during development: expression as a stable maternal mRNA uncoupled from cell division. EMBO (Eur Mol Biol Organ) J 5:3563–3570

Controls of cell proliferation in yeast and animals

Chris Norbury and Paul Nurse

I.C.R.F. Cell Cycle Group, Microbiology Unit, Department of Biochemistry, University of Oxford, OX1 3QU, UK

Abstract. Genetic studies using fission yeast (*Schizosaccharomyces pombe*) have identified a gene, cdc2, whose product (p34[cdc2]) is a protein kinase required for traversal of both the G1 and G2 cell cycle control points. Genetic complementation has been used to demonstrate that p34[cdc2] homologues are functionally and structurally conserved in distantly related eukaryotes, and p34[cdc2]-related proteins are components of both maturation-promoting factor (MPF) and the M phase (growth-associated) histone H1 kinase. The p34[cdc2] homologues of multicellular eukaryotes undergo potentially regulatory phosphorylation changes through the cell cycle. Phosphorylation on serine during late G1 is accompanied by a significant increase in p34[cdc2] kinase activity which, by analogy with fission yeast, may betray a function related to control over entry into S phase. Phosphorylation on threonine and tyrosine in G2 precedes dephosphorylation of these residues during kinase hyperactivation and entry into mitosis. In addition, long-term control of expression of mammalian p34[cdc2] homologues is likely to be exerted at the transcriptional level. These observations provide the framework of a universal model for the control of eukaryotic cell proliferation, in which the p34[cdc2] protein kinase integrates multiple cues to signal the initiation of S phase and, subsequently, mitosis.

1990 Proto-oncogenes in cell development. Wiley, Chichester (Ciba Foundation Symposium 150) p 168–183

The control of eukaryotic cell proliferation can be conceptually divided into two areas; firstly those processes, involving growth factors and proto-oncogene products in higher eukaryotes, which govern the transition from a resting (G0) state to an actively dividing state, and secondly those controls which operate within the cell cycle of the actively dividing cell. It seems likely that connections between elements involved in cell cycle control and proto-oncogene action will be elucidated in the near future; indeed, at least one tentative link of this nature has already been established.

A large number of conditional yeast mutants have been identified which, under restrictive conditions, are unable to proceed beyond specific points in the cell cycle (for review see Nurse 1985). The cdc (cell division cycle) genes defined by these conditional mutants appear to be involved in the regulation of progresssion through the cycle, as cycle-independent growth processes are largely uninhibited by shift to the restrictive conditions. Of the *cdc* genes identified in the fission yeast *Schizosaccharomyces pombe*, the properties of one, *cdc2*, have singled it out for particular attention. This is because temperature-sensitive lesions in the *cdc2* gene cause cell cycle arrest at two important control points; firstly at a point in late G1 called Start, where cells become committed to entry into S phase, and secondly at a point in late G2 where entry into mitosis is controlled (Nurse 1975, Nurse & Bissett 1981). Furthermore, a class of dominant *cdc2* mutants ('wee' mutants) are advanced into mitosis at a cell size smaller than usual (Nurse & Thuriaux 1980), suggesting that the *cdc2* product is centrally involved in the control of mitotic initiation.

The *cdc2* function does not appear to be regulated at the transcriptional level in yeast, since transcripts of $cdc2^+$ do not change in abundance either during the cell cycle or on entry into stationary phase in response to nutrient depletion (Durkacz et al 1986). These invariant transcript levels are reflected in equally invariant levels of the protein product ($p34^{cdc2}$) (Simanis & Nurse 1986). The prediction, based on sequence homologies, that the *cdc2* gene product should have protein kinase activity was validated by performing *in vitro* assays on the immunoprecipitated protein using exogenous casein as a substrate (Simanis & Nurse 1986). Using histone H1 as substrate, the p34 kinase activity has been shown to be periodic in crude extracts of synchronously growing fission yeast cells, despite the constant level of total p34, with maximal activity being found as cells enter mitosis (Moreno et al 1989); kinase activity is lost as cells complete mitosis or leave the cycle in response to nutrient depletion (Moreno et al 1989, Simanis & Nurse 1986). Significantly, the p34 kinase activity isolated from temperature-sensitive *cdc2* mutants grown under permissive conditions was found to be temperature sensitive when assayed *in vitro* (Simanis & Nurse 1986, Moreno et al 1989), suggesting that the kinase activity is, at least in part, the basis of the biological function of the protein. Yeast p34 is itself extensively phosphorylated in actively growing cells, predominantly on threonine residues, and to a lesser extent on tyrosine and serine (Reed et al 1985, Potashkin & Beach 1988, K.L. Gould, unpublished work 1989). The constancy of this phosphorylation is a matter of some controversy; while invariant p34 phosphorylation has been reported for both *S. pombe* and *Saccharomyces cerevisiae* (Potashkin & Beach 1988, Hadwiger & Reed 1988), a separate study of *S. pombe* has shown that $p34^{cdc2}$ becomes dephosphorylated under nutrient depletion conditions concurrently with the loss of kinase activity (Simanis & Nurse 1986). It has been proposed that the $p34^{cdc2}$ protein kinase activity is regulated by changes in the $p34^{cdc2}$ phosphorylation state (Simanis & Nurse 1986). This suggestion is particularly attractive in view of the genetic evidence implicating two additional protein kinases (products of the *wee1* and *nim1* genes) in the mitotic control of *S. pombe* (Russell & Nurse 1987a,b).

The use of yeast genetics to approach an understanding of eukaryotic cell cycle control was validated when Lee & Nurse (1987) described the cloning, by complementation of a temperature-sensitive *cdc2* yeast strain, of a functional human homologue of *cdc2*. Sequence analysis of this human cDNA showed that it encodes a 34 kDa protein which is identical to *S. pombe* p34^{cdc2} at 63% of its amino acid residues; the human gene was termed *CDC2Hs*. Human p34 was also identified immunochemically by using monoclonal antibodies which were raised against yeast p34^{cdc2} (Draetta et al 1987). An antiserum raised against the perfectly conserved *cdc2* peptide EGVPSTAIREISLLKE specifically recognized a 34 kDa protein in extracts prepared from a wide variety of eukaryotic species, including plants and algae (P.C.L. John, personal communication 1989), supporting the notion that a *cdc2*-related p34 might be a cell cycle regulator in all eukaryotes. This view was dramatically substantiated when p34 was shown to be a component of maturation promoting factor (MPF), an apparently universal initiator of the eukaryotic M phase, originally identified by virtue of its ability to cause meiotic maturation upon microinjection into amphibian oocytes. p34 was also found to be a component of the M phase histone H1 kinase, implicated in models of chromosome condensation and essentially identical to MPF (for review see Norbury & Nurse 1989).

Like its yeast counterparts, immunoprecipitated mammalian p34 has *in vitro* protein kinase activity; here too, the kinase activity varies through the cell cycle and appears to be at its maximum level during G2/M (Draetta et al 1987, Draetta & Beach 1988, this study). In addition to exogenous histone H1 or casein, a co-precipitated 62 kDa protein of unknown function or identity (p62) can be used as a substrate in the immunoprecipitate. If p34 is involved in control of progress through the mammalian as well as the yeast cell cycle, then we might expect to detect changes either in the modification state of the mammalian protein or in its association with effectors which reflect its functional status. At the level of total phosphorylation, mammalian p34 behaves much like the fission yeast protein on exit from and re-entry into the cycle; within 18 hours of withdrawal of serum from Swiss mouse 3T3 fibroblasts, p34 becomes completely dephosphorylated, although the dephosphorylated protein persists at a reduced level for long periods of quiescence (Lee et al 1988). In both human and murine fibroblasts p34 is rephosphorylated after cells re-enter the cycle in response to serum stimulation (Lee et al 1988). As in fission yeast, the timing of this rephosphorylation approximately coincides with the late G1/early S period (Simanis & Nurse 1986, Lee et al 1988); this association may be evidence for the involvement of an activating kinase in the execution by *S. pombe* p34 of its genetically defined late G1 (Start) function. The Start control point of yeasts can be considered equivalent to one of the late G1 control points described for mammalian cells, such as the restriction or R point (Pardee et al 1978). Cells which have progressed beyond this point are thought to be committed to the completion of S phase (although not necessarily to the completion of the entire cycle), even if growth factors are removed from the medium. In much the same way, yeast cells become committed on passing Start to enter the mitotic cycle, rather than the alternative developmental pathway of conjugation (Nurse & Bissett 1981). Thus it is possible that phosphorylation of mammalian p34 is involved in the decision made in late G1 to prepare for and enter a new round of DNA synthesis.

Phosphorylation changes have also been shown to accompany passage through G2 in transformed human (HeLa) cells (Draetta & Beach 1988, Draetta et al 1988). In addition to a general increase in total p34 phosphorylation as cells pass from late G1 to late G2, p34 becomes phosphorylated on tyrosine residues during this period in these cells. However, mounting evidence suggests that it is dephosphorylation of p34, rather than phosphorylation on tyrosine or any other residue, which is associated with the activation of its histone H1 kinase activity during transit from G2 to M phase. In both starfish and frog oocytes, activation of the p34 kinase is clearly associated with its dephosphorylation, and the decrease in protein kinase activity on leaving M phase with rephosphorylation (Gautier et al 1989, Labbe et al 1989). More specifically, dephosphorylation of tyrosine residues may activate the p34 kinase in frog oocytes and mouse 3T3 cells (Dunphy & Newport 1989, Morla et al 1989). These results suggesting that removal of phosphate groups from p34 leads to its activation as a protein kinase in mitosis seem to be inconsistent with the behaviour of *S. pombe* p34, which loses kinase activity as it is dephosphorylated on exit from the cycle (Simanis & Nurse 1986). The total level of phosphorylation of p34 is clearly not, therefore, a useful general indicator of its biological activity; more subtle variations in the precise sites of phosphorylation and the association of p34 with substrate and effector molecules probably hold the key to the understanding of its regulation during the cell cycle.

The effect of association with p62 on p34 protein kinase activity is not yet clearly understood. Whereas p62 is apparently complexed with HeLa cell p34 in immunoprecipitates, gel filtration studies suggest that a large proportion of the p34 casein kinase activity in HeLa cell extracts is associated with the monomeric form of p34 (Draetta & Beach 1988). An active form of the p34 kinase in starfish oocytes is also the monomer, rather than a complex between p34 and any other protein (Labbe et al 1989). Clearly, there is evidence to support models for control of the p34 kinase based upon p34 phosphorylation changes and/or multiprotein complex formation; at this stage it remains possible that certain eukaryotic cell types have cell cycle control mechanisms which exploit one regulatory strategy while dispensing with the other.

In addition to the changes in phosphorylation and kinase activity described above, mammalian p34 is subject to further potentially regulatory changes not seen in either budding or fission yeast. Human diploid fibroblasts lost almost all detectable *CDC2Hs* transcripts within 48 hours of serum withdrawal (Lee et al 1988). After serum stimulation of the quiescent cells, a 25-fold increase

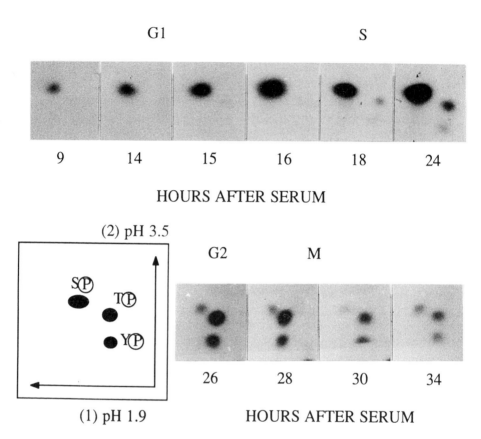

FIG. 1.   Phosphorylation of p34^{CDC2} is biphasic in serum-stimulated Swiss 3T3 cells. Subconfluent dishes of cells were serum starved for 48 hours and restimulated as described by Lee et al (1988). Cells were labelled with [^{32}P] orthophosphate (2.5 mCi per 9 cm dish) for four hours prior to harvesting at the times indicated after re-addition of serum. After immunoprecipitation with anti-peptide antibodies directed against the mammalian p34 C-terminus (LDNQIKKM), phosphate-labelled p34 was subjected to phosphoamino acid analysis by two-dimensional thin layer electrophoresis as described by Morgan et al (1989). The migration of phosphoserine, phosphothreonine and phosphotyrosine standards is indicated diagrammatically in the panel on the lower left. Only part of each autoradiogram is shown. The analysis in the upper panel shows that p34 is phosphorylated exclusively on serine residues between nine and 16 hours after addition of serum. By 18 hours phosphorylation on threonine and tyrosine residues is also detected. In a second experiment (lower right), phosphorylation on threonine, and to a lesser extent tyrosine, residues was seen to predominate by 26 hours after serum addition; this was some two hours before the peak in mitotic index, reached at 28 hours. Though synchrony was beginning to be lost by this stage, an overall decrease in p34 phosphorylation was clearly detected by 30 hours.

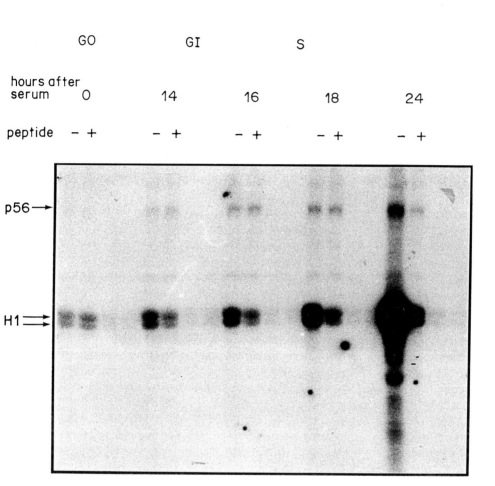

FIG. 2. Protein kinase activity of p34[CDC2] increases before S phase after serum stimulation of Swiss 3T3 cells. Protein kinase activity associated with p34 was measured after serum stimulation as described (Fig. 1) and immunoprecipitation with anti-*CDC2* peptide serum (or with antiserum pre-incubated with the peptide LDNQIKKM), using histone H1 and ATPγ^{32}P as the *in vitro* substrates, essentially as described by Moreno et al (1989). An autoradiogram of the reaction products separated by SDS/PAGE is shown. Tracks labelled ' + ' correspond to peptide-competed immunoprecipitations, tracks labelled ' − ' to uncompeted immunoprecipitations. Peptide-competed p34 kinase activity had increased significantly by 14 hours after serum addition, though the later increase (M phase kinase) was considerably more dramatic. The band labelled p56 is a protein of approximately 56 kDa that was co-precipitated with p34; this may be identical to the p62 described by Draetta et al (1987).

in *CDC2Hs* RNA was seen by the time the cells entered S phase. Total human p34 levels also varied under these conditions, though to a lesser extent; p34 was readily detectable in cells which had been deprived of serum for several days, and the increase in p34 protein level seen after serum stimulation was approximately fivefold (Lee et al 1988). Consequently, while phosphorylation changes are implicated in the control of the biological function of p34 in yeast and mammals, presumably by regulation of kinase activity, long-term modulation of total p34 levels via changes in mRNA availability could influence the proliferative state of cells in multicellular eukaryotes. For example, reduction in mammalian *CDC2* transcript and p34 levels might have a causal role in the initiation of senescence, differentiation or long periods of quiescence *in vivo*.

Despite enormous advances in the elucidation of the pathways involved in vertebrate growth factor signal transduction, we have no real understanding of the presumably essential link or links between growth factor action and elements controlling the cell cycle. In addressing this question, it will be useful to examine closely the route of progression of a vertebrate cell from the quiescent (G0) state to the point at which it becomes committed to enter the mitotic cycle, for it is along this route, rather than within the cycle *per se*, that the majority of known vertebrate growth factors act. With this in mind, we have investigated further the changes which contribute to the increase in total phosphorylated p34 following serum stimulation of 3T3 fibroblasts (Lee et al 1988). Using four-hour labelling periods, we were able to detect p34 phosphorylation as early as nine hours after serum addition; at this time the cells were entering the late G1 phase, as DNA synthesis was first detectable at about 14 hours under these conditions (Lee et al 1988). Late G1/early S phase phosphorylation of p34 (up to 16 hours after serum addition) was exclusively on serine residues (Fig. 1, upper panel). This constitutes the first description of a p34 modification specific to the late G1 phase, and leads us to suggest that mammalian p34 may be activated for a role in late G1 as a result of phosphorylation by a protein serine kinase, as yet unidentified. It is not yet clear to what extent the analogy should be drawn with phosphorylation changes previously described for *S. pombe* p34 during re-feeding of nitrogen starved cells (Simanis & Nurse 1986). In the yeast experiments, as here, p34 became rephosphorylated during late G1 as cells recovered from growth arrest. While it is not yet known if yeast p34 becomes phosphorylated on serine residues under these circumstances, serine phosphorylation of yeast p34 has been detected in asynchronous cultures (Reed et al 1985, Potashkin & Beach 1988).

In *S. pombe* cells recovering from nitrogen starvation, the protein kinase activity of p34, measured in immunoprecipitates using casein as a substrate, was found to increase along with p34 phosphorylation in late G1 (Simanis & Nurse 1986). In order to establish whether any changes in enzymic activity of p34 accompany its late G1-specific serine phosphorylation in 3T3 cells, we performed *in vitro* protein kinase assays on immunoprecipitated p34 after serum stimulation from quiescence (Fig. 2). These assays used histone H1 as exogenous substrate and were performed under conditions optimized for phosphorylation of H1 by the M phase kinase form of p34 (Moreno et al 1989). There was, nonetheless, a significant increase in the p34 protein kinase activity detected as cells passed through late G1. This increased activity could have physiological relevance if the p34 protein kinase does indeed play a role in the G1/S transition in mammals, as it does in yeast. While the G1 increase was small in comparison with the dramatic p34 activation seen in G2/M, histone H1 and the assay conditions used may not be well suited to the G1 form of p34. Alternatively, the small increase in p34 histone H1 kinase activity may be sufficient to allow execution of the putative G1/S function, or p34 may have another function which is undetected by this assay and is associated with its late G1 serine phosphorylation. We note that the kinetics of phosphorylation of a co-precipitated protein of approximately 56 kDa, which may correspond to the p62 species described by Draetta et al (1987), differed from that of H1 phosphorylation in this assay; p34-specific phosphorylation of the co-precipitated protein was not seen until 24 hours after serum addition. This result supports the view that p34 may have different substrate specificities at different stages of the cell cycle.

When 3T3 cells passed through S phase into G2 following serum stimulation from quiescence, p34 was phosphorylated predominantly on threonine residues, and to a lesser extent on tyrosine, with only low levels of phosphoserine detectable (Fig. 1, lower panel). This result agrees with those obtained from both yeast and HeLa cells (Reed et al 1985, Potashkin & Beach 1988, K.L. Gould, unpublished work 1989, Draetta et al 1988). Threonine and tyrosine phosphorylation predominated by 26 hours after serum addition; this was some two hours before the peak in mitotic index, reached at 28 hours. Though synchrony was beginning to be lost by this stage, an overall decrease in p34 phosphorylation was clearly detected by 30 hours. These results support current models of p34 activation with respect to its M phase substrates by dephosphorylation at the G2/M boundary.

In exponentially growing mammalian cells the majority of p34 phosphorylation detected by [^{32}P] phosphate labelling is on threonine residues (Draetta et al 1988, C. J. Norbury, unpublished work 1988). By extrapolation from the results presented here, one might expect such cells in the G1 phase to contain p34 phosphorylated on serine, with G2 cells containing p34 phosphorylated on threonine and tyrosine. Since the majority of exponentially growing mammalian cells are in G1 at any given time, the most abundant p34 phosphoamino acid would be predicted to be phosphoserine. This discrepancy leads us to suggest that the serine phosphorylation found in serum-stimulated 3T3 cells may be at least partly specific to the first G1 period after recovery from G0. One possibility would be that this phosphoserine is rather long lived, and is thus underrepresented when p34 is labelled with phosphate for relatively short periods in continuously cycling cells.

In conclusion, we propose two levels of control of p34 in mammalian cells, one acting in the short term involving phosphorylation of the protein and a second acting in the long term involving regulation of transcript levels. The phosphorylation is biphasic after serum stimulation, with serine phosphorylation potentially activating a late G1 function and phosphorylation in G2 on threonine and tyrosine inhibiting entry into mitosis. Prime candidates for M phase p34 substrates include histone H1, whose phosphorylation may be important for the initiation of chromosome condensation, and the p60 product of the c-*src* gene, which is hyperphosphorylated during mitosis, apparently by a protein complex containing p34 (Morgan et al 1989). The p60[c-src]/p34 interaction is the first described example of a link between a proto-oncogene product and a cell cycle regulator. G1 substrates of p34 have not yet been clearly identified. While p34 may phosphorylate RNA polymerase II in its repeated C-terminal domain (Cisek & Corden 1989), the cell cycle dependence of this interaction is not yet clear. It seems likely that further investigation of the yeast Start control may be informative with respect to the late G1 function of p34, given the high degree of functional and structural conservation of *cdc2* throughout the eukaryotes.

Acknowledgements

This work was supported by the Imperial Cancer Research Fund. We thank Kevin Crawford for help with photography.

References

Cisek LJ, Corden JL 1989 Phosphorylation of RNA polymerase by the murine homologue of the cell-cycle control protein cdc2. Nature (Lond) 339:679–684
Draetta G, Beach D 1988 Activation of cdc2 protein kinase during mitosis in human cells: cell cycle-dependent phosphorylation and subunit rearrangement. Cell 54:17–26
Draetta G, Brizuela L, Potashkin J, Beach D 1987 Identification of p34 and p13, human homologs of the cell cycle regulators of fission yeast encoded by cdc2+ and suc1+. Cell 50:319–325
Draetta G, Piwnica-Worms H, Morrison D, Druker B, Roberts T, Beach D 1988 Human cdc2 protein kinase is a major cell-cycle regulated tyrosine kinase substrate. Nature (Lond) 336:738–744
Dunphy WG, Newport JW 1989 Fission yeast p13 blocks mitotic activation and tyrosine dephosphorylation of the *Xenopus* cdc2 protein kinase. Cell 58:181–191
Durkacz B, Carr A, Nurse P 1986 Transcription of the cdc2 cell cycle control gene of the fission yeast *Schizosaccharomyces pombe*. EMBO (Eur Mol Biol Organ) J 5:369–373
Gautier J, Matsukawa T, Nurse P, Maller J 1989 Dephosphorylation and activation of *Xenopus* p34[cdc2] protein kinase during the cell cycle. Nature (Lond) 339:626–629
Hadwiger JA, Reed S 1988 Invariant phosphorylation of the Saccharomyces cerevisiae Cdc28 protein kinase. Mol Cell Biol 8:2976–2979
Labbe JC, Picard A, Peaucellier G, Cavadore JC, Nurse P, Doree M 1989 Purification of MPF from starfish: identification as the H1 histone kinase p34[cdc2] and a possible mechanism for its periodic activation. Cell 57:253–263

Lee MG, Nurse P 1987 Complementation used to clone a human homologue of the fission yeast cell cycle control gene cdc2. Nature (Lond) 327:31–35

Lee MG, Norbury CJ, Spurr NK, Nurse P 1988 Regulated expression and phosphorylation of a possible mammalian cell-cycle control protein. Nature (Lond) 333:676–679

Moreno S, Hayles J, Nurse P 1989 Regulation of p34^{cdc2} protein kinase during mitosis. Cell 58:361–372

Morgan DO, Kaplan JM, Bishop JM, Varmus HE 1989 Mitosis-specific phosphorylation of p60^{c-src} by p34^{cdc2}-associated protein kinase. Cell 57:775–786

Morla AO, Draetta G, Beach D, Wang JYJ 1989 Reversible tyrosine phosphorylation of cdc2: dephosphorylation accompanies activation during entry into mitosis. Cell 58:193–203

Norbury C, Nurse P 1989 Control of the higher eukaryote cell cycle by p34^{cdc2} homologues. Biochim Biophys Acta 989:85–95

Nurse P 1975 Genetic control of cell size at cell division in yeast. Nature (Lond) 256:547–551

Nurse P 1985 Cell cycle control genes in yeast. Trends Genet 1:51–55

Nurse P, Thuriaux P 1980 Regulatory genes controlling mitosis in the fission yeast *Schizosaccharomyces pombe*. Genetics 96:627–637

Nurse P, Bissett Y 1981 Gene required in G1 for commitment to cell cycle and in G2 for control of mitosis in fission yeast. Nature (Lond) 292:558–560

Pardee AB, Dubrow R, Hamlin JL, Kletzien RF 1978 Animal cell cycle. Annu Rev Biochem 47:715–750

Potashkin JA, Beach DH 1988 Multiple phosphorylated forms of the product of the fission yeast cell division cycle gene cdc2$^+$. Curr Genet 14:235–240

Reed SI, Hadwiger JA, Lorincz AT 1985 Protein kinase activity associated with the product of the yeast cell division cycle gene CDC28. Proc Natl Acad Sci USA 82:4055–4059

Russell P, Nurse P 1987a The mitotic inducer nim1$^+$ functions in a regulatory network of protein kinase homologs controlling the initiation of mitosis. Cell 49:569–576

Russell P, Nurse P 1987b Negative regulation of mitosis by wee1$^+$, a gene encoding a protein kinase homolog. Cell 49:559–567

Simanis V, Nurse P 1986 The cell cycle control gene cdc2$^+$ of fission yeast encodes a protein kinase potentially regulated by phosphorylation. Cell 45:261–268

DISCUSSION

Hunter: Is there an increase in the total amount of p34^{cdc2} at the same time as you see an increase in serine phosphorylation?

Norbury: That depends largely on the experimental system. In serum stimulation of human embryonic fibroblasts, there is about a fivefold increase in total p34 protein by 32 hours after serum (Lee et al 1988). In Swiss 3T3 cells the increase is much less dramatic, only about twofold during that time. Even if we correct for the increase in total p34 in the 3T3 experiment, the increase in *cdc2*-associated kinase in late G1 is still significant. It can't be attributed simply to the increase in total p34 protein.

Hunter: The phosphorylation you detect could be due to crossing a threshold level of protein, which would lead to an increase in the stoichiometry of the phosphorylation. Unfortunately, it's apparently difficult to show a change in the phosphorylation status of this protein on 2-D gels.

Norbury: On a gross level, the amount of *cdc2* kinase per cell increases, by whatever means, in late G1, and there is an apparent increase in phosphoserine content.

Hunter: Have you tried dephosphorylating the *cdc2* protein from late G1 to see whether there's any change in its protein kinase activity?

Norbury: I have tried that experiment using potato acid phosphatase and I wasn't able to change the activity of the kinase. The problem is that in order to retain an active kinase you have to do everything under relatively non-denaturing conditions. It could be that it's very difficult to get access to the appropriate phosphate residues. Jean Wang's group have managed to dephosphorylate the tyrosine residues in the G2 form of p34 without any apparent change in the kinase activity (Morla et al 1989), but there are other phosphorylation events which may be equally important.

Hunter: I was specifically asking about the *cdc2* protein from G1 because your model might predict that serine phosphorylation increases this kinase activity— the converse of what occurs at G2/M when phosphorylation of *cdc2* inhibits its activity and phosphate has to be removed to activate the kinase.

Norbury: I take that point; one possible explanation for finding no change in activity could be that we are inhibiting the G1 form and promoting the G2 form.

Hunter: Obviously you need to show that the phosphate is removed.

Verma: Have you used antisense oligos to *cdc2* to see whether you can block entry into S phase? Or antibodies to p34[cdc2]?

Norbury: I have tried putting antisense *cdc2* oligonucleotides into human fibroblasts during a serum stimulation time course without any effect on entry into S phase. I haven't yet tried it with thio-substituted oligonucleotides. My feeling is that the transcription changes we see might be related to long-term events. For example, transcriptional regulation might be more relevant to senescence or long periods of quiescence *in vivo*, rather than re-entry into the cell cycle. There is p34 protein present at the beginning of that serum stimulation time course, perhaps sufficient to drive entry into S phase.

Verma: As well as a change in the serine phosphorylation of p34, there may be other modifications of the protein.

Norbury: We have only looked at phosphorylation so far. It would seem that there are no other phosphate-containing modifications.

Noble: What happens to *cdc2* or cyclin or any of the components of this system when cells terminally differentiate? Do they disappear?

Norbury: Rosemary Ackhurst has looked at rat myoblasts differentiating to myotubes. There's a gradual disappearance of p34[cdc2] during the terminal differentiation. It is not associated with any obvious differentiative event, but seems to be associated with the gradual loss of the ability to divide.

Hanley: What happens in *S. cerevisiae* with mating factor-induced growth arrest in terms of changes in the level of cyclin or *CDC28*?

Norbury: According to Steve Reed, very little actually happens to budding yeast p34 under a variety of circumstances. He insists there are no phosphorylation changes through the cell cycle (Hadwiger & Reed 1988). I think the situation is going to be complicated by the duplication of gene functions in *S. cerevisiae*.

Land: Could you comment on the role of *suc1* in mitosis and the rest of the cell cycle?

Norbury: The evidence for *suc1* playing a role in the exit from mitosis comes from study of gene deletions. In yeast it is possible to remove one copy of the gene from a diploid, induce sporulation and see what happens to the cells that don't have any *suc1*. Jacky Hayles has found that under those circumstances the cells become blocked late in mitosis (Moreno et al 1989). However, the p13^{suc1} does seem to be associated with p34^{cdc2} at all stages of the cell cycle, so the mechanism for p13 function is not clear.

Heath: You said that the tyrosine phosphorylation was inhibitory. Isn't that a possible mechanism for giving differential control at the S boundary and the M boundary?

Norbury: To ensure that M and S phases occur alternately? Yes, that's possible.

Heath: Is anything known about the role, in yeast, of dephosphorylation?

Norbury: There are protein phosphatases involved in later stages of mitosis, but as far as I am aware they are not involved in dephosphorylating p34 at the onset of mitosis.

Heath: So if you put in phosphatase inhibitors ...

Norbury: We haven't tried that.

Sugimura: Dr Nishimoto at Kyushyu University has shown that okadaic acid (a phosphatase inhibitor) can cause chromosome condensation in cultured mammalian cells.

Hunter: There is a paper by Goris et al (1989) on the effects of okadaic acid in *Xenopus* oocytes showing that it induces MPF formation and maturation.

Norbury: One way to explain that would be to propose inhibition of an inhibitory activity that acts upstream of p34.

Harlow: Have any phosphatases been identified by genetic screens?

Norbury: Yes, several. One is the product of the *dis2* gene of *S. pombe*, mutants in which are defective in chromosome segregation. Another is the product of the *bimG* gene of *Aspergillus*, which is required for the completion of anaphase (reviewed by Cyert & Thorner 1989).

So there are phosphatases involved in the late stages of mitosis. We don't have any evidence for a phosphatase involved in activation of p34. There has been speculation, mainly from David Beach, that *cdc25* may be a tyrosine phosphatase.

Hunter: Dephosphorylation of both threonine and tyrosine may be needed to activate *cdc2* in vertebrate cells.

Verma: What about the concentration of the kinases? Is there a large amount of this kinase? If you over-produce the p34, do you see any effect on the cell? Would it retard the G2/M transition?

Norbury: If p34 is overproduced in yeast, there are pleiotropic effects, best described as a sickening of the cell. There are no dramatic changes in the level of p34 protein phosphorylation, suggesting that the extent of phosphorylation is not limited by availability of the substrate.

Vande Woude: Do these effects occur at a specific stage of the cell cycle? Is there competition for a specific kinase or phosphatase?

Norbury: No, there is a variety of different phenotypes.

Land: Has anybody mutated the tyrosine phosphorylation site in *cdc2*?

Norbury: Kathy Gould has done that for the yeast gene (Gould & Nurse 1989). The tyrosine phosphorylation site in yeast is in the middle of the p34 ATP-binding site. When Kathy mutates that to a phenylalanine residue the protein seems to be constitutively active with respect to the G2/M transition; the cells are constantly trying to enter mitosis.

Land: Are they wee?

Norbury: No, superwee really.

Hunter: Chris, to come back to the issue of what *cdc2* might be associated with, you have immunoprecipitated *cdc2* at different stages of the cell cycle. Do you see any associated phosphoproteins?

Norbury: There is a protein of about 56 kDa that co-precipitates with p34 and that's phosphorylated in the histone H1 kinase assay. It's only phosphorylated in a peptide-competable manner after the cells have gone through the peak of S phase. There is a difference in the kinetics between *cdc2*-specific phosphorylation of histone H1 and phosphorylation of this 56 kDa protein.

Hunter: That is about the size of a cyclin. Does that go away again after S phase?

Norbury: I haven't looked at that. David Beach's lab and others have shown that there is a variety of proteins around 60 kDa which associate with p34 and may be significant for the regulation of its activity during the cell cycle (Draetta & Beach 1988, Pines & Hunter 1989).

Hunter: Do you see this 56 kDa protein after labelling with ^{32}P?

Norbury: No, not after *in vivo* labelling with ^{32}P; however, it does label with [^{35}S]methionine.

Hunter: Have you monitored the size of p34^{cdc2} during this period to see whether it is a monomer or multimer of some sort?

Norbury: I have been leaving the oligomerization to other people! There is evidence to suggest that p34 on its own can act as a kinase. Marcel Doree has purified the growth-associated histone H1 kinase from starfish oocytes and found that the active form can be a monomer (Labbé et al 1989).

Hunter: Given the idea that there may be activation of *cdc2* kinase at G1/S, it will be interesting to know whether that property is intrinsic to the *cdc2* subunit itself or requires the association of another protein.

Verma: Cisek & Corden (1989) showed that the C-terminus of RNA polymerase II was phosphorylated by the *cdc2* kinase. At what time does that occur? Does it occur in yeast?

Norbury: Their paper contains no information on the cell cycle dependence of that phosphorylation.

Hunter: I think Corden said it was high in S phase. There is p34[cdc2] present in his preparation, but they haven't shown strictly that *cdc2* is responsible for the phosphorylation of the RNA polymerase. There is an additional 60 kDa protein. Arno Greenleaf says that his complex from yeast, which phosphorylates the α subunit of RNA polymerase, also has two subunits (Lee & Greenleaf 1989). However, the 60 kDa subunit is the kinase. He has cloned and sequenced it. It has a *cdc2*-like kinase domain and 30 kDa of additional sequence. Corden's complex may contain p34[cdc2], but it is still an open question whether that is the kinase that phosphorylates RNA polymerase.

Vande Woude: Is the phosphorylation site in RNA polymerase II the so-called consensus site for p34[cdc2] kinase?

Hunter: It has the proline but lacks the basic residues that are in the p60[c-src] and histone H1 sites. So you might argue that it isn't the same enzyme, or that the other subunit alters the specificity in some way.

Verma: What is known about the tyrosine phosphatase?

Norbury: Genes for membrane-bound and soluble tyrosine phosphatases have been cloned (reviewed by Hunter 1989). We have checked the sequences against all known yeast *cdc* genes but there is no obvious homology.

Hunter: There is little known about the regulation of tyrosine phosphatases. There are the transmembrane CD45-related phosphatases, which may be regulated by ligand binding—that's not strictly proven yet. There are soluble forms which appear to be monomers of 35 kDa, which are related in sequence to the membrane bound forms that Nick Tonks and Ed Fischer have worked on (Tonks et al 1988a,b). These appear to be constitutively active, which may be why there is so little tyrosine phosphorylation detectable in normal cells. Jean Wang and Alex Morla were able to dephosphorylate p34[cdc2] using a purified soluble tyrosine phosphatase (Morla et al 1989), but whether that's the enzyme that dephosphorylates p34[cdc2] *in vivo* and how that would be regulated just at the G2/M boundary is not known.

Verma: Overproduction of tyrosine phosphatases would be pretty bad for the cell.

Hunter: If you could identify this particular tyrosine phosphatase and overexpress an unregulated form, this might cause premature mitosis.

The experiment described by Morla et al (1989) is subject to a variety of interpretations. If you treat synchronized cells with vanadate they accumulate at the G2/M boundary with highly phosphorylated p34^{cdc2}. If you then wash out the vanadate and release the cells, they rapidly and synchronously enter mitosis. One interpretation is that the phosphatase has been unable to dephosphorylate p34^{cdc2}. Unfortunately, vanadate does many other things to cells.

Brugge: That is in contrast to the ability of some cell types to grow in soft agar in the presence of vanadate (Klarlund 1985). This process obviously requires cell division.

Hunter: I agree the two results are paradoxical. There are also the data from Dunphy & Newport (1989) where p13^{suc1} injected into *Xenopus* oocytes blocks maturation and causes accumulation of the tyrosine phosphorylated form of p34^{cdc2}, again implying that this has to be dephosphorylated for the cell to enter mitosis.

Westermark: In yeast it's clear that there is a point in G2 that is regulated. What is the evidence for a similar thing in higher eukaryotes?

Norbury: The best example would be the arrest of oocytes before meiosis I. There are also restriction points in the lymphocyte cell cycle, one of which would correspond to a G2 block of some description, that Fritz Melchers has described (Melchers & Lernhardt 1985). It is possible to isolate animal cells that are physiologically arrested in G2, for example from the mouse ear (Gelfant 1962, Pedersen & Gelfant 1970).

Brugge: Chris, would you hypothesize that the tyrosine phosphorylation had a positive as well as a negative regulatory effect? The G1/S serine phosphorylated *cdc2* is a poor histone H1 kinase in the absence of tyrosine phosphorylation; therefore you don't really need tyrosine phosphorylation to suppress the histone H1 kinase activity. It seems that tyrosine phosphorylation must have another effect, possibly by bringing some other protein into the complex, or perhaps it suppresses kinase activity only when *cdc2* is bound to cyclin.

Norbury: That's possible. The modifications that are important for cyclin association have not yet been worked out. In my opinion, tyrosine phosphorylation is unlikely to have a positive effect on the kinase activity *per se* because of its position in the ATP binding site.

Heath: Doesn't Kathy Gould's experiment argue against that, at least for yeast, because induction of mitosis is accelerated by removing the tyrosine?

Brugge: Why would you need to add the tyrosine phosphate to what is already a poor histone kinase?

Hunter: The answer is probably that tyrosine phosphorylation is needed only to suppress the cyclin-associated *cdc2*.

Brugge: So as soon as cyclin binds, tyrosine phosphorylation is needed to suppress *cdc2* kinase activity.

References

Cisek L, Corden J 1989 Phosphorylation of RNA polymerase by the murine homologue of the cell-cycle control protein cdc2. Nature (Lond) 339:679–684

Cyert MS, Thorner J 1989 Putting it on and taking it off: phosphoprotein phosphatase involvement in cell cycle regulation. Cell 57:891–893

Draetta G, Beach D 1988 Activation of cdc2 protein kinase during mitosis in human cells: cell cycle-dependent phosphorylation and subunit rearrangement. Cell 54:17–26

Dunphy WG, Newport JW 1989 Fission yeast p13 blocks mitotic activation and tyrosine dephosphorylation of the Xenopus *cdc2* protein kinase. Cell 58:181–191

Gelfant S 1962 Initiation of mitosis in relation to the cell division cycle. Exp Cell Res 26:395–403

Goris J, Hermann J, Hendrix P, Ozon R, Merlevede W 1989 Okadaic acid, a specific protein phosphatase inhibitor induces maturation and MPF formation in Xenopus laevis oocytes. FEBS (Fed Eur Biochem Soc) Lett 245:91–94

Gould K, Nurse P 1989 Tyrosine phosphorylation of the fission yeast *cdc2+* protein kinase regulates entry into mitosis. Nature (Lond) 342:39–45

Hadwiger JA, Reed S 1988 Invariant phosphorylation of the *Saccharomyces cerevisae* Cdc28 protein kinase. Mol Cell Biol 8:2976–2979

Hunter T 1989 Protein-tyrosine phosphates: the other side of the coin. Cell 58:1013–1016

Klarlund JK 1985 Transformation of cells by an inhibitor of phosphatases acting on phosphotyrosine in proteins. Cell 41:707–717

Labbé JC, Picard A, Peaucellier G, Cavadore JC, Nurse P, Doree M 1989 Purification of MPF from starfish: identification as the H1 histone kinase p34[cdc2] and a possible mechanism for its periodic activation. Cell 57:253–263

Lee JM, Greenleaf AL 1989 A protein kinase that phosphorylates the c-terminal repeat domain of the largest subunit of RNA polymerase II. Proc Natl Acad Sci USA 86:3624–3628

Lee MG, Norbury CJ, Spurr NK, Nurse P 1988 Regulated expression and phosphorylation of a possible mammalian cell-cycle control protein. Nature (Lond) 333:676–679

Melchers F, Lernhardt W 1985 Three restriction points in the cell cycle of activated murine B lymphocytes. Proc Natl Acad Sci USA 82:7681–7685

Moreno S, Hayles J, Nurse P 1989 Regulation of p34[cdc2] protein kinase during mitosis. Cell 58:361–372

Morla AO, Draetta G, Beach D, Wang JYJ 1989 Reversible tyrosine phosphorylation of cdc2: dephosphorylation accompanies activation during entry into mitosis. Cell 58:193–203

Pedersen T, Gelfant S 1970 G2-population cells in mouse kidney and duodenum and their behaviour during the cell division cycle. Exp Cell Res 59:32–36

Pines J, Hunter T 1989 Isolation of a human cyclin cDNA: evidence for cyclin mRNA and protein regulation in the cell cycle and for interaction with p34[cdc2]. Cell 58:833–846

Tonks NK, Diltz CD, Fischer EH 1988a Purification of the major protein tyrosine phosphatases of human placenta. J Biol Chem 263:6722–6730

Tonks NK, Diltz CD, Fischer EH 1988b Characterization of the major protein tyrosine phosphatases of human placenta. J Biol Chem 263:6731–6737

General discussion II

Regulation of the eukaryotic cell cycle

Hunter: I shall say something about the work of John Pines in my lab, who has isolated cDNA clones for the human cyclins. Cyclins are proteins that are made periodically during the cell cycle. They accumulate during each cell cycle, then at mitosis they are very abruptly destroyed. During the next cell cycle they reappear and are destroyed again. From this one can deduce that they may be involved in regulating entry into or exit from mitosis.

It has now been shown for *Xenopus* MPF (Gautier et al 1989), the surf clam, *Spisula* (Draetta et al 1989) and yeast (Booher et al 1989) that cyclin homologues are associated with the *cdc2* protein, p34^{cdc2}, during the G2 phase, and that it is the cyclin/*cdc2* complex that is responsible for the mitosis-specific histone H1 kinase activity that is detected at the start of mitosis. It looks as though association of p34^{cdc2} with cyclin may be an important regulatory event in the G2/M decision.

Most of the work on cyclins has looked at what happens during early embryogenesis, particularly in the sea urchin and starfish systems. John Pines thought it would be interesting to identify mammalian cyclin genes and see whether cyclins were involved in the normal somatic cell cycle. Sea urchins are unusual in having only a single cyclin, at least that has been detected so far; other invertebrates and *Xenopus* have at least two cyclin genes, the so-called A type and B type. These both encode proteins of about 60 kDa. There is homology between the central regions of each protein (the cyclin box), but the termini of the two types of cyclin are quite distinct, perhaps indicating that they have discrete functions.

John used a degenerate oligonucleotide to isolate a cyclin B cDNA clone from a HeLa cell library (Pines & Hunter 1989). The properties of this cyclin B are very similar to those observed for invertebrate cyclins; the protein accumulates during each cell cycle, starting somewhere in S phase, peaking in mitosis when it is abruptly destroyed. There is one novel feature of the HeLa cell system, which is probably true of all somatic cells, namely that the RNA level is also regulated during the cell cycle. One sees a low level of cyclin message in G1 phase, then it peaks in the G2 phase. This regulation appears to be at the level of transcription, based on nuclear run-on studies.

There is also regulation of cyclin at the level of degradation. There are hints from the work of Andrew Murray and others that degradation is regulated through an N-terminal region which is on the N-terminal side of the cyclin box. Andrew Murray used cyclin RNAs to drive a cell-free *Xenopus* egg system which undergoes multiple cell cycles *in vitro*. He found that an N-terminally truncated version of cyclin was not degraded and arrested these extracts in mitosis (Murray & Kirschner 1989, Murray et al 1989). So apparently the signal for degradation lies in the N-terminus. There is additional evidence that cyclin degradation may involve phosphorylation.

John Pines has also shown that cyclin B is associated with human p34[cdc2] in HeLa cells. It only associates with the most highly phosphorylated form of p34[cdc2], which is phosphorylated on tyrosine and on threonine. Not all of cyclin B is associated with p34[cdc2]; conversely not all of p34[cdc2] is associated with cyclin. We don't know exactly what that means. p34[cdc2] that is not complexed to cyclin B could be associated with cyclin A. A cyclin A/p34[cdc2] complex is found in *Spisula* (Draetta et al 1989), and therefore is likely to exist in other systems as well. There may also be other proteins associated with p34[cdc2]. The cyclin B/p34[cdc2] complex has histone H1 kinase activity and this is maximal when cells enter mitosis. Activation of the *cdc2* kinase involves dephosphorylation of both tyrosine and threonine residues.

Basically, what we have learned is that the somatic cell cyclins are going to be very similar to the cyclins that are involved in early embryonic cell divisions. We are now interested in what regulates the association of cyclin with p34[cdc2]. Cyclin itself is a substrate for p34[cdc2] kinase in this complex, we would like to know what role that plays in its function. There are additional phosphorylation sites in cyclin that are not apparently phosphorylated by p34[cdc2] itself, so other protein kinases must be involved. We are also interested in systems that degrade cyclin which apparently regulate the metaphase to anaphase transition in mitosis.

It is not clear that there are only two cyclins. There are some hints from Steve Reed and Paul Russell, who have been using PCR cloning with degenerate oligonucleotides corresponding to the cyclin box, that there may be additional cyclins. All of these cyclin family members may associate with p34[cdc2] and may regulate the substrate specificity. There may be cyclins, for instance, that associate with p34[cdc2] in the G1/S phase of the cell cycle and regulate that transition, and other cyclins which associate with it at the G2/M boundary.

This human p34[cdc2] protein is precipitated by anti-PSTAIR peptide antibodies (a sequence conserved in all *cdc2* proteins), although the cyclin/p34[cdc2] complex interestingly is not precipitated by anti-PSTAIR antibodies, implying that the PSTAIR sequence in p34 is masked in the complex. In contrast, you can precipitate this complex with an antibody against the C-terminus of p34[cdc2]. There's every reason to believe that we are dealing with authentic p34[cdc2], although I have heard rumours that there may be more than one *cdc2*-like gene. So it's an open question as to whether all the 34 kDa molecules in this complex are really identical .

Nurse: I would like to say a few words on this subject, and I hope I will not make the waters even muddier than they are already! There is a growing consensus that cyclin proteins, at least partly associated with p34^{cdc2}, accumulate through the cycle and when they reach a critical level the p34 kinase is activated, followed by entry into M phase. There are experiments in *Xenopus* which show that if you add protein synthesis inhibitors quite early in the cycle, the cell still enters M phase essentially on schedule. The interesting point about these experiments is that the requirement for protein synthesis can be fulfilled early in the cycle. The timing varies, depending on the experimental system, but it certainly occurs before 0.5 of a cell cycle, maybe earlier.

The implication of these results is that if cyclin is the only protein that has to be made, then all the cyclin that is necessary is made early in the cycle, suggesting that the level of cyclin that accumulates by entry into M phase does not actually determine the timing of M phase. It could of course contribute to the timing: one possibility is a model where a certain amount of cyclin accumulates, then something happens, then a series of post-translational events occur, leading to entry into M phase. Another possibility is that there are two parallel pathways, one of which consists of accumulation of cyclin and a second (usually longer), which determines entry into M phase. Cyclin is always needed but it may not normally be limiting.

Hunter: We have not shown that cyclin is necessary in the HeLa cell system.

Nurse: Cyclins have been shown to be necessary in the *Xenopus* and yeast systems. There is something about the yeast system that worries me. The *cdc13* gene is a homologue of cyclin, but is it functionally equivalent to cyclins found in eggs and human somatic cells? We know that the human *cdc2* and the mouse *cdc2* genes are functionally equivalent to the yeast *cdc2* gene by complementation. But we and others have tried quite hard to complement *cdc13* mutants in yeast with these other cyclins and these experiments have failed. So, are these cyclins functionally equivalent? It may simply be that some aspect of their function, for example whether they are degraded, is special to each system and otherwise what they do is basically the same.

Therefore, we have to be rather cautious about whether everything we say about yeast *cdc13* also applies to other cyclins. Obviously, we would like them to be the same, because they behave in similar ways. Incidentally, unlike the human cyclin mRNA, the mRNA for *cdc13* is constant through the somatic cell cycle in yeast.

There are a couple of instances where p34 has H1 histone kinase activity when it's not complexed with cyclin. There is a report from Marcel Dorée of purified starfish MPF of 34 kDa native molecular weight, clearly not associated with cyclin, that possesses *in vitro* histone H1 kinase activity. Sergio Moreno in my lab has preliminary evidence that some part of the p34 activity in yeast may also be in only a monomeric form. That doesn't mean that *in vivo* p34 is not in some sort of complex, but it may mean that *in vitro* it can have histone H1 kinase activity when it is monomeric.

Hunter: It would be surprising if a regulatory subunit totally altered the specificity of a protein kinase.

Nurse: So you wouldn't be surprised if p34 did have kinase activity on its own *in vitro* without being complexed with cyclin?

Hunter: No. But with regard to the starfish so-called monomeric p34, it's not clear to me that there couldn't be a fragment of cyclin associated with it.

Nurse: It's possible there is a fragment not detected on SDS gels, but it has to be fairly small because the kinase activity elutes from a sizing column at about 35 kDa.

Hunter: I agree that *cdc13* is not necessarily functionally homologous to the cyclins. In addition, Steve Reed has isolated a series of cyclin-like genes from *S. cerevisiae.*

Verma: Does the product of *cdc13* bind to the mammalian p34^{cdc2}?

Hunter: I don't know.

Nurse: Stuart MacNeil has shown that when the human *cdc2* gene is put into yeast to replace the endogenous yeast *cdc2* gene, the interactions of the human *cdc2* gene with *cdc13* are similar to those seen with the yeast *cdc2* gene. He has not looked at the biochemistry of those interactions yet. Various genetic interactions have been looked at, for example there are alleles of *cdc13* which can be rescued if yeast *cdc2* is overproduced; these can also be rescued by overproduction of human *cdc2.*

Hunter: But have you tried to express *cdc13* in mammalian cells to see whether it acts as a dominant-negative? If it can associate but forms an abortive complex, then you would predict *cdc13* might block the mammalian cell cycle.

Nurse: You mentioned Andrew Murray's deletion mutant of cyclin. Jacqueline Hayles in my lab has made that deletion in *cdc13* and it doesn't have quite the same effect as the deletion in the *Xenopus* cyclin.

Hunter: I think you are right that the signals for degradation may vary between different cyclins, because of the enormous heterogeneity of their N- and C-terminal sequences.

Harlow: There is another complex that's formed in mammalian cells with *cdc2*. The majority of this work was done by Antonio Giordano in my lab and Giulio Draetta in David Beach's group. The original observation was made by Bob Franza looking at 2-D gels. He was immunoprecipitating *cdc2* protein from a mammalian T lymphoblast line. In addition to *cdc2* protein, he saw two spots on the gels that he remembered from an experiment that he and I did together. When you precipitate adenovirus E1A protein, a variety of cellular proteins come down. One of the proteins was about 60 kDa, that we called p60. This was the same protein as Bob saw on the 2-D gel of the *cdc2* immunoprecipitate (Giordano et al 1989).

We decided to look at this complex in more detail. We raised an antibody against p60, a monoclonal antibody called C160. When you precipitate p60 with this antibody, p34^{cdc2} co-precipitates.

Cyclin B also binds to p34, but only to highly phosphorylated tyrosine-containing species of p34^{cdc2}. On the other hand, p60 binds only to the under-phosphorylated form of p34^{cdc2}.

Since we have an antibody against this p60, we can ask whether the p60/p34^{cdc2} protein complex has kinase activity. It does have kinase activity, and this activity is cell cycle regulated. But it doesn't rise and fall at the same times as the cyclin/p34^{cdc2} activity does. It goes up and down earlier in the cell cycle. We know that from three types of experiment.

The two most convincing are, first, we isolated a population of G1 cells using centrifugal elutriation, reinoculated them into culture and took time points. Using an antibody against p34^{cdc2}, kinase activity starts around S phase, continues to mitosis, then drops back down. We then looked at the same samples with the antibody against p60. In this case, kinase activity comes up at the beginning of S phase, but falls about 4–5 hours later, well before mitosis. You can extend this difference by using drugs that block in S phase and in mitosis. If you look at the MPF kinase activity with this strategy, there is about 135-fold higher activity in mitosis than in S phase. If you look at the p60/p34^{cdc2} complex, it's the other way around, the activity is fourfold higher in S phase than in mitosis. So there are distinct physical complexes with p34^{cdc2}.

If the inference from the experiments on cyclin B and *cdc2* is that binding of cyclin to p34^{cdc2} changes not only the temporal activation of the kinase but also the substrate specificity, we have half of that fulfilled with the p60/p34^{cdc2} complex. We don't know yet whether the substrate specificity is different.

What makes the p60/p34^{cdc2} particularly exciting for us is that the adenovirus has tagged it, although we don't know why. It does look like there is different temporal activation of the kinase activity, presumably of p34^{cdc} as the catalytic subunit.

Nurse: Do you know how this p60 relates to the Corden RNA polymerase kinase activity which is also associated with a 60 kDa protein?

Harlow: Our experiments are as unsatisfactory as everybody else's in this regard. If you use the heptapeptide from the C-terminus of RNA polymerase II and ask whether it could be a substrate for the p60/p34^{cdc2} complex, it can be phosphorylated in an *in vitro* kinase assay. In our hands, the cyclin B/p34^{cdc2} complex will also phosphorylate this peptide.

Nurse: Have you ever used your antibodies with Corden's purified protein?

Harlow: No.

Norbury: Has anything been done with microinjection of C160?

Harlow: In early experiments, we injected the antibodies, but at that time we didn't know what to look for. In the original experiment we just checked for changes in morphology. That will have to be done now in more detail.

Vande Woude: Is p60 stable?

Harlow: I don't know. We are now doing experiments across the cell cycle, looking to see when association occurs. We are also trying to establish what adenovirus E1A is doing to this protein: does it turn it on or block interactions or what? One of the major jobs of viral early proteins like E1A is to stimulate DNA synthesis. If the p60/p34^{cdc2} complex is involved in some of those switches, that might explain why E1A is tagging p60. We have preliminary evidence that if you arrest cells by serum deprivation, they stop making p60. It doesn't seem to be made during G0. That was done in normal human diploid fibroblasts.

Méchali: When do you first see p60 during the cell cycle?

Harlow: We can't say yet.

Hunter: Concerning whether p60 and E1A cross-compete, have you looked in a cell expressing E1A to see whether p34^{cdc2} and p60 still form complexes?

Harlow: At the level of resolution we have now, we find E1A bound to p60 but no p34, and p34 bound to p60 but no E1A. We don't see the trimolecular complex. We do see some p60/p34^{cdc2} complexes in cells that contain E1A, but we are looking at a population of cells that is not synchronized.

Nurse: To return to Tony's point about whether there is more than one *cdc2* protein, what antibodies are you using to specify p34^{cdc2}?

Harlow: An antibody against the C-terminal peptide.

Brugge: Have you looked at the differential ability of the p60-bound *cdc2* protein and the mitotic cyclin-bound *cdc2* protein to phosphorylate *Rb* protein? Phosphorylation of p*Rb* seems to correspond temporally with the p60 association.

Harlow: The experiment can't be done very cleanly. It would mean doing double immunoprecipitations.

Brugge: Have you looked at sites of phosphorylation of *Rb* protein?

Harlow: We are just beginning such studies.

Land: To which domain(s) of E1A does p60 bind?

Harlow: Many mutations knock out binding of p60 to E1A. The interaction of p60 with wild-type E1A looks relatively stable like the other complexes that contain E1A, but in mutants the binding is lost very easily. It's not the same sort of interaction that, for example E1A and p*Rb* use, where only a small region of E1A is involved. You can take off the N-terminus of E1A and p60 will still bind, so it's not related to the interactions with the 300 kDa protein. Also, mutations that knock out binding to p*Rb* likewise knock out binding to p60. Therefore, we think it's similar to binding to *Rb* protein, but the interaction of p60 just happens to be weaker or to need more structural information.

Hunter: Does p34^{cdc2} phosphorylate E1A?

Harlow: It doesn't.

Nurse: I am very interested in the kinetics of how the p60-associated kinase activity rises in the cell cycle, because it has enormous significance if *cdc2* is required at the G1/S boundary. In yeast it is, but this is not clear in higher systems. You have said that the p60-associated kinase activity comes up as a sort of shoulder and peaks, but falls away before the peak of M phase. Do you think the p60-associated activity is coming up in late G1, at the middle of S phase or the end of S phase?

Harlow: If you block with hydroxyurea, you get some activity, I don't know whether that is before the peak of kinase activity or not.

Nurse: 3–4-fold?

Harlow: The rise is from zero in early G1, but the timing of that increase is very difficult to assess.

I don't know how many cyclins there are, but p60 might be a cyclin. Suggesting that p60 is the key regulator of the G1/S boundary is premature. I think p60 is likely to be important because the virus has tagged it; viruses in general are pretty smart, they don't do things they don't have to.

Hunter: Jo Milner in Cambridge thinks that p53 is associated with *cdc2*, which would be another complicating factor.

Nurse: John Jenkins has similar data.

Hunter: His genetic evidence from *S. pombe* suggests that p53 and *cdc2* do interact, since expression of *cdc2* rescues cells from the toxic effects of p53 (personal communication).

References

Booher RN, Alfa CE, Hyams JS, Beach DH 1989 The fission yeast cdc2/cdc13/suc1 protein kinase: regulation of catalytic activity and nuclear localization. Cell 58:485–497

Draetta G, Luca F, Westendorf J, Brizvela L, Ruderman J, Beach D 1989 cdc2 protein kinase is complexed with both cyclin A and B: evidence for proteolytic inactivation of MPF. Cell 56:829–838

Gautier J, Minshull J, Lohka M, Glotzer M, Hunt T, Maller JL 1990 Cyclin is a component of MPF from Xenopus. Cell 60:487–494

Giordano A, Whyte P, Harlow E, Franza R, Beach D, Draetta G 1989 A 60 kd cdc2-associated polypeptide complexes with the E1A proteins in adenovirus-infected cells. Cell 58:981–990

Murray AW, Kirschner MW 1989 Cyclin synthesis drives the early embryonic cell cycle. Nature (Lond) 339:275–280

Murray AW, Solomon MJ, Kirschner MW 1989 The role of cyclin synthesis and degradation in the control of maturation promoting factor activity. Nature (Lond) 339:280–286

Pines J, Hunter T 1989 Isolation of a human cyclic cDNA: evidence for cyclin mRNA and protein regulation in the cell cycle and for interaction with p34^{cdc2}. Cell 58:833–846

Role of receptor tyrosine kinases during *Drosophila* development

Ernst Hafen and Konrad Basler

Zoological Institute, University of Zürich, CH-8057 Zürich, Switzerland

Abstract. In vertebrates, a tyrosine kinase activity has been identified as an integral component of growth factor receptors and the products of proto-oncogenes. Many of these receptor tyrosine kinases (RTKs) appear to play a key role in the regulation of cell growth. Recent analyses of several *Drosophila* genes encoding putative RTKs indicate that this class of proteins also plays an important role in decisions about cell fate that depend on cellular interactions during development. The *sevenless* RTK mediates the position-dependent specification of a particular photoreceptor cell type (R7) in the eye. The local specification of R7 cells requires a functional tyrosine kinase domain of the *sevenless* protein but does not depend on the spatially restricted expression of the *sevenless* gene. The *Drosophila* EGF receptor homologue serves multiple functions during development, some of which are clearly unrelated to regulation of cell growth. Finally, the *torso* gene encodes an RTK required for the specification of the terminal regions of the *Drosophila* larva. A number of other genes have been genetically identified that appear to function in the same developmental processes upstream or downstream of these three RTKs. These loci are excellent candidates for genes encoding other components of the signalling pathways, such as ligands or substrates of the RTKs.

1990 Proto-oncogenes in cell development. Wiley, Chichester (Ciba Foundation Symposium 150) p 191–211

Introduction

Protein tyrosine kinases constitute a class of proteins involved in signal transduction. Amongst them, the receptor tyrosine kinases (RTKs) serve a special function in mediating responses of cells to their environment. RTKs are found in both vertebrates and invertebrates, but not in unicellular organisms. It has been suggested that they evolved when cell–cell communication became necessary (Hanks et al 1988). Most vertebrate RTKs for which a function is known are receptors for diffusible growth factors and are associated with the control of cell growth and physiological responses (reviewed in Yarden & Ullrich 1988). A number of RTKs were initially identified as the products of transforming oncogenes of different retroviruses. The characterization of the function of these putative RTKs, as well as the identification of other elements in the signal transduction pathway, has been hindered by the lack of suitable assays.

In *Drosophila*, genetic analysis has led to the identification of genes involved in various decisions during embryonic and pupal development. Recent molecular characterization of some of these genes indicated that the genes *sevenless* (Hafen et al 1987), *torso* (Sprenger et al 1989) and *torpedo* (Price et al 1989, Schejter & Shilo 1989) encode putative RTKs, suggesting that this class of proteins also plays an important role in developmental decisions which depend on cellular interactions. In this article we will summarize what is known about the function of these genes during *Drosophila* development.

The degree of sequence conservation in the catalytic domains of different kinases is a good measure of their phylogenetic relationship and might be used to deduce possible functional properties of a given kinase (Hanks et al 1988). For example, the catalytic domain of the human EGF receptor is more similar to the catalytic domain of the *Drosophila* EGF receptor homologue than to that of most other vertebrate tyrosine kinases (Table 1). The degree of sequence conservation in the tyrosine kinase domain can therefore be used to establish possible cognate relationships between RTKs from vertebrates and invertebrates. The amino acid sequences of the catalytic domains of the RTKs discussed in this review were compared with those of different vertebrate RTKs (Table 1). The percentage of amino acid identity between two unrelated tyrosine kinases is, on average, 30 to 50%. The homology between the catalytic domains of the human EGF receptor and its *Drosophila* homologue is 57%. Similarly, the human and the *Drosophila* insulin receptor are 67% identical in the corresponding region. Based on these criteria, the best candidate for a *sevenless* homologue in vertebrates is c-*ros*, with a homology of 61%. For *torso* the greatest similarity found is 47% amino acid identity with the v-*ret* oncogene. It appears that if a homologue of *torso* exists in vertebrates, it has not been identified yet.

TABLE 1 Comparison of the amino acid identities between the catalytic domains of vertebrate and *Drosophila* receptor tyrosine kinases

	EGF *receptor*	*Insulin* *receptor*	*PDGF* *receptor*	c-*ros*	v-*ret*
DER	**57.4%**	32.4%	36.3%	39.9%	33.3%
DIR	35.0%	**66.8%**	35.2%	49.6%	42.6%
sevenless	41.1%	49.6%	41.8%	**61.1%**	39.5%
torso	37.7%	35.5%	41.5%	39.6%	46.9%

The catalytic domains of the different receptor tyrosine kinases have been aligned according to Hanks et al (1988). For better comparison, the non-conserved insertion sequences between the tyrosine kinase subdomains V and VI (Hanks et al 1988) in the PDGF receptor, insulin receptor, *Drosophila* insulin receptor homologue (DIR), *torso*, and between domains VI and VII in *sevenless* and c-*ros* were deleted before calculating the percentage of amino acid identity. For alignment the algorithm of Needleman & Wunsch (1970) was used with a gap penalty of three and a deletion penalty of nine. The sequence of the *Drosophila* insulin receptor homologue was taken from Nishida et al (1986) and the v-*ret* sequence from Takahashi & Cooper (1987). DER, *Drosophila* EGF receptor homologue.

The *seventless* RTK is a putative receptor for positional information in the developing eye of *Drosophila*

The compound eye of *Drosophila* consists of a repetitive array of several hundred identical units, called ommatidia (Fig. 1 A–D). Each unit is composed of only 20 cells, including eight photoreceptor cells (R1–R8). This highly ordered structure is built in a precise spatial and temporal sequence from an initially unpatterned single-layer epithelium, the eye imaginal disc (reviewed in Ready 1989, Fig. 1E). The regularity of this process has made it possible to study the specification of the different cell types at the single cell level and to identify genes that are involved in this process.

The assembly of each ommatidial unit is initiated by the specification of the R8 photoreceptor to which the other photoreceptors and accessory cells are added in a precise sequence (Fig. 1F). The fate of a cell is determined by the position it occupies in the developing cluster. Tomlinson & Ready (1987) have postulated that a cascade of inductive interactions between differentiating cells and their undetermined neighbours leads to the specification of the different cell types. In wild-type flies, a cell that occupies the position between the developing R1, R8 and R6 cells always develops into an R7 cell (Fig. 2). In *seventless* mutant flies, however, this cell does not develop as a photoreceptor but as a non-neuronal cone cell (Tomlinson & Ready 1986). The use of genetic mosaics, in which mixtures of wild-type and *seventless* mutant cells are generated by mitotic recombination, indicated that an R7 photoreceptor can differentiate only if its precursor contained a wild-type *seventless* gene (Harris et al 1976). Therefore, the *seventless* product is required in the R7 precursor and is presumably involved in receiving and/or transducing the inductive signal rather than producing it. Mutations in another gene, *bride of seventless* (*boss*), have a phenotype that is indistinguishable from the *seventless* phenotype (Reinke & Zipursky 1988). In contrast to *seventless*, the *boss* gene is required in the neighbouring cell R8 for the correct specification of R7 cell fate, demonstrating that *boss* functions on the signal side.

The seventless protein is a receptor tyrosine kinase with an unusual structure

Molecular characterization of the *seventless* gene showed that it encodes a putative receptor tyrosine kinase expressed on the surface of ommatidial precursor cells (Hafen et al 1987). The result of the molecular analysis is consistent with the proposed role for *seventless* and suggests that the *seventless* protein acts as a receptor for an R7-inducing signal, and that signal transduction is mediated through the ligand-induced activation of the tyrosine kinase.

The *seventless* protein differs from other known RTKs both in size and structure (Basler & Hafen 1988, Bowtell et al 1988). With a length of 2554 amino acids corresponding to a calculated molecular weight of 280 kDa, the *seventless*

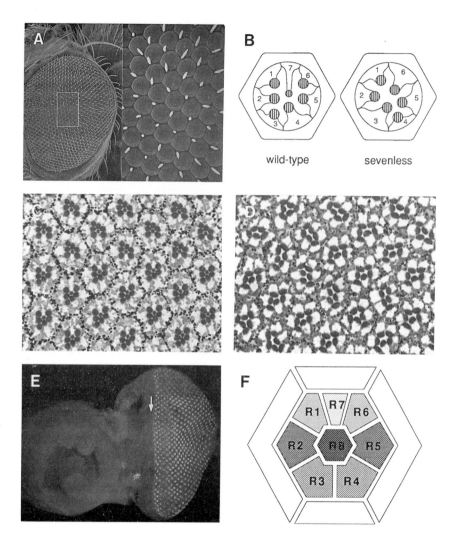

FIG. 1. Structure and development of the compound eye of *Drosophila*. Anterior is to the left. (A) Scanning electron micrograph of a wild-type eye. The enlargement shows the hexagonal arrangement of the individual ommatidia. (B) Schematic representation of cross-sections through the distal part of a wild-type and a *sevenless* mutant ommatidium. The positions of the photoreceptors (R1–R8) are shown. In the *sevenless* mutants the R7 cell is missing. Histological cross-sections through wild-type (C) and *sevenless* mutant eyes (D). (E) Eye imaginal disc stained with an antiserum against the *sevenless* protein visualizing the assembly of the ommatidia. Each of the stained clusters corresponds to a subset of cells in an ommatidium. Ommatidial assembly begins in a wave (arrow) that moves over the disc epithelium in anterior direction; cells anterior to this wave are unlabelled. The different developmental stages of ommatidial assembly are spatially displayed along the anteroposterior axis. (F) Schematic representation of the inner 12 cell unit of an ommatidium at the larval stage. The different grey shades indicate the temporal sequence of assembly. R8 is the first cell in the cluster followed by the pairwise addition of photoreceptors R2/R5, R3/R4, R1/R6 and finally by R7.

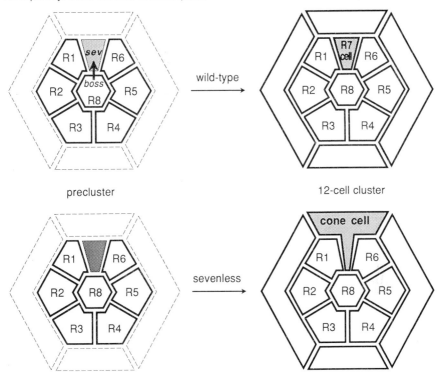

FIG. 2. Fate of R7 precursor cells in wild-type and *sevenless* mutants. A schematic representation of the inner 12 cell unit of a wild-type and a *sevenless* mutant ommatidium is shown. In wild-type, a cell that occupies the position between R1, R8 and R6 develops into an R7 cell. Although in *sevenless* mutants a cell occupies this position, it does not develop into an R7 cell but into a non-neuronal cone cell. An identical mutant phenotype is observed in flies mutant for *boss*. Analysis of genetic mosaics indicates that, for correct R7 development, *sevenless*[+] is required in the R7 precursor whereas *boss*[+] is required in the neighbouring R8 cell.

protein is by far the largest RTK known. In contrast to most RTKs, it contains two putative membrane-spanning regions that divide the protein into three domains: a short N-terminal domain of approximately 100 amino acids located on the cytoplasmic side of the membrane, a large extracellular domain of about 2000 amino acids, and a C-terminal domain that comprises the tyrosine kinase domain, again on the cytoplasmic side. No significant sequence similarities with other known proteins are observed outside the tyrosine kinase domain. In particular, common sequence motifs such as cysteine-rich clusters (Yarden & Ullrich 1988) or immunoglobulin-like domains (Williams 1989), as found in the extracellular domain of other RTKs, are not found in *sevenless*.

Using specific antisera directed against bacterially synthesized *sevenless* proteins, we have demonstrated that the *sevenless* protein is cleaved into

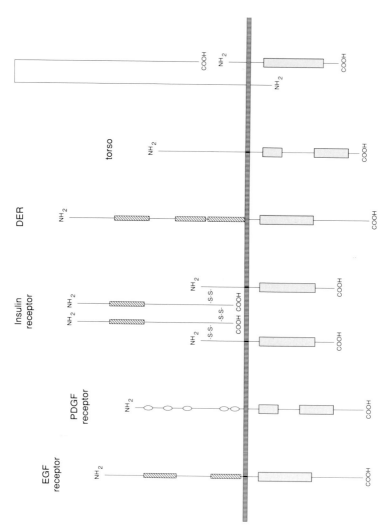

FIG. 3. Structural organization of different receptor tyrosine kinases. The plasma membrane is represented by the horizontal bar. The tyrosine kinase domains are depicted by the shaded boxes. Cross-hatched boxes in the EGF receptor, insulin receptor and *Drosophila* EGF receptor homologue (DER) represent cysteine-rich clusters. Ovals in the PDGF receptor indicate immunoglobulin-like domains. The sizes of the receptors as indicated by their length are drawn to scale.

an N-terminal subunit of 220 kDa and a C-terminal subunit of 60 kDa. The two subunits are tightly associated but not covalently linked (T. Willimann & E. Hafen, unpublished work). The membrane topology of the *sevenless* protein compared with other RTKs is shown in Fig. 3.

sevenless function is critically dependent on a functional tyrosine kinase

To investigate the role of the *sevenless* tyrosine kinase domain in the specification of R7 development, we have used site-directed mutagenesis to substitute a conserved residue (Lys^{2242}) in the ATP-binding site of the *sevenless* tyrosine kinase domain with a methionine (Basler & Hafen 1988). Mutations at analogous positions in other tyrosine kinases have been shown to abolish completely kinase activity (Yarden & Ullrich 1988). The mutated gene is unable to rescue the *sevenless* mutant phenotype when introduced into the germline of *sevenless* mutant flies. Therefore, the tyrosine kinase domain plays an essential role in the transduction of the R7-inducing signal.

Role of the spatial and temporal expression pattern of sevenless

The *sevenless* gene is expressed in a complex spatial and temporal pattern (Tomlinson et al 1987). During the third instar larval period, it is almost exclusively expressed in the developing eye imaginal disc. Within the eye disc it is expressed transiently in only a subset of the ommatidial precursors (Fig. 1E). To test whether the restricted expression pattern of *sevenless* contributes to the spatially restricted specification of R7 cells, we have placed the *sevenless* gene under the control of an inducible ubiquitous promoter, that for the heat shock gene, *hsp70*, and introduced this construct into the germline of *sevenless* mutant flies (Basler & Hafen 1989, Basler et al 1989, Bowtell et al 1989). Ubiquitous expression of *sevenless* during development specifies R7 cells in correct positions but does not interfere with the development of other cells where *sevenless* is not normally expressed. This result suggests that the complex spatial and temporal regulation of *sevenless* gene expression does not contribute to the spatially restricted specification of R7 cells. Specificity of R7 selection may therefore be controlled by the local presentation of the *sevenless* ligand rather than by the restricted expression of the receptor. The idea of a localized R7-inducing signal is supported by the recent genetic analysis of the *boss* gene which may encode the *sevenless* ligand or a product involved in its production (Reinke & Zipursky 1988). The *boss* gene has been shown to be exclusively required in R8, one of the three cells that contact the R7 precursor. It is likely that the ligand for *sevenless* is a membrane-bound molecule, in contrast to ligands of other RTKs that are small, diffusible molecules such as EGF and insulin. It is not known whether the different nature of the *sevenless* ligand accounts for the unusual structure of the *sevenless* RTK.

The *Drosophila* EGF receptor homologue plays multiple roles during development

Genetic and molecular characterization of the mutations *torpedo*, *faint little ball* and *Ellipse* has shown that they all correspond to the gene encoding the EGF receptor homologue of *Drosophila* (DER). The three mutations cause defects at different stages of development, suggesting a complex role for DER during development.

The DER gene was cloned by virtue of its homology to the human EGF receptor (Wadsworth et al 1985, Schejter et al 1986). Localization of DER transcripts during embryonic and larval development indicated a widespread distribution of transcripts at different stages of development (Schejter et al 1986). Loss-of-function mutations in the DER gene cause embryonic lethality (Schejter & Shilo 1989). Embryos lacking DER function fail to undergo germ band retraction, a process that precedes organogenesis. In addition, they exhibit abnormal segregation and death of cells in the head region (Schejter & Shilo 1989, Price et al 1989). Complementation tests indicated that mutations in the DER gene are allelic to the embryonic lethal mutation *faint little ball*, which was isolated in a screen for embryonic lethal mutations on the second chromosome (Nüsslein-Volhard et al 1984).

Independently, Schüpbach (1987) has characterized the maternal effect mutation *torpedo*, which is required in the mother for the correct formation of the dorsoventral axis of both the egg shell and the embryo. Mothers that are homozygous for *torpedo* produce eggs with abnormal eggshells, and embryos developing from these eggs exhibit a ventralized phenotype. Complete loss-of-function alleles of *torpedo* are embryonic lethal mutations which have been shown to be allelic to *faint little ball* and, hence, to DER (Price et al 1989).

Two other genes (*fs(1)K10* and *gurken*) have been identified that are required for the establishment of the dorsoventral axis of both embryo and egg shell (Schüpbach 1987). Mutations in the gene *gurken* cause homozygous females to produce eggs and embryos with a phenotype very similar to the *torpedo* phenotype. Interestingly, genetic mosaic analysis has indicated that *gurken* and *torpedo* are required in different tissues for the correct specification of the dorsoventral axis. *torpedo* is required in the somatically derived follicle cells that form the egg shell and *gurken* is required in the germline-derived oocyte (Schüpbach 1987). Therefore, it appears that the germline and the somatic components of the ovary communicate with each other in a cooperative fashion to determine the pattern of the somatically derived egg shell and the embryo. DER might function as a receptor on follicle cells for a germline-derived signal that depends on *gurken*.

In addition to the involvement of DER in oogenesis and embryogenesis, DER is also required for the correct development of some adult structures, such as

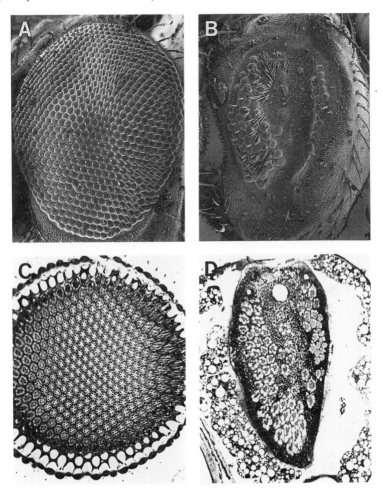

FIG. 4. Effects of *Ellipse*—a gain-of-function mutation of the *Drosophila* EGF receptor homologue—on eye development. Scanning electron micrographs of a wild-type (A) and an *Ellipse/Ellipse* mutant eye (B). Histological cross-sections through a wild-type eye (C) and a *Ellipse/Ellipse* mutant eye (D). The number of ommatidia in *Ellipse* mutant eyes is greatly reduced. (From Baker & Rubin 1989.)

the compound eye and the wing. Flies homozygous for the original *torpedo* allele not only produce defective eggs but also contain rough eyes and defective wings (Price et al 1989). Recently, Baker & Rubin (1989) have demonstrated that the dominant mutation, *Ellipse*, is allelic to DER. Flies homozygous or heterozygous for *Ellipse* have rough eyes with a reduced number of ommatidia (Fig. 4). The dominant effect of *Ellipse* can be suppressed by reducing the number of wild-type

DER gene copies. Whereas in *Ellipse/+* flies the number of ommatidia is reduced, mutant flies that carry the *Ellipse* allele over a deficiency for DER have almost the normal number of ommatidia. This suggests that the *Ellipse* phenotype is caused by the increased gene activity of the *Ellipse* mutant allele. Whether the elevated gene activity is caused by an increased rate of transcription or by an increased basal activity of the mutant DER protein is not known.

In normal development, ommatidial clusters are specified in the disc epithelium at regular intervals. In *Ellipse* mutants, however, only very few ommatidia begin to develop and do so at irregular intervals. The initial steps in the formation of ommatidia involve the regular spatial specification of R8 photoreceptor cells, which are the founder cells for all clusters. Although little is known about this process whereby R8 cells prevent their undetermined neighbours from also entering the R8 pathway, it is likely that cellular interactions play an important role. The DER protein might function as a receptor on undetermined cells that receive an inhibitory signal from R8 cells. In the dominant DER mutant allele *Ellipse*, DER activity might be elevated even in the absence of a signal, thereby preventing most cells from entering the R8 pathway (Baker & Rubin 1989). Subsequent ommatidial assembly, however, occurs normally in *Ellipse* mutants. Therefore, the cascade of inductive interactions that results in the specification of the different cell types appears to be unaffected by DER gene activity.

The human EGF receptor is thought to mediate a signal for cell proliferation in response to diffusible growth factors. It is not clear whether DER serves analogous functions during *Drosophila* development. Cells affected by the lack of DER function are often post-mitotic. On the other hand, the fact that clones of homozygous mutant DER cells do not survive to adulthood (Baker & Rubin 1989) suggests that DER is also involved in certain basic cellular processes.

torso—a receptor tyrosine kinase involved in the specification of the termini of the *Drosophila* larva

The *torso* gene is the third example of an RTK that plays an essential role during *Drosophila* development (Sprenger et al 1989). The *torso* gene belongs to a group of maternal and zygotic genes that control the formation of the unsegmented termini—the acron and the telson–of the *Drosophila* larva (Klingler et al 1988). Lack-of-function mutations in any of these genes lead to the absence of acron and telson (Fig. 5e). *torso* occupies a special position in this group because gain-of-function alleles have been identified that exhibit the opposite phenotype. In *torso* gain-of-function mutations, the central region that normally gives rise to the segmented thorax and abdomen does not form, the acron forms normally and the telson is enlarged (Klingler et al 1988, Fig. 5a–c).

Localization of *torso* transcripts in early embryos indicates a uniform distribution along the anteroposterior axis (Sprenger et al 1989). Therefore it appears that *torso* product is present along the entire body axis but functions only in the terminal regions. Since *torso* is a putative receptor kinase, Sprenger

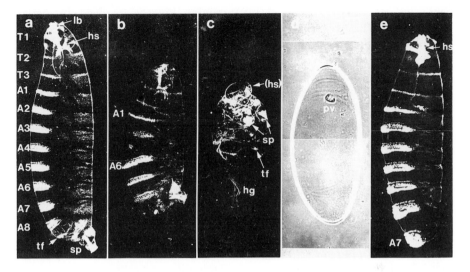

FIG. 5. Phenotype of *torso* alleles. Dark-field and phase-contrast photographs of cuticular preparations of embryos are shown. (a) Wild-type. The denticle belts mark the thoracic (T1 to T3) and abdominal segments (A1 to A8). Whereas most of the involuted head skeleton (hs) derives from segmental anlagen, the labrum (lb) is thought to be part of the unsegmented acron. The telson includes the anal pads, the tuft (tf), and the spiracles (sp) with the prominent filzkörper. (b–d) *torso* gain-of-function phenotypes. (b) Embryo with a weak phenotype. Head and telson are unaffected, whereas the segments A2–A4 are partially or completely deleted. (c) In the intermediate phenotype, all derivatives of the segmental anlagen of head, thorax and abdomen are defective. (d) In the strong gain-of-function phenotype no cuticule is formed. This embryo has not been removed from its egg shell. (e) *torso* lack-of-function phenotype. Labrum and telson are missing and the head skeleton is collapsed. The thoracic and most of the abdominal segments are not affected. (From Klingler et al 1988.)

et al (1989) suggested that *torso* is activated by a ligand localized or released in the terminal region but not in the central region. In the dominant gain-of-function mutants, *torso* appears to be active without ligand stimulation throughout the entire embryo.

Since mutations in several other genes cause a phenotype very similar to that of *torso*, it is likely that they function in the same developmental pathway. By making double mutant combinations of *torso* gain-of-function alleles and loss-of-function alleles of the other genes, it is possible to determine whether these genes act upstream or downstream of *torso*. For example, the dominant phenotype of *torso* is suppressed by loss-of-function mutations in the *tailless* gene. This suggests that the *torso* dominant phenotype is caused by the ectopic activation of *tailless* and that *tailless* acts downstream of *torso* in the same signal transduction cascade (Klingler et al 1988). Interestingly another gene, *lethal(1) polehole*, also functions downstream of *torso* (Sprenger et al 1989). This gene has been cloned recently and shown to be the *Drosophila* homologue of a member of the *raf* protein family in vertebrates (Nishida et al 1988). The *raf* genes in vertebrates encode serine-threonine kinases involved in signal transduction. They were first isolated in altered form as oncogenes. It has been proposed that the *raf* proteins function as signal transducers in pathways that involve RTKs such as the receptors for platelet-derived growth factor and colony-stimulating factor 1 (Rapp et al 1988). The genetic analysis of *torso* and the *Drosophila raf* homologue (D-*raf*) strongly suggests that D-*raf* plays an essential role in the transduction of the signal mediated by *torso*.

A candidate for a gene acting upstream of *torso* is *torsolike*. Genetic mosaic analysis has indicated that in contrast to the other *torso*-group genes, *torsolike* is required in the somatic rather than the germline-derived cells of the ovary. *torsolike* could encode a ligand that is secreted locally by the somatically derived follicle cells that surround the embryo during oogenesis (Sprenger et al 1989).

Conclusions

The genetic and molecular characterization of *sevenless*, DER and *torso* indicated that RTKs play an important role in the position-dependent specification of cell fate. In vertebrates it is difficult to investigate the role of these genes in similar processes during development because loss-of-function mutations are not known. The function of RTKs has mostly been characterized in tissue culture systems, which precludes the analysis of complex interactions between cell and tissue types during development. The relatively high sequence conservation between vertebrate and invertebrate cognate genes makes it likely that at least some of these genes serve a similar function in vertebrates. With the advent of homologous recombination it should be possible to generate null mutations in these genes and to assess directly their role in vertebrate development (Capecchi 1989).

The *sevenless* and *torso* RTKs are required and active in only a subset of the cells in which they are expressed. Furthermore, ectopic expression of *sevenless* does not cause a detectable phenotype. The local activation of the RTK-mediated signalling pathway appears to be controlled not by the spatially restricted expression of the receptor but more probably by the local presentation of the ligand. The receptor renders cells competent to respond to the inductive signals, but only those cells that come in contact with the localized ligand will be determined. In contrast, the ligands for growth factor receptors, such as CSF-1 which controls growth and differentiation of certain haemopoietic precursors (Sherr et al 1985), are diffusible and widely spread. In this case, the specificity is controlled by the expression of the receptor: only cells containing the CSF-1 receptor can respond to the mitogenic signal.

Although the nature of the ligands and substrates for RTKs discussed above is not known, genes with similar or identical mutant phenotypes are likely to function in the same pathway and are therefore excellent candidates for genes encoding ligands or substrates. Furthermore, whether a gene functions upstream or downstream of the receptor can be determined by using genetic mosaic analysis (as in the case of *sevenless*, DER and *torsolike*), or by double mutant combinations with gain-of-function alleles of the RTK gene (as in the case of *torso* and *tailless*). The molecular characterization of these genes will help to elucidate the mechanisms involved in cell–cell interactions during development. Using the combination of genetic and molecular genetic techniques available in *Drosophila*, it is conceivable that most or all of the products that participate in a given RTK-mediated signalling pathway can be identified in the near future.

Acknowledgements

We thank N. Baker and G. Rubin as well as F. Sprenger and C. Nüsslein-Volhard for providing photographs for Figs 4 and 5. We also thank A. Fritz and D. Yen for comments on the manuscript. This work was supported by grants from the Swiss National Science Foundation and the Sandoz Foundation.

References

Baker NE, Rubin GM 1989 Effect on eye development of dominant mutations in the Drosophila homologue of the EGF receptor. Nature (Lond) 340:150–153

Basler K, Hafen E 1988 Control of photoreceptor cell fate by the *sevenless* protein requires a functional tyrosine kinase domain. Cell 54:299–311

Basler K, Hafen E 1989 Ubiquitous expression of *sevenless*: position-dependent specification of cell fate. Science (Wash DC) 243:931–934

Basler K, Siegrist P, Hafen E 1989 The spatial and temporal expression pattern of *sevenless* is exclusively controlled by gene-internal elements. EMBO (Eur Mol Biol Organ) J 8:2381–2386

Bowtell DDL, Simon MA, Rubin GM 1988 Nucleotide sequence and structure of the *sevenless* gene of Drosophila melanogaster. Genes Dev 2:620–634

Bowtell DDL, Simon MA, Rubin FM 1989 Ommatidia in the developing Drosophila eye require and can respond to *sevenless* for only a restricted period. Cell 56:931–936

Capecchi MR 1989 The new mouse genetics: altering the genome by gene targeting. Trends Genet 5:70–76

Hafen E, Basler K, Edstroem JE, Rubin GM 1987 *Sevenless*, a cell-specific homeotic gene of Drosophila, encodes a putative transmembrane receptor with a tyrosine kinase domain. Science (Wash DC) 236:55–63

Hanks SK, Quinn AM, Hunter T 1988 The protein kinase family: conserved features and deduced phylogeny of the catalytic domains. Science (Wash DC) 241:42–52

Harris WA, Stark WS, Walker JA 1976 Genetic dissection of the photoreceptor system in the compound eye of Drosophila melanogaster. J Physiol (Lond) 256:415–439

Klingler M, Erdélyi M, Szabad J, Nüsslein-Volhard C 1988 Function of *torso* in determining the terminal anlagen of the Drosophila embryo. Nature (Lond) 335:275–277

Needleman SB, Wunsch CD 1970 A general method applicable to the search for similarities in the amino acid sequences of two proteins. J Mol Biol 48:443–453

Nishida Y, Hata M, Nishizuka Y, Rutter WJ, Ebina Y 1986 Cloning of a Drosophila cDNA encoding a polypeptide similar to the human insulin receptor precursor. Biochem Biophys Res Commun 141:474–481

Nishida Y, Hata M, Ayaki T et al 1988 Proliferation of both somatic and germ cells is affected in Drosophila mutants of the raf proto-oncogene. EMBO (Eur Mol Biol Organ) J 7:775–781

Nüsslein-Volhard C, Wieschaus E, Kluding H 1984 Mutations affecting the pattern of the larval cuticle in Drosophila melanogaster I. Zygotic loci on the second chromosome. Roux's Arch Dev Biol 193:267–282

Price JV, Clifford RJ, Schüpbach T 1989 The maternal ventralizing locus *torpedo* is allelic to *faint little ball*, an embryonic lethal, and encodes the Drosophila EGF receptor homolog. Cell 56:1085–1092

Rapp UP, Heidecker G, Huleihel M et al 1988 raf family serine/threonine kinases in mitogen signal transduction. Cold Spring Harbor Symp Quant Biol 53:173–184

Ready DF 1989 A multifaceted approach to neural development. Trends Neurosci 12:102–110

Reinke R, Zipursky SL 1988 Cell–cell interaction in the Drosophila retina: the *bride-of-sevenless* gene is required in photoreceptor cell R8 for R7 cell development. Cell 55:321–330

Schejter ED, Shilo B-Z 1989 The Drosophila EGF receptor homolog (DER) gene is allelic to *faint little ball*, a locus essential for embryonic development. Cell 56:1093–1104

Schejter ED, Segal D, Glazer L, Shilo B-Z 1986 Alternative 5' exons and tissue-specific expression of the Drosophila EGF receptor homolog transcripts. Cell 46:1091–1101

Schüpbach T 1987 Germline and soma cooperate during oogenesis to establish the dorsoventral pattern of egg shell and embryo in Drosophila melanogaster. Cell 49:699–707

Sherr CJ, Rettenmier CW, Sacca R, Roussel MF, Look AT, Stanley ER 1985 The c-fms proto-oncogene product is related to the receptor for the mononuclear phagocyte growth factor, CSF-1. Cell 41:665–676

Sprenger F, Stevens LM, Nüsslein-Volhard C 1989 The Drosophila gene *torso* encodes a putative receptor tyrosine kinase. Nature (Lond) 338:478–483

Takahashi M, Cooper GM 1987 ret transforming gene encodes a fusion protein homologous to tyrosine kinases. Mol Cell Biol 7:1378–1385

Tomlinson A, Ready DF 1986 *sevenless*: a cell-specific homeotic mutation of the Drosophila eye. Science (Wash DC) 231:400–402

Tomlinson A, Ready DF 1987 Neuronal differentiation in the Drosophila ommatidium. Dev Biol 120:366–376

Tomlinson A, Bowtell DDL, Hafen E, Rubin GM 1987 Localization of the *sevenless* protein, a putative receptor for positional information in the eye imaginal disc of Drosophila. Cell 51:143–150

Wadsworth SC, Vincent WS, Bilodeau-Wentworth D 1985 A Drosophila genomic sequence with homology to human epidermal growth factor receptor. Nature (Lond) 314:178–180

Williams LT 1989 Signal transduction by the platelet-derived growth factor receptor. Science (Wash DC) 243:1564–1570

Yarden Y, Ullrich A 1988 Growth factor receptor tyrosine kinases. Annu Rev Biochem 57:443–478

DISCUSSION

Heath: Since the cells R3 and R4, which express *sevenless*, are recruited before R7, could one explanation be that specification as R7 has more to do with the timing of recruitment rather than the expression of *sevenless*?

Hafen: Yes, you have to explain why R3 and R4 which also contact the ligand-presenting R8 cell do not become R7 cells and why they are not affected by the absence of *sevenless* product in mutants. There are many explanations: one is that to form an R7 cell you need another signal that may come from R1 and R6. The one we favour is that it's the timing of events; it is not only local presentation of the ligand, but the ligand is also presented at only one stage and not the other. There is a time difference of about 10 hours between the determination of R3 and R4, and of R7. During this period the R8 cell, which produces the R7-inducing signal, undergoes its normal process of differentiation. The ligand for *sevenless* might be expressed only at a later stage of this differentiation, after R3 and R4 have already been committed to their fate.

Heath: *boss* is the putative ligand, but that doesn't affect the other cells that express *sevenless*, which argues against the ligand presentation model.

Hafen: No, you just have to propose that *boss*, if it is the ligand, is expressed only relatively late in the differentiation of R8. Then the only cell that sees the *boss* product is the R7 cell, R3 and R4 have already been determined.

Verma: What's the phenotype of the *boss* mutants?

Hafen: The same as that of *sevenless* mutants, the ommatidium just lacks this one cell.

Noble: Are there data on the localization of *boss* expression?

Hafen: No, that's what we are all waiting for, *boss* hasn't been cloned yet. Larry Zipursky and Rosemary Reinke are trying but they haven't identified it yet.

McMahon: Is anything known about what's involved in the localization of *sevenless* protein in the membrane? In the hsp70 constructs, what does the localization look like in the other ommatidia?

Hafen: It is not known why *sevenless* protein is located in the membrane. It may be that a ligand–receptor interaction leads to clustering, which would contradict the hypothesis that R3 and R4 never see the ligand. It is also possible that there is a special interface between these differentiating cells and other cells or matrices that causes localization of the protein there.

From the transformed lines containing the hsp-*sevenless* construct, we know that *sevenless* protein is expressed at the cell surface. We know that it can functionally interact with the ligand because it can rescue the mutant phenotype, but we haven't studied this by electron microscopy to see whether it was the same kind of localization.

Nusse: Has anybody looked at the localization of the *sevenless* protein in *boss* mutants?

Hafen: That's a very good experiment but it has not been done yet.

Hunter: There is certainly room for two ligand-binding sites on a protein of 200 kDa. One ligand could localize the protein and the other could give the signal.

Noble: In the vertebrate systems that I know about there are very few growth factor-like interactions that cause developmental specification. If you take the regions of non-homology from, for example, the EGF receptor, and use those to probe a human library, do you isolate genes that resemble *Drosophila* genes in other regions? Is it possible that you are getting part of another family, that is not really the classical EGF receptor?

Hafen: I haven't done the experiment. We have looked extensively for cross-homologous clones in *Drosophila* libraries using sequences from the extracellular domain as probes, but we didn't find any. That might be because the sensitivity of the hybridization is not high enough.

Heath: I disagree with Mark that there are no known growth factors that play a role in development. Mesoderm induction is one good example (Slack et al 1987).

Hanley: Has anyone explored whether there's a specific relationship or function that can be predicted from the homology of *sevenless* with the mammalian tyrosine kinase c-*ros*? Is this homology only relevant to the structural or evolutionary relationships of tyrosine kinases, or can, for example, a chimaeric protein containing the N-terminus of *sevenless* with the *ros* tyrosine kinase domain rescue the *sevenless* mutant phenotype?

Hafen: We have made the construct but we haven't injected it yet.

Jessell: In order to approach downstream functions of *sevenless*, is anything known about other genes that give a *sevenless* phenotype, albeit not in a totally specific manner?

Hafen: I only know of one example that Richard Carthew in Gerry Rubin's lab has isolated in a search for other eye mutations. This is not as specific as *sevenless*, but also leads primarily to the absence of the R7 cell. By genetic mosaic analysis, this has been shown to be specific for the R7 cell (G. Rubin, personal communication). It appears to act downstream of *sevenless*, but I don't know any more about that. To my knowledge no other such gene has been identified.

We are using *boss* mutants to look for second site revertants of the *boss* phenotype, to get mutations that cause either constitutive activation of the *sevenless* receptor or any gene that acts downstream in this pathway. We have nothing to report yet.

Hunter: It would be interesting to replace the kinase domain with a kinase domain known to be constitutively active and see whether it specifies R7 in a *boss* mutant.

Hafen: We have made a series of hybrid receptors with the extracellular domain of the human EGF receptor and the intracellular tyrosine kinase domain of *sevenless* to get around the problem of lacking a ligand. Such hybrid receptors have been reported to work (Yarden & Ullrich 1988). We have introduced this hybrid receptor into flies; we have also crossed in a construct containing an EGF cDNA clone under the control of the heat shock promoter (K. Basler & E. Hafen, unpublished). It should therefore be possible to produce EGF by heat shock in all the cells and thereby stimulate the hybrid receptor and activate the *sevenless* kinase. Similar chimaeras were also made with the human insulin receptor and *sevenless*. So far we have, however, no positive evidence that these chimaeric receptors work. It is possible that the different structure of the *sevenless* receptor also involves a different kind of activation.

Hunter: The kinase minus *sevenless* protein doesn't restore function, yet presumably the ligand–*sevenless* interaction is still occurring. There's no reason to think that a point mutation will have affected binding of ligand in any way.

Hafen: We have addressed this using one of the C-terminal deletions that makes a partially active product—partially active means that only some of the cells become R7 cells. If there is one copy of this gene, only 40% of the ommatidia are rescued. If there are two copies, about 90% of the ommatidia are rescued. Into this background, we crossed the ATP-binding site mutant protein. If there are two copies of this non-functional gene, the 90% rescue is reduced to 40%. It looks as if the mutant protein can compete for either ligand or substrate.

Harlow: My impression of developmental mutants is that when the gene products are identified, those involved in downstream events normally turn out to be transcription factors and those involved in upstream events are some sort of membrane receptor. Are there any examples of developmental mutations that affect events in the middle stages of signal transduction? I would guess that anything central to the signalling pathway might be a lethal mutation, and *lethal(1) polehole* fits that description.

Hafen: In this case it is like that. The fact that despite extensive screens for visible mutations that affect eye development, only a limited number of viable mutations has been identified suggests that some components of this system are used elsewhere and mutations are therefore lethal.

Harlow: Is that a general rule? Would one say that is the way things work in other systems?

Hafen: I don't think you can say that intermediate steps are used in all cells and transcription factors are not. There are transcription factors that are highly specific, for example the *rough* protein, a homeobox-containing protein that is potentially a transcription factor, only functions in the eye. But other transcription factors are used at more than one stage of development, it has not been tested whether they also play a role in eye development or in other later stages of development.

Maness: Could the products of genes like *torso* or *sevenless* be involved in cellular migration? If so, they could bring the cell into proximity with another cell which produces a locally diffusible factor or a protein on its surface with which it might interact.

Hafen: At the time when *torso* activity is required and *sevenless* has to be active, there is no morphological evidence for any cell migration. This is just recruitment of cells that are already in physical contact.

Maness: By electron microscopy, can you see specific types of cell–cell interactions between, for example R7 and R8?

Hafen: EM shows that *sevenless* protein is concentrated at the zona adherens, the tight junctions, although it is expressed on other parts of the membrane. It is also expressed at the microvillar tips of the same cells but no one really has an idea of what it does there.

Vande Woude: During the formation of a cone cell, how many cell divisions are there?

Hafen: All these recruitments in the eye occur in post-mitotic cells.

Vande Woude: So all you see is a phenotypic change in the cell. If you could correct *sevenless* in the cone cell would it then become an R7 receptor?

Hafen: In *sevenless* mutants the presumptive R7 cell becomes a cone cell because at the stage where it should have the receptor, it doesn't, so it follows a different pathway. If we produce the *sevenless* protein later using the heat inducible *sevenless* construct, we can no longer rescue the cell, presumably because it is already committed to the cone cell fate. There appears to be very tight temporal control of when cells can switch from one fate to another. This might be partly due to the fact that they are only presented with ligand at a certain time.

Sherr: When you express *sevenless* under the control of the heat shock promoter, is it expressed in the R8 cell?

Hafen: It is expressed in the R8 cell.

Sherr: And that has no effect on R8?

Hafen: No.

Hunter: So there is no autocrine signal to that cell?

Hafen: Well, again you could say that by the time R8 expresses the ligand it is already committed to its own developmental pathway, so stimulation has no effect. It is the same argument as for R3 and R4, which have probably been committed prior to seeing the ligand.

Hunter: The eye mutations were identified by a screening procedure that monitored the response to UV light. What cell types actually respond to UV? If you don't have an R7, it seems that you can't test whether *sevenless* functions in the mature cell as well as in the development of that cell.

Hafen: That is the point. In the *sevenless* mutant, you cannot address the question of whether the protein plays an additional role in the R7 cell later in development. Our experiment was to allow R7 cells to form, then remove *sevenless* protein to ask whether it has an effect later in the function of those cells.

Hunter: Have you shown that the protein is not stable for days?

Hafen: In order to completely rescue the eye in a *sevenless* mutant that contains our heat-inducible *sev* construct, we have to heat-induce every four or six hours. If we heat induce every 12 hours, we get stripes of wild-type and mutant ommatidia, showing that over 12 hours the level of *sevenless* protein has dropped below a functional threshold.

Hunter: But that's during development, what about in the fully mature cell? How do you know the protein is not infinitely stable, such that once an R7 cell has been generated it always has *sevenless* protein in it?

Hafen: The basic argument we have is the stability during development. By Western blot analysis we have shown that at the larval stage after ubiquitous expression of *sevenless*, there is a decay after a few hours.

Hunter: You could stain the eye for *sevenless* protein.

Hafen: But you don't know what levels are functional; if it is active at very low concentrations, you might not see it by antibody staining.

Sherr: You co-expressed two different receptors, one of which was ATP-binding site-negative, and got dominant suppression?

Hafen: Only under very special circumstances, such as using this partially functional receptor. If you do it in the wild-type, you do not get suppression.

Sherr: Would that predict that if you were studying the *Ellipse* mutation and you put in an ATP binding site from DER, those cells would then differentiate into neurons?

Hafen: I guess you could compete it down.

Sherr: Here you have an EGF receptor that, when expressed, prevents cells from committing to a neuronal pathway. If you introduced a mutant receptor which competed for ligand, would the logical assumption be that you would get neuronal development?

Hafen: That's basically the experiment Baker & Rubin (1989) have done except that instead of introducing a defective gene they removed the wild-type gene, and that reversed the dominant effect.

Sherr: And the pattern formation of R8 cells is normal?

Hafen: Yes. There is a graded series, despite the fact that *Ellipse* is a dominant mutation in genetic terms. The phenotype I showed was homozygous for *Ellipse*. Increasing the levels of DER activity prevents more and more cells from entering the photoreceptor cell pathway.

Sherr: I am interested in this idea of dominant suppression by inactive kinases. There is another situation involving *kit* where this appears to be the mechanism by which the mutant form exerts its dominant effects. *kit* mutants, as I understand, are kinase negative mutants but are dominant and give dramatic developmental phenotypes.

The white (W) locus in the mouse gives a pleiotropic phenotype, including white specks in the coat colour. There are different versions of the gene, some are lethal, some are tolerated. It has been found that the *kit* locus, which encodes a tyrosine kinase of the PDGF receptor, CSF-1 receptor subfamily, is deleted or mutated in mice with W phenotype. The coat colour defect has something to do with migration of melanocytes during development. There is also a haemopoietic anomaly in which the primary defect is in stem cell migration. If you inoculate bone marrow from W/W^v mice into syngeneic wild-type animals, you do not get colony formation in the recipient spleen. One mutation, W^{42}, is a point mutation in the kinase domain which knocks out the kinase activity. This is a most interesting case, because the point mutation gives a dominant phenotype. The loss of kinase activity associated with a dominant phenotype is an example of what I was driving at with *Ellipse*.

Hafen: Can't that just be haploinsufficiency? If you delete one copy of the gene or you knock one out by a mutation in the ATP binding site, you reduce the amount of gene product. If that product acts at a rate-limiting step, it would produce a phenotype.

Heath: It's also misleading to describe the heterozygous W^{42} allele of the mouse as dominant. The homozygotes are completely white, whereas the heterozygotes are more or less coloured with small patches of white. The erythroid defect is also a mild form of the severe anaemia seen in homozygotes.

Sherr: There is partial penetrance in the heterozygotes.

Heath: That's not the same as complete dominance.

Hunter: If a point mutation (W^{42}) gives a stronger phenotype than deletion of the gene (W^{44}), that would argue that the protein that's still around does something.

Sherr: By analogy with the *boss* and *sevenless* situation, there is another mutation in the mouse, called *Steel*. It gives rise to mice that are phenotypically identical to those with defects of the W locus. *Steel* appears not to affect the homing ability of haemopoietic stem cells but rather the microenvironment in which those cells mature. In a reconstitution experiment with a *Steel* mouse, if you inject stem cells from a normal donor, you get no colony-forming units. Therefore it's been suggested that *Steel* encodes the ligand for *kit*, a similar reasoning to that for *boss* and *sevenless*.

Maness: By the same token, *Steel* could affect migration of a cell towards a site where the ligand for the c-*kit* protein is being produced. A cell defective in *Steel* would not be found in the proximity of the requisite ligand, and would therefore be unable to respond.

Harlow: boss could also act indirectly.

Hafen: Yes. There is no evidence that it directly encodes the ligand; it could encode a transcription factor that turns on expression of the gene for the ligand. It is upstream and in the neighbouring cell, that's all one can say from this genetic analysis.

Hunter: What's known about the molecular details of the *Ellipse* mutation?

Hafen: I don't know how much Baker & Rubin have looked at that. It might just be overexpression.

Hunter: In which case the gene would look normal.

McMahon: I am fascinated by this morphogenetic wave that sweeps across the eye. Is anything known about the nature of this? Is there any way of manipulating this or affecting eye development to get a handle on it?

Hafen: The drawback of this system is that you can't do any sort of manipulation. The morphogenetic furrow is a morphological depression that sweeps across the disc epithelium from posterior to anterior. Cells anterior to the furrow are unpatterned and dividing, whereas cells posterior to the furrow become integrated into the ommatidial clusters (see Fig. 1). How and why the morphogenetic furrow moves across the eye disc is not known, although it is clearly related to synchronous cell division and the cell cycle in some respect. The furrow is caused by the simultaneous downward movement of the majority of the cells, which then undergo a last synchronous division and generate all the precursors except for the five original photoreceptor cells. Those post-mitotic cells are then incorporated into the cluster. Why this occurs with such synchrony, I don't know.

McMahon: As the cells move downwards are they moving to somewhere where they could see a signal which would set them off on the first step?

Hafen: No, I think recruitment occurs, morphologically speaking, in the top layers. They come into the cluster in the apical region of the epithelium.

References

Baker NE, Rubin GM 1989 Effect on eye development of dominant mutations in the Drosophila homologue of the EGF receptor. Nature (Lond) 340:150–153

Slack J, Darlington B, Heath JK, Godsave S 1987 Mesoderm induction in xenopus by heparin binding growth factors. Nature (Lond) 326:197–200

Yarden Y, Ullrich A 1988 Growth factor receptor tyrosine kinases. Annu Rev Biochem 57:443–478

The *int* genes in mouse mammary tumorigenesis and in normal development

Roel Nusse

Division of Molecular Biology, Netherlands Cancer Institute, Plesmanlaan 121, 1066 CX Amsterdam, The Netherlands

Abstract. In mice, the mouse mammary tumour virus causes tumours by insertional activation of host cell oncogenes. By the application of transposon tagging techniques, several cellular oncogenes, called *int*, have been discovered. The *int-1* gene encodes a cysteine-rich protein with a signal peptide, suggesting that it may act as an extracellular growth or differentiation factor. Normally, the int-1 gene is expressed in early embryogenesis of the mouse, in particular in the developing nervous system. The essential role of *int-1* in embryogenesis is underscored by its high degree of homology with the *Drosophila* segment polarity gene *wingless*, a gene involved in pattern formation in segments of the developing fly. In *Drosophila*, the *int-1/wingless* gene appears to encode a secreted factor, as concluded from antibody staining experiments. The *int-4* gene is not yet fully characterized at the molecular level. From its expression pattern, however, we have concluded that *int-4* may also act in the control of embryogenesis: the gene is expressed only during specific time intervals in mouse embryos and it is highly conserved in evolution.

1990 Proto-oncogenes in cell development. Wiley, Chichester (Ciba Foundation Symposium 150) p 212–226

The work done in our laboratory is concerned with the molecular mechanism of viral mammary tumorigenesis in mice. The virus causing the tumours, the mouse mammary tumour virus (MMTV), is a retrovirus lacking a host cell-derived viral oncogene. Proviral insertion and activation of nearby host cell oncogenes is an important step in mammary tumorigenesis caused by MMTV. Peculiarly, the known genes at MMTV insertion sites, collectively called *int*, have never been identified as activated oncogenes in other tumours or as viral oncogenes; they were all originally isolated by provirus tagging, using the inserted MMTV provirus as a starting point for molecular cloning. Two of these genes, *int-1* and *int-2*, have been analysed in great detail. Although *int-1* and *int-2* are structurally not related to each other, they share intriguing properties: temporal expression in a very restricted pattern during early development of the mouse, and protein products with characteristics of secreted factors (reviewed in Nusse 1988).

Provirus tagging using the mouse mammary tumour virus

By inbreeding and selection, several strains of mice have been obtained that are prone to mammary tumorigenesis, even without the administration of carcinogenic chemicals. These mice carry MMTV, a B-type retrovirus, in the milk. In some strains of mice MMTV is present as an endogenous but highly active provirus, still capable of inducing tumours when the milk-transmitted virus is removed. The tumour-inducing properties of MMTV are now generally thought to be intrinsically related to an obligatory step in the retroviral life cycle: insertion of proviral DNA into the host cell DNA. Integration is a mutagenic event for the host cells and one consequence may be the activation of proto-oncogenes whose expression is normally tightly regulated. This is not to say that the whole process of tumorigenesis is caused by a single integration: additional events, not necessarily initiated by MMTV, may contribute essential steps.

Provirus tagging exercises are most conveniently done in tumours with a single acquired MMTV integration, followed by chromosome walking and scoring for independent insertions in the same area. This strategy has led to the identification of four such common integration domains (Fig. 1; Table 1). The typical orientation and the large distance of some proviruses from the *int-1* or *int-2* promoters indicate that the transcriptional activation of the *int* genes has been mediated by enhancers in the MMTV genome acting on the *int* promoters (Nusse et al 1984). For some enhancers it has been documented that they can act only on the first promoter that is encountered in *cis*, which would explain the typical orientation of MMTV proviruses near *int-1* and *int-2*. The opposite orientation would interpose the promoter on the MMTV long terminal repeat (LTR) between the *int* promoter and the viral enhancer.

By operational definition, all genes identified through insertions of MMTV genes have been called *int*, with the exception of *hst* (Peters et al 1989). Since the tumours where the different *int* genes are activated are morphologically and biologically similar if not identical, one would expect the *int* genes to share other properties. They do; as a rule they encode proteins with hallmarks of secretory proteins. In addition, the *int* genes are all expressed in early embryogenesis. Most members of the *int* group are, however, not related by sequence to each other, nor do they have an obvious identical role in development.

The *int-1* gene

int-1 is quite frequently activated by MMTV, regardless of the strain of mice or the MMTV variant (Nusse & Varmus 1982). The gene is unusually highly

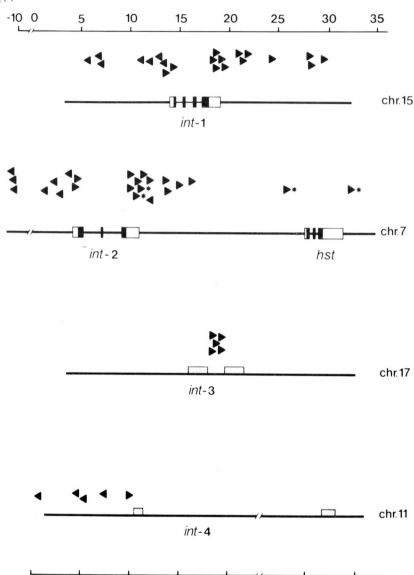

FIG. 1. Organization of the various *int* genes. Shown are four chromosomal areas (number of chromosome indicated) where MMTV proviral insertion (arrows indicate site and orientation) is found in different mammary tumours. Structure of the genes, if known, is indicated by boxes: filled area corresponds to protein-encoding domain; stippled boxes at *int-3* and *int-4* indicate position of probes hybridizing to tumour-specific RNA. Top scale applies to *int-1*, *int-2* and *int-3*, bottom scale to *int-4*. Asterisks at arrows in the *int-2/hst* domain are at MMTV inserts in those tumours where *hst* is expressed.

TABLE 1 Properties of the different *int* genes

Gene	Mouse chromosome	Human chromosome	Size of protein	Gene to which related	Homologues in other species (% identity)	Expression in embryonal carcinoma cells	Frequency of activation in mouse strains
int-1	15	12	44 kDa	*irp*	Human (99%) Xenopus (68%) Drosophila (54%) C. elegans Zebrafish	differentiated P19	80% C3H 70% BR6 30% GR
int-2	7	11	27 kDa	FGF	Human (89%)	differentiated F9, PCC4, PSA-1	65% BR6 5% C3H 20% GR
int-3	17						40% CzII 8% BR6
int-4	11	17				differentiated P19	10% GR
hst	7	11	22 kDa	FGF	Human (82%)	undifferentiated F9, PCC4	10% BR6

FGF, fibroblast growth factor; *irp*, *int*-related protein.

conserved in evolution, having a virtually identical human homologue and easily recognizable amphibian and insect counterparts. *int-1* encodes a protein with many cysteine residues and a signal peptide but without a transmembrane domain (Van Ooyen & Nusse 1984). These are properties of secretory proteins, and *int-1* protein species, if overexpressed in various cell lines, do enter the secretory pathway (Papkoff et al 1987, Papkoff 1989). Secreted forms of the *int-1* protein have been difficult to detect, however, due to the low affinity of the available antibodies. A solution to the problem of weak antigenicity of *int-1* has been found in *Drosophila*, where *int-1* is identical to the segment polarity gene *wingless* (Rijsewijk et al 1987a). The mouse and *Drosophila int-1* proteins are identical in 54% of their amino acids, but the *Drosophila* gene has an insert encoding an additional 85 amino acids. Antibodies raised to the *dint-1/wingless* protein are able to detect the protein in whole mount embryos and in individual cells, and the insert probably provides the highly antigenic determinant. Secretion of the *wingless* protein has been observed using the electron microscope, where the protein is seen in small membrane-bound vesicles, in multivesicular bodies and in the intercellular space (van den Heuvel et al 1989). Movement of the protein is suggested by a double staining of wild-type embryos with both *wingless* and *engrailed* antibodies. The *engrailed* protein is located in the nucleus of cells next to the *wingless* cells. The multivesicular bodies containing the wingless protein are occasionally found in *engrailed*-positive cells, suggesting that the *wingless* protein behaves as a paracrine signal.

int-1 expression pattern and role in embryogenesis

In adult mice, *int-1* is not expressed in any tissue, except the mature testis (Shackleford & Varmus 1987). Expression is localized in post-meiotic, round spermatids, but the biological significance of this is unclear. In mouse embryos, *int-1* is expressed between Day 8 and Day 14, specifically in the developing nervous system (Wilkinson et al 1987). Areas in the neural plate, the anterior head folds, the folding neural tube and the spinal cord show a highly localized expression of *int-1*. During folding of the neural tube, expression moves from anterior to posterior. In *Xenopus* embryos also, *int-1* expression is associated with neural development (Noordermeer et al 1989). It has been suggested that the gene is involved in positional signalling in the developing neural system, providing information specifying the midline axis. In agreement with this theory are the dramatic effects of ectopic expression of *int-1* in *Xenopus* embryos, leading to duplication of the neural tube and the anterior head folds (McMahon & Moon 1989). Most tested embryonal carcinoma cell lines do not express *int-1*, except for P19 cells when induced by retinoic acid to differentiate along the neural pathway (Schuuring et al 1989).

From clonal analysis of *wingless* cells, it appears that the phenotype is non-autonomous in mosaics, which has been interpreted as suggesting a role in intercellular signalling for the wild-type protein (Morata & Lawrence 1977). Various alleles of *wingless* are known: a viable allele (which gave the gene its name) that leads to homeotic transformation of the wing into a notum, and embryonic lethal alleles that cause severe effects. Most conspicuous in the lethal mutations are the absence of segment boundaries and an inverted polarity of the posterior zone of each segment. Expression of *wingless* has been localized by both *in situ* RNA hybridization and immunostaining. The gene is expressed in regions which are aberrantly developed in *wingless* mutants: different regions of the head, the hindgut/anal region and in 15 stripes in the trunk region (Rijsewijk et al 1987a, van den Heuvel et al 1989). After gastrulation, expression is seen in each segment, in a few cells just anterior to the parasegment boundary.

Assays for *int-1* activity

There are now several biological assays for *int-1* activity. The most dramatic effects of the gene are seen in intact animals. In the mouse, a transgene consisting of *int-1* with an MMTV LTR inserted upstream and in the opposite orientation to the gene, causes massive proliferation of mammary epithelial cells, resulting in hyperplastic glands (Tsukamoto et al 1988). Those mice are not able to nurse their offspring. After a period of latency, mammary tumours emerge in a stochastic pattern, presumably by somatic mutations. The hyperplasia in *int-1* transgenic mice is generally not seen in mice carrying other oncogenes linked to the MMTV LTR, which could mean that the gene has a mammary cell-specific growth stimulatory effect.

Transgenic flies with the *wingless* gene under the control of a heat shock promoter display interesting phenotypes during embryogenesis. The most dramatic phenotype is a naked cuticle, which is, not unexpectedly, the opposite of the *wingless*⁻ phenotype (J. Noordermeer, R. Nusse & P. Lawrence, unpublished).

The *in vitro* effects of *int-1* are less spectacular. By retroviral or direct gene transfer, two cell lines of mammary origin can be morphologically transformed (Brown et al 1986, Rijsewijk et al 1987b). These cells, called C57MG and RAC, normally have a cuboidal phenotype, grow to a low density and are not tumorigenic. Upon uptake and overexpression of *int-1* constructs, the cells become elongated, form foci, and, in case of the RAC cells, grow as tumours in suitable recipient mice. These effects are not seen when fibroblastic lines are used, which suggests a cell type specificity of *int-1* transforming ability. A problem with these assays is that the transformed cells are quite unlike mammary tumour cells with activated *int-1*; they lack most markers of mammary cells and the tumours growing out of the transfected RAC cells have a different pathology to primary mouse mammary tumours. The full spectrum of *int-1* action may thus not be completely represented by these *in vitro* assays.

An *int-1* receptor?

Experimental evidence for an *int-1* receptor has not been obtained, but given the secretion of the protein, we can expect that an *int-1* receptor would behave like other growth factor receptors: binding its ligand on the cell surface then being internalized. Physical evidence for a receptor has to come from biochemical experiments with purified *int-1* protein, which is difficult to isolate in a biologically active form. The genetics of *Drosophila* segmentation may offer an alternative approach; the phenotype of a *wingless* receptor mutant may be similar to *wingless* itself, and eight genes with such a phenotype have been characterized (reviewed in Ingham 1988). Some of those mutants behave in a cell autonomous way, which would be expected of a receptor gene. As many of these genes are now being cloned, more information will soon become available. Further genetic and biochemical analysis of *wingless* in the process of segmentation and in outgrowth of imaginal discs will undoubtedly elucidate the signal transduction pathway in which the gene is involved, with obvious relevance to the mechanism of action of *int-1* in the mouse.

The *int*-related protein, *irp*

The gene for *int*-related protein; *irp*, was found by Wainwright et al (1988) who were searching for a gene predisposing to cystic fibrosis (CF) on human chromosome 7 by a chromosome walk. *irp* is not the CF gene, but shows 38% identity with *int-1*, including conservation of all cysteines, hence the name.

In view of the homology with *int-1*, we have recently examined whether *irp* can also behave as an oncogene in mouse mammary tumours. No cases of tumours with MMTV insertions near the gene were found, but two tumours appeared to have amplified copies of *irp*. In those tumours, which were negative for *int-1* activation, considerable overexpression of *irp* was detected, suggesting that amplification of *irp* had been instrumental in tumorigenesis.

The *int-4* oncogene

This gene is rather infrequently activated by MMTV; it has been observed in only five individual tumours, all from GR mice. The size of the transcriptional unit, which appears to cover more than 50 kb, has complicated its structural analysis. *int-4* is normally expressed in mouse embryos as a 3.8 kb mRNA, probably in the developing brain, and in differentiating P19 embryonal carcinoma cells. The pattern of expression is rather similar to that of *int-1*, but whether *int-1* and *int-4* have other properties in common is not known (H. Roelink et al, unpublished observations).

Cooperation between different *int* genes

With the identification of more *int* genes, many examples have now been found where more than one gene is activated in a single tumour. In a set of 35 BR6 tumours, Peters et al (1986) found 50% of the samples to have *int-1* plus *int-2* rearranged, most likely in a clonal fashion. Another example is those tumours where *int-2* and *hst* are both activated, albeit by a single proviral insertion. In our laboratory we have seen tumours where *int-4* is activated which also have amplified copies of the *irp* gene, with corresponding high levels of *irp* RNA. Intriguing are some early arising tumours in the GR strain, which are still completely dependent on pregnancy hormones for growth. These tumours may have multiple *int* genes activated, but in different clones within the same tumour. The clonal balance of these subpopulations is maintained when the tumour is transplanted, suggesting mutual interactions of the clones. Such interactions may well be mediated directly by the *int* proteins as paracrine growth factors.

These observations suggest that different *int* genes can cooperate in tumorigenesis. A single *int* activation is probably not sufficient for a fully malignant tumour: transgenic mice carrying a germline copy of *int-1* that is transcriptionally active in many of the mammary cells develop tumours in a stochastic way. This indicates that additional mutations, not caused by MMTV, which is absent from these mice, cause clonal outgrowth of the tumours. The nature of these events is unknown; transcriptional activity of *int-2* has not been detected in these tumours. Unexpectedly, the tumours in transgenic animals emerge after approximately the same period of latency as do tumours where *int-1* is somatically activated by MMTV. Possibly the virus does assist in tumorigenesis by providing a gene product which by itself is not sufficient to cause a tumour (otherwise tumours would not be clonal), but which can in cooperation with an *int* gene product.

Speculations and conclusions

The *int* genes form one of the best examples of a link between cancer and the control of normal early development. Cancer in this case would be the consequence of erroneous expression of genes whose normal function is confined to decision-making during embryogenesis. In addition to what is summarized here on *int-1*, essential roles in development for the *int-2* gene product, which is related to the fibroblast growth factor family (Dickson & Peters 1987), are apparent from its expression pattern in mouse embryos. By *in situ* RNA hybridizations, transcription has been detected as early as Day 7.5 of the developing mouse embryo. Most intriguing is the expression seen in presumptive mesodermal cells, migrating through the primitive streak, during gastrulation, and in migrating parietal endodermal cells (Wilkinson et al 1989). These findings strongly indicate that the *int-2* protein may play a decisive role in embryonic development as a local cell–cell interaction molecule.

In reviewing all that is known on the *int* genes, one wonders why these virally induced mouse mammary tumours arise by activation of developmental genes, rather than activation of a more classical oncogene. In other words, why is *myc* or c-*erbB2* never activated, when they and many other other oncogenes are capable of inducing mouse mammary tumours if present as a transgene? Apparently, the mouse mammary gland is extremely sensitive to the inadvertent expression of an *int* gene. The mammary cells must have receptors for the *int* genes, but it is unlikely that they are the only cells equipped with such receptors, nor can we assume that mammary cells would not have receptors for other oncogene products.

It is possible that the *int* genes act as developmental genes in tumorigenesis: developmental genes may be potent mammary oncogenes because the tumours arise in a developing organ, namely mammary glands cycling through multiple rounds of pregnancy. In this aspect, the mammary gland is unique among the adult organs, which may explain the specific association of abnormal mammary growth with ectopic activation of developmental genes. The mitogenic stimulus comes from pregnancy hormones, and aberrant growth may result from activation of genes controlling normal growth at an abnormal point in life.

Another mechanism suggested by the specific biology of the mammary gland is delay of cell death after lactation, when the gland regresses. During a subsequent round of gland development, cells with a prolonged lifespan would still be present, have a growth advantage and be prone to secondary tumorigenic events. This hypothesis is supported by the observation that *int-1* as a transgene leads to mammary hyperplasia in male mice. The male mammary gland develops initially as in the female, but the gland regresses under the influence of hormones acting on the surrounding mesenchyme, which in turn secretes substances destroying the gland. The *int-1* transgene apparently interferes with this regression of the male gland, and may delay cell death.

It is anticipated that new insight into the mechanism of action of the *int* genes will be generated by inactivating the genes in the germline of mice and examining the phenotype of the resulting mutant mice. The possibility of manipulating embryonic stem cells *in vitro* to introduce mutations at will by homologous recombination has been explored for both *int-1* and *int-2*. For *int-2*, such mutations have been obtained at relatively high frequency (Mansour et al 1988), whereas very few homologous recombinants at *int-1* were generated. This contrast may be due to differences in expression levels of the genes in embryonic stem cells, since the selection procedures used depend in part on the transcriptional capacity of the target locus. To date, however, no mice bearing a germline mutation have been generated. Once obtained, such animals will be invaluable in examining what these intriguing genes do in the developing animal and, by extrapolation, in the process of mammary tumorigenesis.

Acknowledgement

Studies in my laboratory are supported by the Netherlands Cancer Foundation.

References

Brown AMC, Wildin RS, Prendergast TJ, Varmus HE 1986 A retrovirus vector expressing the putative mammary oncogene int-1 causes partial transformation of a mammary epithelial cell line. Cell 46:1001–1009

Dickson C, Peters G 1987 Potential oncogene product related to growth factors. Nature (Lond) 326:833

Ingham PW 1988 The molecular genetics of embryonic pattern formation in Drosophila. Nature (Lond) 335:25–34

Mansour SL, Thomas KR, Capecchi MR 1988 Disruption of the proto-oncogene int-2 in mouse embryo-derived stem cells: a general strategy for targeting mutations to non-selectable genes. Nature (Lond) 336:348–352

McMahon AP, Moon RT 1989 Ectopic expression of the proto-oncogene int-1 in Xenopus embryos leads to duplication of the embryonic axis. Cell 58:1075–1084

Morata G, Lawrence PA 1977 The development of wingless, a homeotic mutation of Drosophila. Dev Biol 56:227–240

Noordermeer J, Meijlink F, Verrijzer P, Rijsewijk F, Destree O 1989 Isolation of the Xenopus homolog of int-1/wingless and expression during neurula stages of early development. Nucleic Acids Res 17:11–18

Nusse R 1988 The int genes in mammary tumorigenesis and in normal development. Trends Genet 4:291–295

Nusse R, Van Ooygen A, Cox D, Fung YKT, Varmus HE 1984 Mode of proviral activation of a putative mammary oncogene (int-1) on mouse chromosome 15. Nature (Lond) 307:131–136

Nusse R, Varmus HE 1982 Many tumors induced by the mouse mammary tumor virus contain a provirus integrated in the same region of the host genome. Cell 31:99–109

Papkoff J 1989 Inducible overexpression and secretion of int-1 protein. Mol Cell Biol 9:3377–3384

Papkoff J, Brown AMC, Varmus HE, 1987 The int-1 proto-oncogene products are glycoproteins that appear to enter the secretory pathway. Mol Cell Biol 7:3978–3984

Peters G, Lee AE, Dickson C 1986 Concerted activation of two potential proto-oncogenes in carcinomas induced by mouse mammary tumor virus. Nature (Lond) 320:628–631

Peters G, Brookes S, Smith R, Placzek M, Dickson C 1989 The mouse homolog of the hst/k-FGF gene is adjacent to int-2 and activated by proviral insertion in some virally induced mammary tumors. Proc Natl Acad Sci USA 86:5678–5682

Rijsewijk F, Schuermann M, Wagenaar E, Parren P, Weigel D, Nusse R 1987a The Drosophila homologue of the mammary oncogene int-1 is identical to the segment polarity gene wingless. Cell 50:649–657

Rijsewijk F, Van Deemter L, Wagenaar E, Sonnenberg A, Nusse R 1987b Transfection of the int-1 mammary oncogene in cuboidal RAC mammary cell line results in morphological transformation and tumorigenicity. EMBO (Eur Mol Biol Organ) J 6:127–131

Schuuring E, van Deemter E, Roelink H, Nusse R 1989 Expression of the int-1 proto-oncogene during differentiation of P19 embryonal carcinoma cells. Mol Cell Biol 9:1357–1361

Shackleford GM, Varmus HE 1987 Expression of the proto-oncogene int-1 is restricted to postmeiotic male germ cells and the neural tube of mid-gestational embryos. Cell 50:89–95

Tsukamoto AS, Grosschedl R, Guzman RC, Parslow T, Varmus HE 1988 Expression of the int-1 gene in transgenic mice is associated with mammary gland hyperplasia and adenocarcinomas in male and female mice. Cell 55:619–625

van den Heuvel M, Nusse R, Johnston P, Lawrence PA 1989 Distribution of the *wingless* gene product in Drosophila embryos: a protein involved in cell–cell communication. Cell 59:739–749

Van Ooyen A, Nusse R 1984 Structure and nucleotide sequence of the putative mammary oncogene int-1: proviral insertions leave the protein-encoding domain intact. Cell 39:233–240

Wainwright BJ, Scambler PJ, Stanier P et al 1988 Isolation of a human gene with protein sequence similarity to human and murine int-1 and the Drosophila segment polarity mutant wingless. EMBO (Eur Mol Biol Organ) J 7:1743–1748

Wilkinson DG, Bailes JA, McMahon AP 1987 Expression of the proto-oncogene int-1 is restricted to specific neural cells in the developing mouse embryo. Cell 50:79–88

Wilkinson DG, Bhatt S, McMahon AP 1989 Expression pattern of the FGF-related proto-oncogene int-2 suggests multiple roles in fetal development. Development 105:131–136

DISCUSSION

Hunter: wingless protein is highly localized but you don't know whether that is because it's bound to receptor on the neighbouring cells or to the matrix, is that correct?

Nusse: It could be that the protein is immediately caught by the receptor and therefore doesn't travel far. There is evidence that in *Drosophila* the ligand of *torso* is able to travel away from the poles of the embryo when the *torso* gene product is absent. If you inject *torso* mRNA into the middle of a *torso* mutant embryo, you get formation of the terminal structures in the middle of the embryo. This shows that the *torso* ligand is able to diffuse all over the embryo when the receptor is absent. When the receptor is present the ligand, which is made at the poles, is caught and shows a localized effect.

Hunter: What about the flies in which *wingless* is under the control of the heat shock promoter? What is the distribution of the *wingless* protein in regions where it is not normally expressed?

Nusse: We haven't looked at that by electron microscopy yet. I would predict that the receptor is everywhere.

Hafen: What about in *porcupine⁻* embryos, have you looked at the distribution of *wingless* by electron microscopy?

Nusse: No, not yet.

McMahon: Have you been able to characterize the *Drosophila* protein with your antibodies? For instance, is the *wingless* protein a dimer?

Nusse: We have done Western blots of a reducing gel and the protein does not appear to be proteolytically processed. On a non-reducing gel there are bands in the 100–150 kDa region, so the protein may be a multimer. The monomer is 52 kDa.

McMahon: The prediction is that there are dominant mutations; are you trying to isolate new mutations?

Nusse: Yes. That is one approach. The other approach is making transgenic flies with P element vectors carrying the mutant *wingless* gene constructs to see whether they have a dominant-negative effect on the expression of the normal *wingless* allele.

Wagner: In the embryos in which *wingless* is ectopically expressed, have you measured the expression of other developmental genes? How does *wingless* affect their expression?

Nusse: We are looking at expression of *engrailed* but we have no results yet. We are also looking at the expression of *armadillo*. Eric Wieschaus has shown that the *armadillo* RNA is expressed all over the *Drosophila* embryo. The protein is seen only in the zone of cells that make *wingless* protein. In the *wingless*⁻ mutant, that staining is absent, so it appears that the *armadillo* protein is also dependent on *wingless*. How that works is not known. It would be interesting to look for expression of *armadillo* protein in these embryos expressing *wingless* ectopically.

Hafen: Can you see any genetic interaction by making double heterozygotes between the different segment polarity genes? Would that show which ones might interact? If you reduce the level of potential receptors, do you see a phenotype?

Nusse: We haven't done that. I am surprised that it hasn't been done by anyone yet, although it does require good mutants. Most of these mutants are lethal. *wingless*[1] is an interesting candidate to look for effects in double heterozygotes. It could be done with *wingless*-lethal but *wingless*[1] is better because it is viable. It would be interesting to see whether there is any effect of crossing in a second polarity gene.

McMahon: We have looked at the expression of *int-1* in mice and the patterns are relatively simple (Wilkinson et al 1987). At early stages of neurulation, prior to neural tube closure and any recognizable neural differentiation, the *int-1* mRNA is distributed fairly broadly over the anterior head folds. It is not found in more posterior regions of presumptive brain tissue or in the spinal cord, that's at about 8.5 days of mouse embryonic development. By about 9.5 days, what looks like the final pattern of expression has been established. *int-1* has predominantly localized to the dorsal midline of the mid-brain, the equivalent region in the hind brain and right the way down the presumptive spinal cord. So it's localized symmetrically about the dorsal midline, in an area called the roof plate. These cells are thought to be a population of radial glia.

That's how it stays until about 16.5 days of development, which is the last stage at which we have definitely seen expression. It's difficult to tell whether it is expressed in those cells from then onwards because they form a non-mitotic population that is undergoing rather dramatic changes in shape. All our studies have been done at the RNA level, and not at the protein level.

Noble: If you do *in situ* hybridization in newborn mice, do you see *int-1* mRNA?

McMahon: We haven't gone beyond 16.5 days, because by then it's pretty difficult for us to see expression.

Noble: Do you have a useful antibody?

McMahon: We don't.

Nusse: We were lucky that *wingless* protein has this large insert that is not in the mouse *int-1* protein. We have evidence that the insert is the antigenic determinant. The remainder of the protein is rich in cysteine residues and it may be folded such that most of the antibodies raised against peptide fragments from the protein do not see their antigen. The insert domain may be on the outside of the protein, and it contains charged amino acids, so it is a good antigenic component.

Brugge: In normal mouse mammary tissue, is *int-1* induced by oestrogen?

Nusse: No, we have looked at that, it's not.

Brugge: Presumably, the receptor is expressed in those cells.

Hunter: This is a bit of a paradox—if *int-1* is never expressed in the adult, why should the receptor be there?

Nusse: One explanation could be that there is a related gene product that acts on the receptor. Another is that the receptor is ubiquitous, although I find that more difficult to believe.

McMahon: A third possibility is that *int-1* is expressed in the early developing mammary gland. As far as I know, that has never really been looked at.

Heath: Can you complement the *Drosophila wingless* mutation with the mammalian *int-1* gene?

Nusse: We haven't tried this yet, because we have not even been able to complement the *wingless* mutation by *wingless*. It appears that some of the mutations in *wingless* are at sites 5kb downstream of the gene and that there are extensive regulatory elements surrounding the gene. We have to make larger constructs to include all the regulatory sequences before we can do the complementation experiments.

Norbury: Is there any reason to expect that there might be mouse genes other than *int-1* that resemble *wingless* more closely?

McMahon: The *Drosophila wingless* gene was pulled out by Roel Nusse by cross-hybridization with the mouse *int-1* and it didn't pull out any other genes. So at the nucleotide level, *wingless* is definitely the most closely related gene in *Drosophila*. Whether there are other related genes remains to be seen. The gene for *int*-related protein does not appear to have a *Drosophila* counterpart, as judged by low stringency screening.

Nusse: In the fly we have looked for other *wingless*-like genes and we haven't seen them yet.

Westermark: With the *int-1*-transformed mouse cells, you don't find a growth factor in the medium, but is there any activity associated with the cell surface or any growth factor activity in the matrix?

Nusse: Tony Brown has shown by Western blotting that there is a lot of *int-1* protein in the matrix. The matrix is the material that remains on the dish after you scrape off the cells or lyse them mildly. He has not shown that that protein is biologically active.

Westermark: If you seed cells in that matrix, do they not grow?

Nusse: There is no effect. However, if Tony co-cultivates cells that make *int-1* with cells that are able to respond to the *int-1* signals by morphological transformation, the co-cultivated cells become transformed. Whether that is due to the *int-1* protein is not known, because the antibodies that are available do not inhibit that kind of interaction.

Hunter: I am not sure that Jackie Papkoff would necessarily agree with Tony Brown that much *int-1* is associated with the matrix.

Vande Woude: Has the *Drosophila* gene been tried in the mouse cell transformation assay?

Nusse: Yes, we saw marginal morphological transformation.

Vande Woude: Does that require high levels of expression?

Nusse: We did it with a *wingless* construct driven by an MSV LTR.

Hunter: What about the effects of *int-1* in *Xenopus*?

McMahon: We injected mouse *int-1* mRNA into *Xenopus* embryos, hoping that it would tell us what *int-1* might be doing in the neural tube, assuming that *Xenopus int-1* is expressed in the same way as the murine form. We have no direct evidence, but Olivier Destrée's group in Holland have shown that *Xenopus* has an *int-1* gene which is activated at the neurula stage of development (Nordermeer et al 1989), consistent with the mouse.

When we inject murine *int-1* into fertilized eggs or 2-cell embryos, or even at the 8-cell stage, it leads to duplication of the embryonic axis (McMahon & Moon 1989). The duplication is quite profound; it is a duplication of the notochordal tissue which then induces extra neural plate. There is also duplication of somitic tissue. It requires intact *int-1* protein. If one mutates one of the conserved cysteine residues, one knocks out the duplication. We removed the signal sequence and the truncated protein had no effect whatsoever. When we added back a completely unrelated signal peptide sequence, from chick lysozyme, it restored the effect. This suggests the protein operates through some stage of a secretory pathway.

Hunter: That's interesting, because there is no evidence that the *int-1* signal sequence is cleaved, whereas the lysosome signal peptide would be.

Nusse: It is, if you add microsomes to *in vitro* translations of *int-1* RNA, the signal peptide is cleaved from the translated *int-1* protein.

McMahon: We can't reconcile the phenotype that we see in *Xenopus* with what we expected from the expression pattern in the mouse. However, we can make some statements as to how we think it operates. Our best guess is that the effect is in some way caused by an interaction with the organizing system of the frog embryonic axis. Spemann & Mangold (1924) defined a structure, that they called the dorsal organizer, which is responsible for setting up the embryonic axis. The sorts of phenotypes we see are what one sees if one perturbs the organizer, so we are looking for a role of *int-1* in the dorsal organizer.

Méchali: Do you get a complete duplication?

McMahon: No. There are only anterior duplications. They resemble embryos produced in some experiments that Jonathan Cooke performed several years ago in which organizers were grafted into embryos close to the existing organizer (Cooke 1972). If you decrease the angle between the organizers, they cause progressively more anterior duplications. So the *int-1*-injected *Xenopus* embryos look like they are caused by a local split in the organizer.

Hunter: But in the transgenic mice, there's no evidence for developmental abnormality is there?

Nusse: It probably depends on the promoter. The experiments of Harold Varmus' group were done with an MMTV promoter (Tsukamoto et al 1988).

McMahon: We are trying to do experiments with other promoters in mice.

References

Cooke J 1972 Properties of the primary organization field in the embryo of *Xenopus laevis*. II. Positional information for axial organization in embryos with two head organizers. J Embryol Exp Morphol 28:27–46

McMahon AP, Moon RT 1989 Ectopic expression of the proto-oncogene *int-1* in *Xenopus* embryos leads to duplication of the embryonic axis. Cell 58:1075–1084

Nordermeer J, Meijlink F, Verrijzer P, Rijsewijk F, Destrée O 1989 Isolation of the *Xenopus* homologue of *int-1 wingless* and expression during neurula stages of early development. Nucl Acid Res 17:11–18

Spemann H, Mangold H 1924 Uber Induktion von Embryonenanlagen durch Implantation artfremder Organisatonen. Wilhelm Roux Arch Entwicklungsmech Org 100:599–638

Tsukamoto AS, Grosschedl R, Guzman RC, Parslow T, Varmus HE 1988 Expression of the *int*-1 gene in transgenic mice is associated with mammary gland hyperplasia and adenocarcinomas in male and female mice. Cell 55:619–625

Wilkinson DG, Bailes JA, McMahon AP 1987 Expression of the proto-oncogene *int-1* is restricted to specific neural cells in the developing mouse embryo. Cell 50:79–88

Control of division and differentiation in oligodendrocyte-type-2 astrocyte progenitor cells

Mark Noble* Susan C. Barnett* Oliver Bögler*
Hartmut Land° Guus Wolswijk* † and Damian Wren†

*Ludwig Institute for Cancer Research, 91 Riding House Street, London W1P 8BT, UK;
†Department of Clinical Neurology, Institute of Neurology, Queen Square, London
WC1N 3BG, UK; °Imperial Cancer Research Fund, Lincolns Inn Field, London WC2R 3LF, UK

Abstract. Oligodendrocyte-type-2 astrocyte (O-2A) progenitor cells give rise to
oligodendrocytes and type-2 astrocytes in cultures of rat optic nerve. These
progenitors are one of the few cell types in which most aspects of proliferation
and differentiation can be manipulated in a defined *in vitro* environment. When
exposed to platelet-derived growth factor (PDGF), O-2A progenitors divide a
limited number of times before clonally related cells differentiate into
oligodendrocytes with a timing similar to that seen *in vivo*. In contrast, O-2A
progenitors grown in the absence of mitogen do not divide but differentiate
prematurely into oligodendrocytes, and progenitors exposed to appropriate
inducing factors differentiate into type-2 astrocytes. O-2A progenitors can become
immortalized through at least two different mechanisms. First, when O-2A
progenitors are exposed to a combination of PDGF and basic fibroblast growth
factor (bFGF) these cells undergo continuous self-renewal in the absence of
differentiation. In contrast, the application of bFGF alone is associated with
premature oligodendrocytic differentiation of dividing O-2A lineage cells. Thus,
cooperation between growth factors can modulate O-2A progenitor self-renewal
in a defined chemical environment by eliciting a novel programme of division and
differentiation which cannot be predicted from the effects of either factor examined
in isolation.
A further mechanism which allows prolonged self-renewal in the O-2A lineage
is the generation of a stem cell. O-2A progenitors isolated from optic nerves of
perinatal rats also have the capacity to give rise to a population of cells called
O-2A^{adult} progenitors, which differ from their perinatal counterparts in many
characteristics. Most importantly, O-2A^{adult} progenitors have a slow cell cycle,
divide and differentiate asymmetrically and appear to have the capacity for
prolonged self-renewal. Thus, immortalization in this lineage can also be achieved
by the generation of a cell with stem cell-like characteristics from a rapidly dividing
progenitor population.

*1990 Proto-oncogenes in cell development. Wiley, Chichester (Ciba Foundation
Symposium 150) p 227–249*

Our attempts to understand the association between division and differentiation in normal development have been focused on the optic nerve, the simplest part of the central nervous system (CNS). This tissue is composed of a small number of cell types, of which just three are derived from the neuroectoderm. Type-1 astrocytes, and their precursors, seem to contribute to the morphogenetic development of the optic nerve by offering a preferred substrate for growing axons (Noble et al 1984, Silver & Sapiro 1981). These cells also appear to interact with endothelial cells to induce formation of the blood–brain barrier (Janzer & Raff 1987), and (as discussed below) are a source of mitogen for other cells in the nerve. Oligodendrocytes enwrap large axons of the CNS with myelin sheaths, while type-2 astrocytes extend processes which are associated with axons at the nodes of Ranvier (ffrench-Constant & Raff 1986, Miller et al 1989), the regions between consecutive myelin sheaths where ion fluxes occur during transmission of impulses along the myelinated axon. Thus, oligodendrocytes and type-2 astrocytes appear to be specialized to create the anatomical specializations which characterize the myelinated tracts of the CNS.

The three differentiated glial cell types of the optic nerve are generated at specific developmental periods from two distinct cellular lineages. The neuro-epithelial cells which form the optic stalk, the embryonic anlage of the optic nerve, seem to give rise only to type-1 astrocytes (Small et al 1987). The initial differentiation of type-1 astrocytes, at Day 16 of embryogenesis (E16), is followed developmentally by the appearance within the nerve of oligo-dendrocyte-type-2 astrocyte (O-2A) progenitor cells, which can be induced to differentiate *in vitro* into either oligodendrocytes or type-2 astrocytes (Raff et al 1983b). The O-2A progenitors arise in a germinal zone located in or near the optic chiasm, and first migrate into the optic nerve at E17 (Small et al 1987). Several days later, at the time of birth of the rat (E21 or P0), oligodendrocytes are first seen in the nerve (Miller et al 1985). Type-2 astrocytes are the third glial cell type to differentiate within the nerve, and do not begin to develop until around eight days after birth (P8; Miller et al 1985).

O-2A progenitors can also be identified in the optic nerves of adult rats, but we defined these cells separately as O-2A[adult] progenitors because they differ from their perinatal counterparts in antigen expression, morphology, cell cycle length, motility, time-course of differentiation and the manner in which they generate oligodendrocytes (Wolswijk & Noble 1989, submitted, Wren et al, submitted; see Fig. 1 for summary). As discussed later in detail, the O-2A[adult] progenitors appear to be derived from a subset of O-2A[perinatal] progenitors (Wren et al, submitted). Thus, O-2A[perinatal] progenitors represent tripotential, rather than bipotential, cells.

All of the diverse cell types of the optic nerve can be readily distinguished from each other *in vitro* on the basis of antigen expression and morphology. The identifying characteristics of the cells of the O-2A lineage are summarized in Fig. 1. As also shown in Fig. 1, O-2A progenitors grown in the presence of appropriate inducing factors (e.g. fetal calf serum or ciliary neurotrophic factor) differentiate into type-2 astrocytes, while growth in the absence of inducing agents is associated with differentiation of O-2A progenitors into oligodendrocytes (Raff et al 1983b, Hughes & Raff 1987, Lillien et al 1988). We shall not discuss the type-2 astrocyte pathway of differentiation in this paper, and the reader is referred to recent reviews (Anderson 1989, Raff 1989) for further information about this particular topic.

The role of type-1 astrocytes in modulating division and oligodendrocytic differentiation of O-2A^{perinatal} progenitors

When we originally identified the O-2A^{perinatal} progenitor cell we were puzzled by two observations. First, these cells were not dividing in our tissue culture conditions (Raff et al 1983b), even though they were being removed from the rat optic nerve at seven days after birth, a time previously identified as a peak period for division of oligodendroglial precursors (Skoff et al 1976a,b). Second, virtually all of these progenitor cells differentiated within three days *in vitro* (Raff et al 1983b), even though generation of oligodendrocytes normally continues *in vivo* for at least three weeks beyond the time of our dissection (Skoff et al 1976a,b). The failure of O-2A^{perinatal} progenitor cells to divide or differentiate *in vitro* according to their normal *in vivo* schedule suggested that a necessary component of the *in vivo* environment was missing in the tissue culture dish.

In subsequent experiments, we found that purified type-1 astrocytes appear to be the vital component which promotes the normal division and differentiation of O-2A^{perinatal} progenitors and that the division of O-2A^{perinatal} progenitors *in vitro* was caused by a soluble mitogenic activity secreted by type-1 astrocytes. We initially found that purified type-1 astrocytes secreted a soluble mitogenic activity capable of causing extensive division of O-2A^{perinatal} progenitor cells *in vitro* (Noble & Murray 1984). Dividing O-2A^{perinatal} progenitors grown in the presence of type-1 astrocytes continued to produce increasing numbers of progenitors, and oligodendrocytes, for an extended period in tissue culture. Thus, just as *in vivo*, growth of O-2A^{perinatal} progenitors in the presence of type-1 astrocytes was associated with maintenance of a dividing population of progenitors for extended periods, and also with the continued production of new oligodendrocytes (Noble & Murray 1984). In contrast, removal of the eye of newborn rats, which leads to degeneration of the axons in the nerve, had no effect on the division, or differentiation, of progenitors in the optic nerve (David et al 1984), suggesting that type-1 astrocytes may be primarily responsible for promoting division of O-2A^{perinatal} progenitor cells.

As well as promoting the *in vitro* division of O-2A^{perinatal} progenitors, type-1 astrocytes restored the normal timing of oligodendrocytic differentiation. As mentioned earlier, oligodendrocytes first appear in the rat optic nerve at about

THE OLIGODENDROCYTE-TYPE-2 ASTROCYTE LINEAGE

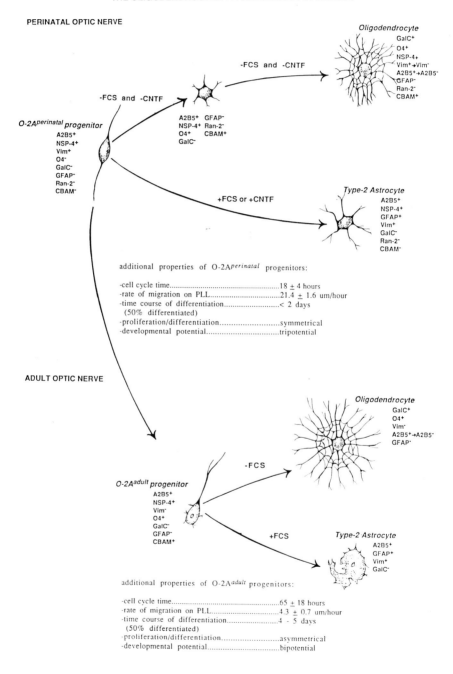

PERINATAL OPTIC NERVE

Oligodendrocyte
GalC⁺
O4⁺
NSP-4₊
Vim⁺→Vim⁻
A2B5⁺→A2B5⁻
GFAP⁻
Ran-2⁻
CBAM⁺

-FCS and -CNTF

-FCS and -CNTF

A2B5⁺ GFAP⁻
NSP-4⁺ Ran-2⁻
O4⁺ CBAM⁺
GalC⁻

O-2Aperinatal progenitor
A2B5⁺
NSP-4⁺
Vim⁺
O4⁻
GalC⁻
GFAP⁻
Ran-2⁻
CBAM⁻

+FCS or +CNTF

Type-2 Astrocyte
A2B5⁺
NSP-4⁺
GFAP⁺
Vim⁺
GalC⁻
Ran-2⁻
CBAM⁻

additional properties of O-2Aperinatal progenitors:

-cell cycle time...18 ± 4 hours
-rate of migration on PLL.................................21.4 ± 1.6 um/hour
-time course of differentiation..........................< 2 days
 (50% differentiated)
-proliferation/differentiation..........................symmetrical
-developmental potential................................tripotential

ADULT OPTIC NERVE

Oligodendrocyte
GalC⁺
O4⁺
Vim⁻
A2B5⁺→A2B5⁻
GFAP⁻

-FCS

O-2Aadult progenitor
A2B5⁺
NSP-4⁺
Vim⁻
O4⁺
GalC⁻
GFAP⁻
CBAM⁺

+FCS

Type-2 Astrocyte
A2B5⁺
GFAP⁺
Vim⁺
GalC⁻

additional properties of O-2Aadult progenitors:

-cell cycle time...65 ± 18 hours
-rate of migration on PLL.................................4.3 ± 0.7 um/hour
-time course of differentiation..........................4 - 5 days
 (50% differentiated)
-proliferation/differentiation..........................asymmetrical
-developmental potential................................bipotential

E21, the time of birth of the rat. However, if O-2A progenitors derived from optic nerves of embryonic rats of any age are grown in chemically defined medium, these cells always give rise to oligodendrocytes within 48 hours of plating (Raff et al 1985). In contrast, embryonic progenitors induced to divide *in vitro* by growth in the presence of type-1 astrocytes give rise to oligodendrocytes

---

FIG. 1. This figure summarizes the defining characteristics of O-2A*perinatal* and O-2A*adult* progenitor cells and their derivatives. O-2A*perinatal* progenitors are bipolar cells which divide with a cell cycle of about 18 hours and migrate with an average speed of 21 μm/h when grown in the presence of type-1 astrocytes or platelet-derived growth factor (Small et al 1987, Noble et al 1988). These cells can be distinguished antigenically by labelling with monoclonal antibodies A2B5 and NSP-4, by their expression of vimentin intermediate filaments (IFs) and by their lack of labelling with the O4 monoclonal antibody when dividing *in vitro* (Raff et al 1983b, ffrench-Constant & Raff 1986, Wolswijk & Noble 1989). O-2A*perinatal* progenitors divide predominantly in a symmetrical fashion, followed by differentiation of all cells within a single family into oligodendrocytes after 1–8 divisions. O-2A*perinatal* progenitors differentiating into oligodendrocytes first pass through an intermediate stage of differentiation which can be identified by labelling with the O4 monoclonal antibody (Sommer et al, submitted).

O-2A*adult* progenitors have a characteristic unipolar morphology, divide with a cell cycle time of 65 hours and migrate with an average speed of only 4 μm/h when grown in the presence of type-1 astrocytes (Wolswijk & Noble 1989). Unlike O-2A*perinatal* progenitors, the O-2A*adult* progenitors divide as O4$^+$ cells *in vitro*. Moreover, these cells appear to divide and differentiate asymmetrically (Wren et al, submitted), such that a single clone of cells continues to generate more progenitors for at least several divisions after the first generation of oligodendrocytes within the clone.

Both O-2A*perinatal* progenitors and O-2A*adult* progenitors can differentiate into either oligodendrocytes or type-2 astrocytes. Astrocytic differentiation is induced by exposure to fetal calf serum (FCS) or ciliary neurotrophic factor (CNTF; Raff et al 1983a,b, Hughes et al 1988, Lillien et al 1988). As CNTF-like molecules are present in extracts prepared from optic nerves of three week old rats (Hughes et al 1988), it appears that CNTF may play an important role in type-2 astrocyte differentiation *in vivo*. It is not yet known whether O-2A*adult* progenitors are induced to differentiate into type-2 astrocytes in the presence of CNTF.

Oligodendrocytes can be identified *in vitro* by their multipolar morphology, their expression of galactocerobroside (GalC), by O4 labelling and by their lack of intermediate filaments (Raff et al 1978, Sommer & Schachner 1981, Raff et al 1985). Type-2 astrocytes generated from either O-2A*perinatal* progenitors or O-2A*adult* progenitors contain both vimentin and glial fibrillary acidic protein (GFAP) intermediate filaments, and are A2B5$^+$ and NSP-4$^+$ (Raff et al 1983a,b, Raff et al 1985, ffrench-Constant & Raff 1986, Wolswijk & Noble 1989). While type-2 astrocytes generated from O-2A*perinatal* progenitors are process-bearing cells (Raff et al 1983b, Lillien et al 1988), type-2 astrocytes generated by O-2A*adult* progenitors usually have a more flattened morphology (Wolswijk & Noble 1989).

Oligodendrocytes and O-2A*adult* progenitors are capable of binding and activating complement in the absence of antibody, resulting in their lysis. O-2A*perinatal* progenitors are resistant to the effects of complement, and acquire the capacity to bind and activate complement while progressing through stages of differentiation intermediate between the progenitor cell and the GalC$^+$ oligodendrocyte (Wren & Noble 1989). CBAM, complement binding and activating molecule(s); Vim, vimentin.

after a period of time equivalent to the age of the embryo plus the number of days that would have been required for the embryo to reach E21 (e.g. E17 + 4 days, E18 + 3 days). These results suggest that the onset of oligodendrocytic differentiation is controlled by a biological clock, and that type-1 astrocytes promote the normal function of this clock *in vitro*.

The biological clock which promotes the normally timed differentiation of oligodendrocytes appears to function by causing all members of a clonally related family of O-2A*perinatal* progenitor cells to differentiate synchronously into oligodendrocytes within a limited number of cell divisions. Although we do not yet understand the molecular nature of this clock, it seems most likely that the mechanism resides within the O-2A*perinatal* progenitors themselves. For example, different families of O-2A*perinatal* progenitors in a culture dish can divide dissimilar numbers of times before differentiating, suggesting that induction of differentiation is not mediated by a soluble factor. Moreover, sister O-2A*perinatal* progenitor cells grown in separate dishes divide an equal number of times before differentiating (Temple & Raff 1986), further suggesting that the mechanism which controls timing of the clock is internal to the progenitor cells themselves.

Growth of O-2A*perinatal* progenitors in the presence of type-1 astrocytes also has specific effects on the morphology and motility of these cells. O-2A*perinatal* progenitors grown in chemically defined medium (DMEM-BS; Raff et al 1983b, Bottenstein & Sato 1979) are non-motile cells which rapidly develop a multipolar morphology and differentiate into oligodendrocytes. In contrast, O-2A*perinatal* progenitors induced to divide by type-1 astrocytes express a characteristic bipolar morphology and are extremely motile cells, capable of migrating at speeds of up to 100 μm/h, and with average speeds of 21 μm/h (Small et al 1987).

Platelet-derived growth factor is the astrocyte-derived mitogen which modulates division and differentiation of O-2A*perinatal* progenitors *in vitro*

The observation that type-1 astrocytes had dramatic effects on O-2A*perinatal* progenitors led us to investigate the molecular basis for this cell–cell interaction. In screening a variety of known mitogens, we found that only platelet-derived growth factor (PDGF) was capable of mimicking the effects of type-1 astrocytes on O-2A*perinatal* progenitor cells (Noble et al 1988). PDGF promoted DNA synthesis in O-2A*perinatal* progenitors as effectively as type-1 astrocytes or astrocyte-conditioned medium (Astro-CM). With the best PDGF preparations, DNA synthesis was induced at < 5 pM and the response plateaued at < 100 pM. With a similar dose–response curve, PDGF also inhibited the premature differentiation of progenitors into oligodendrocytes seen when these cells are grown in the absence of type-1 astrocytes. After 72 hours of growth in chemically defined medium virtually all O-2A*perinatal* progenitors differentiate into oligo-dendrocytes that express galactocerebroside. In contrast, after 48 or 72 hours of growth in the presence of PDGF or Astro-CM, over half of the O-2A^{perinatal} lineage cells still expressed the antigenic and morphological phenotype of O-2A^{perinatal} progenitors. Moreover, O-2A^{perinatal} progenitors grown in the presence of PDGF and analysed by time-lapse microcinematography had similar rates of migration and cell cycle lengths to those grown in Astro-CM (24.6 ± 5.4 µm/h versus 21.4 ± 1.6 µm/h; 19.7 ± 6.4 h versus 17.7 ± 4.0 h in PDGF and Astro-CM, respectively).

PDGF was also capable of driving the clock which modulates the timing of oligodendrocytic differentiation *in vitro* (Raff et al 1988). If PDGF was added to cultures of embryonic rat optic nerve cells grown in chemically defined medium, then the correct timing of oligodendrocyte differentiation was restored. In addition, O-2A^{perinatal} progenitors grown in the presence of PDGF generated families of cells in which all members of the family differentiated synchronously after a limited number of divisions. In contrast, PDGF neither promoted differentiation of O-2A^{perinatal} progenitors into type-2 astrocytes nor inhibited such differentiation when fetal calf serum was present in the medium (Noble et al 1988).

Several lines of evidence indicate strongly that the effects of type-1 astrocytes on O-2A^{perinatal} progenitors are indeed mediated by PDGF: (i) antibodies to PDGF block the effects of type-1 astrocytes on O-2A^{perinatal} progenitors derived from either embryonic or perinatal rats (Noble et al 1988, Raff et al 1988, Richardson et al 1988); (ii) type-1 astrocytes make mRNA for PDGF A-chain, and (iii) the mitogenic activity secreted by type-1 astrocytes *in vitro* co-migrates with PDGF on chromatography columns (Richardson et al 1988). That this factor is physiologically relevant is supported by observations that (i) PDGF is effective at picomolar concentrations (Noble et al 1988, Raff et al 1988, Richardson et al 1988), (ii) mitogenic activity for O-2A^{perinatal} progenitors found in extracts of optic nerve is largely blocked by anti-PDGF antibodies (Raff et al 1988), and (iii) *in situ* hybridization suggests that optic nerve cells express mRNA for PDGF A-chain *in vivo* (Pringle et al 1989). Moreover, O-2A^{perinatal} progenitor cells express α type PDGF receptors, in agreement with their ability to respond to A-chain homodimers (Hart et al 1989).

The programme of behaviour induced in O-2A^{perinatal} progenitors by PDGF is mitogen specific

One of the next problems we addressed was the extent to which individual aspects of the behavioural programme of O-2A^{perinatal} progenitor cells can be separately regulated. The ability of PDGF to replace completely type-1 astrocytes in modulating the *in vitro* development of O-2A^{perinatal} progenitors suggested that these cells have a complex and constitutive behavioural phenotype, controlled by processes internal to the progenitors themselves, and that initiation of this programme requires interaction of progenitors with only a single polypeptide mitogen. But, does induction of cell division necessarily cause expression of the entire programme of behaviour seen in O-2A^{perinatal} progenitor cells grown in PDGF or the presence of type-1 astrocytes, or can this programme be uncoupled from cell division?

Our studies on the effects of basic fibroblast growth factor (bFGF) on O-2A^{perinatal} progenitor cells have shown that it is possible to separate the promotion of progenitor division from other possible behaviours of dividing O-2A^{perinatal} progenitor cells (Bögler et al, submitted). For example, although bFGF is an effective mitogen for O-2A^{perinatal} lineage cells, O-2A^{perinatal} progenitors dividing in the presence of bFGF rapidly develop a multipolar morphology and show little or no migratory behaviour. In addition, most O-2A^{perinatal} progenitor cells grown in the presence of bFGF differentiate into oligodendrocytes within 72 hours. Thus, O-2A^{perinatal} progenitors can be induced to divide without necessarily expressing other properties associated with progenitors grown in the presence of type-1 astrocytes or PDGF.

Another difference between PDGF and bFGF is that bFGF can induce division of oligodendrocytes (Eccleston & Silberberg 1985, Saneto & De Vellis 1985); thus, even though oligodendrocytes are not induced to divide by PDGF they are nonetheless competent to do so. This finding suggests that the functioning of the biological clock which modulates the timing of oligodendrocyte differentiation, although associated with a withdrawal from division prior to oligodendrocytic differentiation, is not associated with terminal loss of the ability to divide. In addition, the observation that oligodendrocytes can be induced to divide by bFGF focuses attention on the paradoxical observation that these cells express PDGF receptors (Hart et al 1989), but are not induced to divide by PDGF (Noble et al 1988).

Growth factors can cooperate to cause continuous self-renewal, and inhibit differentiation, of progenitor cells

In striking contrast to the effects of exposure to either bFGF or PDGF alone, oligodendrocytic differentiation of O-2A progenitors was strongly inhibited by simultaneous exposure to both of these mitogens (Bögler et al, submitted). A dramatic demonstration of the ability of PDGF + bFGF to inhibit oligo-dendrocytic differentiation was obtained by adding these mitogens to cultures of optic nerve cells from rats. In these cultures, O-2A progenitor cells exposed to PDGF, or grown in the presence of PDGF-secreting type-1 astrocytes, normally begin to generate oligodendrocytes after two days of *in vitro* growth (Raff et al 1985, 1988), a timing which mimics that seen *in vivo* (Miller et al 1985). In contrast, identical cells grown in the presence of PDGF + bFGF did not differentiate into oligodendrocytes even after 10 days of *in vitro* growth.

The inhibition of differentiation seen in embryonic cultures exposed to PDGF + bFGF could be maintained for extended periods *in vitro*, but was conditional upon the continued application of these mitogens. For example, when E19 cultures grown in the presence of PDGF + bFGF were switched after 10 days to chemically defined medium or to defined medium supplemented with 10 ng/ml of PDGF, oligodendrocytes began to appear within 48 hours, and steadily increased in number in the days following. In contrast, we have been able to passage O-2A progenitors for over five months in the absence of differentiation by continually growing cells in the presence of PDGF + bFGF (S. Barnett, M. Noble, unpublished observations).

The effects of bFGF and PDGF on O-2A progenitor differentiation lead to the following conclusions in regards to the O-2A lineage: (i) The biological clock which promotes the correctly timed differentiation of oligodendrocytes *in vitro* can be regulated by the environment. Thus, the clock does not function simply through the counting of cell divisions. In this regard, the clock which promotes the correctly timed appearance of oligodendrocytes is apparently very different from the one which regulates the switch from production of fetal to adult haemoglobin in the erythroid lineage, as all experiments to date have indicated that the timing of the haemoglobin switch is not regulated by environmental factors (Wood et al 1985, 1988). (ii) Induction of oligodendrocytic differentiation of O-2A progenitors can be uncoupled from division in two opposite directions. Progenitors can undergo premature differentiation while being induced to divide (by exposure to bFGF alone), and can also be induced to divide without differentiating (by co-application of PDGF + bFGF). Our results therefore demonstrate that the correctly timed production of oligodendrocytes *in vitro* is dependent upon the stimulation of particular signalling pathways in O-2A progenitor cells, and indicate that PDGF is acting in this lineage both as a mitogen and as a regulatory agent which elicits a specific programme of differentiation.

Studies of other cellular lineages suggest that the ability to induced prolonged progenitor division in the absence of differentiation represents a phenomenon which is not specific to the O-2A lineage, although our experiments represent the first reduction of this phenomenon to a cooperative interaction between two well-defined growth factors. For example, growth of murine embryonic stem cells in medium containing differentiation-inducing activity and fetal calf serum similarly inhibits differentiation (Smith et al 1988). Our results suggest that cell–cell interactions which inhibit progenitor differentiation may be mediated by the interaction between a small number of definable molecules, and moreover, demonstrate that cooperation between growth factors can elicit effects on differentiation which are completely unpredictable from the effects of either growth factor analysed in isolation. It is also of interest that the inhibition of differentiation induced by co-application of PDGF + bFGF to cultures of O-2A progenitors is strikingly similar to that induced by the expression of activated nuclear oncogenes in these cells after retroviral-mediated gene insertion (M. Noble, H. Land, in preparation). Such observations raise obvious questions about the molecular mechanisms which underlly the inhibition of differentiation in normal cells.

The O-2A^{adult} progenitor cell

As mentioned above, O-2A^{perinatal} progenitors also give rise to a second progenitor population, the O-2A^{adult} progenitors, which have properties very different from those of their perinatal counterparts (summarized in Fig. 1). For example, O-2A^{adult} progenitors have a cell cycle time of about 65 hours, over three times as long as that of O-2A^{perinatal} progenitors. O-2A^{adult} progenitors also migrate more slowly than their perinatal counterparts, express a different morphology and have a different antigenic phenotype. These properties are all expressed by O-2A^{adult} progenitors grown in the presence of astrocyte-derived mitogen(s). More recent experiments suggest that the effects of type-1 astrocytes on O-2A^{adult} progenitors are likely to be mediated by PDGF. In addition, when grown in chemically defined medium, or in the presence of fetal calf serum, O-2A^{adult} progenitors differentiate into oligodendrocytes or type-2 astrocytes several times more slowly than O-2A^{perinatal} progenitors (Wolswijk & Noble 1989).

The differences between *perinatal* and *adult* O-2A progenitor cells appear to be intrinsic to the cells themselves, as the progenitor populations can coexist *in vivo* and in tissue culture for extended periods (Wolswijk & Noble, submitted). For example, during normal development, O-2A^{adult} progenitors first appear in the rat optic nerve at about seven days after birth, when the majority of progenitors in the nerve are of the perinatal phenotype. O-2A^{perinatal} progenitors are gradually lost from the population of O-2A lineage cells during the subsequent 3–4 weeks, and the ratio of perinatal:adult progenitors in the optic nerve switches in favour of the adult phenotype during this time. Finally, by one month after birth, many progenitors in the nerve are of the adult phenotype. However, O-2A^{perinatal} progenitors and O-2A^{adult} progenitors isolated from the optic nerve during this period of transition each express their characteristic behaviour, and can be readily distinguished from each other. Their characteristic distinguishing features are also expressed when these cells are grown in the presence of PDGF.

Asymmetry and self-renewal

A well recognized means of achieving immortalization of normal cells is through asymmetric division and differentiation, such that dividing precursor cells generate both more precursors and differentiated end-stage cells, but where there is at least a slightly greater likelihood of generating a new precursor (i.e. the self-renewal probability is >0.5). This is the mechanism which is thought to underly the immortality of stem cell populations. In this regard, one of the most striking differences between O-2A^{adult} progenitors and O-2A^{perinatal} progenitors is that the adult cells divide and differentiate asymmetrically, rather than in the symmetrical manner expressed by most perinatal cells (Wren et al, submitted). For example, when we examined colonies derived from dividing O-2A^{adult} progenitors, growing on monolayers of type-1 astrocytes, we found that over 75% of colonies contained both oligodendrocytes (which do not divide in these tissue culture conditions) and progenitors that were engaged in DNA synthesis (as determined by [3H]thymidine labelling) after 15 or 25 days *in vitro* (Wren et al, submitted). Only 10% of the colonies visualized on Day 25 (a length of time equal to <10 divisions) consisted entirely of oligodendrocytes, and the remaining 14% contained oligodendrocytes and progenitors which were unlabelled by [3H]thymidine. Although colony size increased, the proportion of colonies which contained both oligodendrocytes and radiolabelled O-2A progenitor cells remained as high on Day 25 as on Day 15. This result suggests that asymmetric generation of oligodendrocytes and O-2A^{adult} progenitors represented the dominant pattern of division and differentiation in these colonies.

The size of individual colonies derived from O-2A^{adult} progenitors was also consistent with the view that the adult cells differentiated in a constitutively asymmetric manner. Symmetrically differentiating O-2A^{perinatal} progenitor cells isolated from perinatal rats generate colonies which cluster at sizes of 2, 4, 8, 16, 32 cells and so on in these culture conditions (Temple & Raff 1986; see also below). In contrast, the sizes of colonies derived from O-2A^{adult} progenitors showed no clustering at factors of two on either Day 15 or Day 25 of *in vitro* growth (Wren et al, submitted).

A further demonstration of the asymmetric behavior of O-2A^{adult} progenitors was obtained by using time-lapse microcinematography to follow the behaviour of dividing cells. With such films we have directly observed that cells within a single clonal family of O-2A^{adult} progenitors continued to generate more progenitor cells for at least four divisions after the first generation of an oligodendrocyte (G. Wolswijk, unpublished observations). Such behaviour is very different from that seen in most families of O-2A^{perinatal} progenitors, where clonally related cells generally all differentiate within one cell division of each other (Temple & Raff 1986, Raff et al 1988).

As mentioned earlier, the composition of colonies derived from O-2A^{perinatal} progenitors is generally consistent with symmetrical division and differentiation of the perinatal cells (Wren et al, submitted). However, it is noteworthy, in our experiments, that a small proportion (14%) of colonies derived from cells contained in the optic nerves of newborn rats were apparently asymmetric and contained both oligodendrocytes and radiolabelled O-2A progenitor cells after a length of time equivalent to >10 cell divisions. This figure is consistent with other observations indicating that a minority population of O-2A^{perinatal} progenitor cells divide and differentiate asymmetrically *in vitro* (Wren et al submitted, Temple & Raff 1986, Dubois-Dalcq 1987).

The origin of O-2A*adult* progenitors:
derivation from a subpopulation of O-2A*perinatal* progenitors

What happens to the asymmetric O-2A*perinatal* progenitors during later stages of development? We have previously observed that the optic nerves of adult rats do not appear to contain any O-2A*perinatal* progenitors (Wolswijk & Noble 1989). Although many of these cells are likely to disappear due to differentiation into type-2 astrocytes, our data also suggest another reason for the disappearance of O-2A*perinatal* progenitors during development: they differentiate into O-2A*adult* progenitors.

Time-lapse microcinematographic analysis of cells derived from optic nerves of three week old rats (the time when the relative proportion of perinatal to adult progenitors changes most rapidly; Wolswijk & Noble, submitted), has offered us clear examples of cell families that meet the following three stringent criteria: (i) the founder cell expressed the morphology, cell cycle length and motility characteristic of an O-2A*perinatal* progenitor, (ii) the first division observed gave rise to more O-2A*perinatal* progenitor-like cells, and (iii) cells derived from subsequent divisions expressed the unipolar morphology, slow division time and slow migration rate typical of O-2A*adult* progenitors (Wren et al, submitted). Previous studies by Dubois-Dalcq (1987) have also suggested that some families of O-2A*perinatal* progenitors can give rise to cells with the antigenic phenotype of O-2A*adult* progenitors, although characterization of the resultant cells was insufficient to allow certainty about their identity. The only other cell of origin we have ever seen by time-lapse microcinematography for O-2A*adult* progenitors is other O-2A*adult* progenitors, and we have never seen the birth of O-2A*perinatal* progenitors from anything other than O-2A*perinatal* progenitors.

Further support for the view that O-2A*adult* progenitor cells are directly derived from a subpopulation of O-2A*perinatal* progenitors comes from experiments in which we serially passaged perinatal optic nerve cells through six passages on monolayers of purified and irradiated type-1 astrocytes over three months. In two experiments we passaged optic nerve cells from newborn rats, which contain no detectable adult-like cells (Wolswijk & Noble, submitted). To allow for the possibility that O-2A*adult* progenitor cells are derived from a subset of O-2A*perinatal* progenitors not yet present in the optic nerves of newborn rats, we also separately passaged cells from optic nerves of seven day old rats, in which <2% of the O-2A progenitors appear adult like, as defined by antigen expression and morphology (Wolswijk & Noble, submitted); in the latter case, the suspensions of optic nerve cells were first treated with the O4 monoclonal antibody (Sommer & Schacher 1981) and complement to lyse adult-like progenitors (which are O4$^+$, see Fig. 1) and also to lyse cells progressing through intermediate stages of differentiation which lie between the earliest O-2A*perinatal* progenitors and terminally committed oligodendrocytes (Sommer et al, submitted; see Fig. 1). Thus, these passaging experiments were focused on the most immature members of the O-2A lineage, growing in the absence of neurons and in conditions which, in these experiments, did not generally promote differentiation of progenitors into type-2 astrocytes (D. Wren, unpublished observations).

Serial passaging of O-2A^{perinatal} progenitors from newborn and seven day old rats yielded similar results and was associated with the generation of O-2A^{adult} progenitors and the loss of O-2A^{perinatal} progenitors, as judged by antigenic criteria, morphological criteria and changes in the population doubling times (Wren et al, submitted). These experiments confirm that cells in the O-2A lineage are capable of extensive self-renewal *in vitro*, but that the expression of this ability is associated with replacement of the O-2A^{perinatal} progenitor cells with O-2A^{adult} progenitors.

What is the biological significance of having a perinatal-to-adult transition in O-2A progenitor populations?

In these penultimate sections of our review, we would like to present some suggestions on the relationship between our *in vitro* observations on the biology of perinatal and adult progenitors and the probable roles of these cells in normal development and in regeneration.

In order to achieve rapid myelination of CNS axon tracts during early development, it is necessary to generate large numbers of progenitors and oligodendrocytes within a relatively short period. Thus, a rapid cell cycle time and the capacity for exponential expansion in cell number are requisite properties for an O-2A progenitor cell in the optic nerve of perinatal rats. Moreover, it appears that O-2A^{perinatal} progenitors originate from a germinal zone outside the optic nerve and populate the optic nerve in a wave of migration during perinatal development (Small et al 1987). As oligodendrocytes are non-motile cells (Small et al 1987, Noble et al 1988), asymmetric division and oligodendrocytic differentiation of O-2A^{perinatal} progenitors would reduce the production of the large population of motile O-2A lineage cells necessary for the successful colonization of optic nerves during early development. A further property of O-2A^{perinatal} progenitors which is likely to be well suited to the physiological requirements of the developing nervous system is the biological clock expressed by these cells, which induces symmetrical differentiation of clonally related progenitor cells within a limited number of cell divisions (Temple & Raff 1986, Raff et al 1988). Regulation of differentiation with such a clock offers a simple means of first generating large numbers of progenitors, and then generating large numbers of oligodendrocytes, at precisely the times when these cells are needed during early development.

The presently available evidence suggests that O-2A^{perinatal} progenitors are committed to differentiating within a limited number of cell divisions, and should thus be considered as *bona fide* progenitor cells. This view is supported by the observations that O-2A *perinatal* progenitors disappear from the optic nerve *in vivo*, and are at least greatly reduced in number during serial passage *in vitro*. The surprising observation is that some of these perinatal cells are apparently able to differentiate into oligodendrocytes, type-2 astrocytes or O-2A*adult* progenitors, and thus have three, and not two, potential pathways of differentiation.

Although the adult CNS does not have sufficient space to accommodate the large numbers of new cells produced by rapid exponential growth and symmetrical differentiation, O-2A progenitor cells are still likely to be required in order to replace oligodendrocytes and type-2 astrocytes lost through normal turnover, disease or injury. Asymmetric division and differentiation of O-2A*adult* progenitors, coupled with greatly lengthened cell cycles, would allow maintenance of functional progenitor populations in adult tissue for extended periods without the generation of large numbers of unrequired cells. If we are correct in suggesting that a pre-progenitor cell is not involved in self-renewal in the O-2A lineage, then the capacity for self-renewal of O-2A*adult* progenitors must indeed be extensive, for the O-2A*adult* progenitors which develop *in vivo* during the first month after birth would then seem likely to be the direct ancestors of the O-2A*adult* progenitors in adult rats.

The generation of cells with the stem cell-like characteristics of O-2A*adult* progenitors from rapidly dividing 0.2A*perinatal* progenitors offers a previously unrecognized means of halting the rapid cell division of embryogenesis, by generating a population of cells which respond to particular cell–cell interactions in a manner different from their ancestors and more in keeping with the physiological requirements of adult tissues. Moreover, the pattern of development seen in the O-2A lineage suggests that precursor cells are developmentally nested, such that precursors with properties appropriate for early development give rise to both terminally committed differentiated cells and also a new group of precursors with properties appropriate to the physiological needs of more mature animals.

The apparent development of O-2A*adult* progenitors, with stem cell-like characteristics, from a rapidly dividing perinatal population differs significantly from the pattern of development seen in previously studied lineages, where slowly dividing stem cells have been seen to give rise to rapidly dividing progenitors. However, no other studies have focused on the origin of potential stem cell populations in the manner in which we have; for example, the phenomenon we have observed may be more closely analogous to the generation of haemopoietic stem cells from fetal liver cells than to the generation of mature lymphoid populations from haemopoietic stem cells.

Implications for understanding the repair of CNS myelin and the disease of multiple sclerosis

Our results indicate that the O-2A*adult* progenitor cells likely to be responsible for the regenerative events of remyelination, at least in the rat optic nerve, are not only significantly different in their biology from the cells relevant to developmental processes, but are likely to be much less effective than their perinatal counterparts in remyelinating large areas of damaged tissue. Although we do not yet know whether human glial progenitors are similar to those of the rat optic nerve, there is a striking congruence between the comparative properties of O-2A^{perinatal} and O-2A^{adult} progenitor cells and observations made on individuals afflicted with multiple sclerosis, a disease characterized by destruction of CNS myelin. Demyelinated lesions appear to be efficiently repaired in young patients presenting with optic neuritis (Kriss et al 1988), a demyelinating disease of the optic nerve, but are not effectively repaired in adult patients with either optic neuritis or systemic multiple sclerosis (Kriss et al 1988, Prineas & Connell 1979). The number of oligodendrocytes required for repair of a demyelinated lesion can be calculated from observations (by magnetic resonance imaging; personal communication from W. I. McDonald) that demyelinated plaques in patients with optic neuritis generally involve up to 25% of the axons in the optic nerve (i.e. 250 000 axons) and extend a distance of 1 cm (i.e. about 50 internodes). As previous estimates indicate that one oligodendrocyte can myelinate up to 25 internodes (Peters & Vaughn 1970), 5×10^5 oligodendrocytes would be required to repair an average lesion. Using the optimal figures for oligodendrocyte production indicated by current data, 2000 O-2A^{perinatal} progenitors could produce 5×10^5 oligodendrocytes in eight days, while 2000 O-2A^{adult} progenitors would require two months to produce the same number of oligodendrocytes. During these two months glial scars will be forming within the demyelinated plaque, and these may prevent migration of O-2A progenitors into the lesion (Raff et al 1987). Moreover, other of our recent studies indicate that O-2A^{adult} progenitor cells and oligodendrocytes are specifically able to bind and activate complement in the absence of specific antibody, thus leading to their own destruction; this property is not shared by 0-2A^{perinatal} progenitors, or any other CNS glial cells (Wren & Noble 1989). As breakdown of the blood–brain barrier is a consistent and early feature of multiple sclerosis lesions (W. I. McDonald, personal communication), it may be that entry of complement components into the white matter further contributes to the failure of repair by reducing the available pool of O-2A^{adult} progenitors. Thus, the specific cellular biological properties of the O-2A^{adult} progenitors characterized in the rat optic nerve are consistent with, and may contribute significantly to, the general failure of myelin repair in adults with multiple sclerosis. However, it must be noted that as yet unknown mechanisms may exist which would allow the production of larger numbers of oligodendrocytes or type-2 astrocytes following CNS damage.

References

Anderson D 1989 New roles for PDGF and CNTF in development of the nervous system. Trends Neurosci 12:83–85

Bögler O, Wren D, Barnett SC, Land H, Noble M 1990 Cooperation between two growth factors promotes extended self-renewal, and inhibits differentiation, of O-2A progenitor cells. (submitted)

Bottenstein JE, Sato GH 1979 Growth of a rat neuroblastoma cell line in a serum-free supplemented medium. Proc Natl Acad Sci USA 76:514–517

David S, Miller RH, Patel R, Raff MC 1984 Effect of neonatal transection of glial cell development in the rat optic nerve: evidence that the oligodendrocyte-type-2 astrocyte lineage depends on axons for its survival. J Neurocytol 13:961–973

Dubois-Dalcq M 1987 Characterization of a slowly proliferative cell along the oligodendrocyte differentiation pathway. EMBO (Eur Mol Biol Organ) J 6: 2587–2595

Eccleston A, Silberberg DR 1985 Fibroblast growth factor is a mitogen for oligodendrocytes in vitro. Dev Brain Res 21:315–318

ffrench-Constant C, Raff MC 1986 The oligodendrocyte-type-2 astrocyte cell lineage is specialized for myelination. Nature (Lond) 323:335–338

Hart IK, Richardson WD, Heldin CH, Westermark B, Raff MC 1989 PDGF receptors on cells of the oligodendrocyte-type-2 astrocyte (O-2A) cell lineage. Development 105:595–604

Hughes S, Raff MC 1987 An inducer protein may control the timing of fate switching in a bipotential glial progenitor cell in rat optic nerve. Development 101:157–167

Hughes S, Lillien LE, Raff MC, Rohrer H, Sendtner M 1988 Ciliary neurotrophic factor induces type-2 astrocyte differentiation in culture. Nature (Lond) 335:70–73

Janzer R, Raff MC 1987 Astrocytes induce blood–brain barrier properties in endothelial cells. Nature (Lond) 325:253–257

Kriss A, Francis DA, Cuendet F et al 1988 Recovery from optic neuritis in childhood. J Neurol Neurosurg Psychiatry 51:1253–1258

Lillien LE, Sendtner M, Rohrer H, Hughes SM, Raff MC 1988 Type-2 astrocyte development in rat brain cultures is initiated by a CNTF-like protein produced by type-1 astrocytes. Neuron 1:485–494

Miller RH, David S, Patel ER, Raff MC 1985 A quantitative immunohistochemical study of macroglial cell development in the rat optic nerve: in vivo evidence for two distinct astrocyte lineages. Dev Biol 111:35–43

Miller RH, Fulton B, Raff MC 1989 A novel type of glial cell associated with nodes of Ranvier in rat optic nerve. Eur J Neurosci 1:172–180

Noble M, Murray K 1984 Purified astrocytes promote the in vitro division of a bipotential glial progenitor cell. EMBO (Eur Mol Biol Organ) J 3:2243–2247

Noble M, Fok-Seang J, Cohen J 1984 Glia are a unique substrate for the in vitro growth of central nervous system neurons. J Neurosci 4:1892–1903

Noble M, Murray K, Stroobant P, Waterfield M, Riddle P 1988 Platelet-derived growth factor promotes division and motility and inhibits premature differentiation of the oligodendrocyte-type-2 astrocyte progenitor cell. Nature (Lond) 333:560–562

Peters A, Vaughn JE 1970 Morphology and development of myelin sheaths. In: Davision AN, Peters A (eds) Myelination. Charles C Thomas, p 3–7

Prineas JW, Connell F 1979 Remyelination in multiple sclerosis. Ann Neurol 5:22–31

Pringle N, Collarini EJ, Mosley MJ, Heldin C-H, Westermark B, Richardson WD 1989 PDGF A chain homodimers drive proliferation of bipotential (O-2A) glial progenitor cells in the developing rat optic nerve. EMBO (Eur Mol Biol Organ) J 8:1049–1056

Raff MC 1989 Glial cell diversification in the rat optic nerve. Science (Wash DC) 243:1450–1455

Raff MC, Mirsky R, Fields KL et al 1978 Galactocerobroside, a specific cell surface antigenic marker for oligodendrocytes in culture. Nature (Lond) 274:813–816

Raff MC, Abney ER, Cohen J, Lindsay R, Noble M 1983a Two types of astrocytes in cultures of developing rat white matter: differences in morphology, surface gangliosides and growth characteristics. J Neurosci 3:1289–1300

Raff MC, Miller RH, Noble M 1983b A glial progenitor cell that develops into an astrocyte or an oligodendrocyte depending on the culture medium. Nature (Lond) 303: 390–396

Raff MC, Abney ER, Fok-Seang J 1985 Reconstitution of a developmental clock *in vitro*: a critical role for astrocytes in the timing of oligodendrocyte differentiation. Cell 42:61–69

Raff MC, ffrench-Constant C, Miller RH 1987 Glial cells in the rat optic nerve and some thoughts on remyelination in the mammalian CNS. J Exp Biol 132:35–41

Raff MC, Lillien LE, Richardson WD, Burne JF, Noble MD 1988 Platelet-derived growth factor from astrocytes drives the clock that times oligodendrocyte development in culture. Nature (Lond) 333:562–565

Richardson W, Pringle N, Mosley M, Westermark B, Dubois-Dalcq M 1988 A role for platelet-derived growth factor in normal gliogenesis in the central nervous system. Cell 53:309–319

Saneto RP, De Vellis J 1985 Characterization of cultured rat oligodendrocytes proliferating in a serum-free chemically defined medium. Proc Natl Acad Sci USA 82:3509–3513

Silver J, Sapiro J 1981 Axonal guidance during development of the optic nerve: the role of pigmented epithelia and other intrinsic factors. J Comp Neurol 202:521–538

Skoff RP, Price DL, Stocks A 1976a Electron microscopic autoradiographic studies of gliogenesis in rat optic nerve. 1. Cell proliferation. J Comp Anat 169:291–311

Skoff RP, Price DL, Stocks A 1976b Electron microscopic autoradiographic studies of gliogenesis in rat optic nerve. 2. Time of origin. J Comp Anat 169:313–323

Small RK, Riddle P, Noble M 1987 Evidence for migration of oligodendrocyte-type-2 astrocyte progenitor cells into the developing rat optic nerve. Nature (Lond) 328:155–157

Smith AB, Heath JK, Donaldson DD et al 1988 Inhibition of pluripotential embryonic stem cell differentiation by purified polypeptides. Nature (Lond) 336:688–690

Sommer I, Schachner M 1981 Monoclonal antibodies (O1 to O4) to oligodendrocyte cell surfaces: an immunocytological study in the central nervous system. Dev Biol 83:311–327

Sommer I, Barnett S, Hutchins A-M, Wolswijk G, Wren D, Noble M 1990 Plasticity and commitment in oligodendrocytic differentiation. (submitted)

Temple S, Raff MC 1986 Clonal analysis of oligodendrocyte development in culture: evidence for a developmental clock that counts cell divisions. Cell 44:773–779

Wolswijk G, Noble M 1989 Identification of an adult-specific glial progenitor cell. Development 105:387–400

Wolswijk G, Noble M 1990 Co-existence during development of the rat optic nerve of perinatal and adult forms of a glial progenitor cell. Development, in press

Wood WG, Bunch C, Kelly S, Gunn Y, Breckon G 1985 Control of haemoglobin switching by a developmental clock? Nature (Lond) 313:320–323

Wood WG, Hawes S, Bunch C 1988 Developmental clocks and haemoglobin switching. In: Stamatoyannopoulos G (ed) Hemoglobin switching. Vol 5, in press

Wren D, Noble M 1989 Oligodendrocytes and oligodendrocyte/type 2 astrocyte progenitor cells of adult rats are specifically susceptible to the lytic effects of complement in the absence of antibody. Proc Natl Acad Sci USA 86:9025–9029

Wren D, Wolswijk G, Noble M 1990 Developmental nesting of glial progenitor populations. (submitted)

DISCUSSION

Maness: Could you explain more about the migration of the O-2A progenitors? Where do the cells come from? What causes them to migrate and how do they stop?

Noble: In studies that Peter Riddle, Rochelle Small and I did we found that the O-2A progenitors are very migratory. They are the only migratory cell type in that lineage; oligodendrocytes and type-2 astrocytes don't move. We looked at the optic nerve during early embryogenesis, which in this tissue means embryonic day 15 (E15) to E19. Up to E16 there were no O-2A progenitors in the nerve. At E17 we saw a few close to the optic chiasm; on E18 there were a few in the next part of the nerve moving towards the eye and by E19 there were a few in the region adjacent to the eye. It takes about 12 to 14 days for the numbers of O-2A progenitors to equalize across the nerve.

At E16 there were progenitor cells in the optic chiasm. This is very close to one of the germinal zones in the CNS. We do not yet know whether there is a specific portion of germinal zone in the optic chiasm that gives rise to these cells or whether they migrate in from a larger germinal zone. There are studies which suggest there may be a very specific germinal cell zone. One of the most interesting pieces of evidence supporting that view is that although there is movement of cells towards the eye during this embryonic period, it is not until somewhat later that there is any movement of cells back along the optic tract towards the superior colliculus. Rochelle Small has evidence from chick/quail chimaeras for migration of appropriate oligodendrocyte-related cells into the optic nerve, which would also indicate that migration plays a role in the generation of this tissue.

Hunter: Are there specific surface markers for these progenitor cells?

Noble: We have lots of markers but unfortunately there are as yet no markers good enough for staining sections.

Hunter: Can you raise monoclonal antibodies?

Noble: We certainly can. It is the eternal choice of what's the best thing to do. In looking for genes involved in lineage specification we may obtain a new set of markers. There are other markers which may be suitable for O-2A progenitor cells, for instance the PDGF α-receptor, but so far nothing has been satisfactory.

If PDGF is the major mitogen, then for the mechanism of migration one wonders whether there is a gradient of PDGF along the nerve. That's very hard to demonstrate. Bill Richards has done *in situ* hybridization and sees no obvious sign of a gradient. However, the progenitor cell is intrinsically migratory and it looks like these cells organize themselves in chains along substrates that may be axons or may be something else. It may be that, as in neural crest migration, the cells don't have the opportunity to go backwards because there are other cells behind them. When the progenitor cells become oligodendrocytes, either because of intrinsic biological clocks or as they enter a mitogen-deficient zone, they stop migrating. Even dividing oligodendrocytes are non-migratory cells in our experiments.

Another factor that limits migration is that close to the retina, from studies done in the 1930s, there is a tissue called the lamina cribrosa, which appears to be composed of specialized type-1 astrocytes. The lamina cribrosa looks very like a glial scar. Charles ffrench-Constant and Martin Raff have suggested that that structure is a barrier to migration. In rabbits there is no lamina cribrosa and O-2A progenitors enter the eye and myelinate axons in the retina, which presumably grossly interferes with visual function—I am sure that's why rabbits don't read very well!

Jessell: In the embryonic O-2A cells, can you override these maintenance effects of FGF and PDGF by putting in factors that promote terminal differentiation, such as CNTF?

Noble: Certainly the two pathways interplay but the complexities are considerable. The CNTF that Tom is referring to is from work done by Martin Raff, Simon Hughes, Laura Lillien, Michael Sendtner and Herman Rohrer. Their work indicates that one way by which O2-A progenitors go along the type-2 pathway involves exposure to a molecule called ciliary neurotrophic factor (CNTF), which was originally studied as a survival factor for the ciliary ganglior neurons that innervate the skeletal muscle of the eye. This factor can induce expression of glial fibrillary acidic protein (GFAP). However, this factor on its own is not sufficient to give type-2 astrocyte differentiation— the cells transiently express GFAP and then all become oligodendrocytes. CNTF plus PDGF plus possibly an extracellular matrix factor seem all to be necessary for terminal differentiation along this pathway.

There is another way to achieve terminal differentiation, and this is where we see some of the complexities. For a variety of reasons, we decided to look at the effects of vascular cells on these progenitor cells. Endothelial cells are very potent inducers of differentiation along the type-2 astrocyte pathway. They produce a factor that we think is not CNTF, which we are trying to purify. Endothelial cells are also potent producers of TGF-β and FGF. When you add medium conditioned by endothelial cells to progenitors, initially there is an enormous division of the progenitors, characteristic of the response to PDGF plus FGF. But by three to four days in culture the GFAP type-2 astrocyte expression becomes dominant and the cells become terminally committed non-dividing type-2 astrocytes. So, it appears that this collection of factors overrides the immortalizing effects of co-application of PDGF and FGF.

However, if you mix endothelial cells with type-1 astrocytes there is continued generation of new progenitors and also constant generation of new type-2 astrocytes. Somehow the astrocytes override the terminal differentiation pathway induced by endothelial cells. These interactions are very complicated and obviously until we get this endothelial-derived factor purified it is very difficult to study satisfactorily.

Heath: Presumably endothelial cell conditioned medium has the same effect as live endothelial cells?

Noble: Yes.

Heath: What happens if you add exogenous PDGF and FGF to endothelial cell conditioned medium?

Noble: Very little.

Heath: So it's not a trivial matter that the cells just don't produce enough of these factors?

Noble: We don't think so, but until we have purified material to work with I am very unhappy with the experiments. We know nothing about degradation or incorporation into matrix, *etcetera*.

Alemà: Type-1 astrocytes produce PDGF and CNTF—have you any idea what determines the timing of expression of these two growth factors?

Noble: PDGF seems to be constitutively expressed, in that we have not yet found anything which regulates it. CNTF is controlled by a timing mechanism that remains mysterious. One hypothesis was that the progenitors start making oligodendrocytes and oligodendrocytes cause type-1 astrocytes to make CNTF. Laura Lillien has found that this is entirely wrong. So we don't know what controls timing of CNTF expression. We won't be able study that until we have the molecular probes.

Alemà: Can you, for example, artificially increase the population of differentiated oligodendrocytes in the same culture and see whether this accelerates the appearance of type-2 astrocytes.

Noble: There is no effect.

Hunter: As was mentioned earlier in the symposium, there are clearly multiple FGFs and potentially multiple FGF receptors. Do you think the difference in response between the perinatal and adult progenitors could be due to a difference in the FGF receptor? Have you tried different FGFs in your system?

Noble: No, not yet. We will as soon as we can get the probes. All our experiments are done with basic FGF. We have done a little with acidic FGF, but nothing with keratinocyte growth factor or any of the others.

Hunter: Presumably endothelial cells make acidic FGF.

Heath: But it's not secreted.

Noble: X. Klagsburn and colleagues say it is secreted into the extracellular matrix. The O-2A progenitors make lots of proteases which could break down extracellular matrix and release FGF. However, the initial biological response of the O-2A progenitors to endothelial cell conditioned medium looks like it was elicited by pure PDGF and pure FGF.

Heath: Tom Maciag and others have described glial cells and glial blastoma cell lines that produce increased amounts of acidic FGF. Is there anything known about the phenotype of these cells?

Noble: The problem is that gliomas are not like lymphoid tumours. At the moment our understanding of the relationship between the glioma phenotypes and the phenotypes we see in the normal CNS is almost as close as that of the relationship between what I have described and knowledge of how this directly relates to topoisomerase II function! Most gliomas antigenically don't even look like CNS cells.

Heath: When you film the perinatal to adult transition, have you ever seen an adult type progenitor revert to the perinatal phenotype?

Noble: This has bothered us. If the adult progenitor is going to behave in the way that we feel it should in respect to wound repair, it should be able to generate burst-forming units. There should be a way of making this cell divide rapidly without differentiation. The results from the application of FGF and PDGF together point in the direction of such a pathway. However, adult progenitors, dividing in response to type-1 astrocytes, never do that—they always remain as an adult progenitor and never express the perinatal phenotype. So the FGF work is very interesting but always faces the problem of what could it possibly mean *in vivo*.

Heath: The problem is whether what you are seeing is an actual differentiation event or whether it's some kind of senescence.

Noble: These cells continue to divide.

Heath: I don't mean senescence in the proliferative sense. I mean in the sense of maturation of phenotype over a period in culture.

Noble: I think I am saying exactly the same thing— maturation of phenotype, where the phenotype is the adult progenitor. My prejudice, and it is entirely a prejudice at the moment, is that the type-2 astrocyte pathway is an induced pathway, the oligodendrocyte pathway is a preferred pathway in that dividing cells will preferentially become oligodendrocytes, but if neither of these two events has occurred by the time the cell has undergone a certain number of divisions, the cell enters the adult progenitor pathway. The adult progenitor pathway is essential for proper function of the adult nervous system, because you can't have a rapidly dividing cell that is exponentially increasing in number in this crowded tissue. This is a type of escape valve and we are working on the hypothesis that this is controlled by yet another clock.

Hunter: So you have found no exogenous factor that alters the frequency of conversion into adult progenitors?

Noble: Not yet.

Hunter: If there were a clock, you might expect to be able to influence it.

Noble: That's an interesting difference between this clock and the fetal to adult haemoglobin clock. It looks like the fetal to adult haemoglobin clock is hard wired. If you put fetal precursors in the adult, 60 days later you get adult haemoglobin. If you use precursors from different aged fetuses, the timing is appropriately altered. That makes sense physiologically because if you don't have adult haemoglobin at the right time, you die. In the CNS, where presumably there is a greater need for flexibility, you can override the clock in two different directions. You can get division and differentiation, or you can get division and inhibition of differentiation. It is a very different sort of clock from the haemoglobin clock and it is clearly not hard wired.

Hunter: But you have lost all the perinatal progenitors by one month after birth.

Noble: In vivo in the optic nerve, yes. We are looking at whether in specific germinal zones of the adult brain, perinatal progenitor populations are retained.

Norbury: You checked various candidates for the mechanism of the biological clock that regulates the synchrony of the terminal differentiation. What sorts of things do you look at?

Noble: The first things were diffusible factors and contact; we are ready to rule those out. Because of the difference between FGF and PDGF we thought we would do some of the classical experiments that one does in those situations. We got very excited early on with our work on phorbol esters because it looked like FGF might be being switched to a PDGF-like response by application of phorbol esters. In fact, phorbol esters combined with either growth factor seem to inhibit differentiation and act as one would expect for tumour promoters.

cAMP inhibits division. Another simple idea is that the cell makes x divisions and the PDGF receptors, for instance, are gradually diluted out. We have shown in unpublished work that PDGF is a survival factor for oligodendrocytes. Ian Hart, Bill Richardson and Martin Raff have shown very clearly that the PDGF receptor is still on oligodendrocytes. So we thought perhaps the cell division machinery had run down. But you can give oligodendrocytes FGF and they will divide. So why the PDGF receptors are non- functional we do not know.

There could be an intrinsic limitation on the number of divisions the cells can make. But with PDGF plus FGF, the cells will divide forever, so we have no indication of senescence. These are the sorts of things we are ruling out.

We have not yet ruled out an hypothesis that I like, because it's trendy, namely that PDGF plus FGF hyperinduces *myc* in the generic sense, which prevents differentiation. We know from the work Hucky Land and I have done that nuclear oncogenes inhibit differentiation in this pathway. But one might also say that the altered phosphorylation of p*Rb* or p60, or changing the levels of expression of these proteins could be responsible. That is the problem, there are so many hypotheses and all we do at the moment is keep eliminating them.

Norbury: As far as you see, it's not at either end of the pathway, it is something in the vast undefined middle?

Noble: Unfortunately, I think that's right.

Wagner: You talked about multiple sclerosis, how do you hope to achieve this remyelination?

Noble: One thing that is clear from our studies is that if one is interested in repair applied to a clinical setting, multiple sclerosis is not the most promising disease to work on. We can explore mechanisms of remyelination and migration and how certain genes affect functions and do all the basic biology, but I am not wholly confident that this will prove relevant to the patient with multiple sclerosis, except over the course of many years. Moreover, if our ideas about complement fixation are correct, then this is really bad news for the patient with multiple sclerosis, because that's a very non-specific lesion.

Heath: Doesn't this system contain the seeds of its own destruction? In multiple sclerosis you can't re-evoke the perinatal type of progenitor. If you graft in perinatal progenitors, they will mature into adult types and be re-susceptible to breakdown of the blood/brain barrier.

Noble: I began work on multiple sclerosis with the belief that a biological understanding of the disease was necessary for any hopes of progress. The problem with the biological understanding is that the connection with treatment may take many years to achieve. Thus, in terms of the broad interest that we have in organ regeneration, we are beginning also to work on tissues where regeneration is a more tractable problem.

Differential control of muscle-specific gene expression specified by *src* and *myc* oncogenes in myogenic cells

Germana Falcone*, M. Cristina Gauzzi, Franco Tatò* and Stefano Alemà

*Istituto di Biologia Cellulare, C.N.R., Viale Marx 43, 00137 Roma and *Dipartimento di Biologia Cellulare e dello Sviluppo, Università 'La Sapienza', Via degli Apuli 1, 00185 Roma, Italy*

Abstract. Myogenic cells can be transformed *in vitro* by the introduction of several exogenous viral oncogenes. Transformed myoblasts are prevented from terminal differentiation into myotubes by the continuous expression of oncogenes such as *myc* and *src*, chosen as prototypes of nuclear and cytoplasmic oncogenes. A comparative analysis of the relationship between transformation and differentiation in myoblasts and cells belonging to other lineages has led to the proposal that terminal differentiation of *myc*-transformed quail myoblasts is indirectly prevented by the loss of growth control and that *myc*-bearing cells remain susceptible to growth regulation by interaction with adjacent normal cells. On the contrary, the *src* oncogene appears to affect expression of the myogenic programme via a direct mechanism, independent from abnormal growth control. There is increasing evidence for the existence of master regulatory genes that govern and influence muscle development *in vivo* and myogenic differentiation *in vitro*. Expression of cytoplasmic oncogenes such as *src*, *ras* and polyoma middle T in the mouse myogenic cell line, C2, results in inhibition of biochemical differentiation and a marked down-regulation of the *MyoD1* and *myogenin* genes.

1990 Proto-oncogenes in cell development. Wiley, Chichester (Ciba Foundation Symposium 150) p 250–261

It is often surmised that proto-oncogenes are crucially involved in growth regulation and differentiation. For obvious reasons, it is important to know how proto-oncogenes function in normal differentiated cells during development. One possibility would be to study how they are expressed in specific tissues and at defined developmental stages and indeed this approach has revealed significant clues to their function. Alternatively, conclusions on the possible role of oncogenes in cell differentiation can be inferred from experiments in which oncogenes are introduced into well-defined populations of *in vitro* differentiating cells. Myogenic cells expressing exogenous oncogenes represent a valuable model to address questions about the influence of genetic and epigenetic messages in

250

myogenesis and their interplay with oncogenes, growth control and expression of differentiated functions.

Effects of retroviral oncogenes on myogenesis *in vitro*

All the principal features of myogenesis can be faithfully reproduced by cultures of myoblasts derived from embryonic muscles. Replicating myogenic cells do not express skeletal muscle-specific genes, with the possible exception of desmin, the muscle-specific subunit of intermediate filaments. It has been reported that several types of cycling avian and mammalian myogenic cells express low levels of desmin, which can therefore be considered as a specific marker of the penultimate stage of the myogenic lineage (Kaufman & Foster 1988). After a proliferative phase myoblasts irreversibly withdraw from the cell cycle (Okazaki & Holtzer 1966) and terminally differentiate into definitive post-mitotic myoblasts that up-regulate the coordinate expression of a vast array of muscle-specific genes (encoding sarcomeric proteins, surface receptors and enzymes) and acquire the ability to fuse into multinucleated myotubes (for a review, see Pearson & Epstein 1982).

Terminal commitment events in myogenic cells appear to be under the control of serum mitogens and specific growth factors such as fibroblast growth factor (FGF) and TGF-β, which are potent reversible inhibitors of muscle differentiation (Massague et al 1986, Spizz et al 1987).

The orderly sequence of events outlined above is prevented by the expression of retroviral oncogenes in quail embryo myoblasts. Several oncogenes belonging to different functional groups share the ability to induce anchorage independence and to block terminal differentiation (Table 1). This block, however, is unstable and sensitive to environmental cues. Single cell analysis of myoblasts transformed by *src* (and other cytoplasmic oncoproteins) has shown that a variable proportion of the transformed population may escape this constraint and initiate the expression of a qualitatively similar programme by forming 'revertant' myotubes that do not progress to acquire the functional capacity of mature myotubes (Alemà & Tatò 1987).

Although primary cultures of myoblasts are likely to be more representative of the normal regulation of differentiated functions, continuous lines of mammalian myogenic cells also provide a convenient experimental system in which the molecular control of myogenesis can be examined in some detail. The most salient characteristics of phenotypes induced by three cytoplasmic oncogenes in the mouse C2 myogenic cell line are summarized in Table 2.

v-*myc* blocks differentiation via uncontrolled cell proliferation

Quail embryo myoblasts are efficiently transformed by the v-*myc* oncogene, as assessed by colony formation in semi-solid media and suppression of

TABLE 1 Phenotypes of quail myotubes transformed by different viral oncogenes

Oncogene	Localization	Anchorage independence	Desmin in cycling cells	Terminal differentiation
none		−	+	+ + +
v-*myc*	nuclear	+	+	− [a]
v-*src*	cytoplasmic	+	−	− / + [b]
v-*fps*	cytoplasmic	+	−	− / +
v-*erbB*	cytoplasmic	+	−	− / +
v-*ras*	cytoplasmic	+	−	− / +
v-*erbA*	nuclear	−	ND	+ + +
v-*ski*	nuclear	−	+	+ + +

[a]Incidence of differentiated cells less than 10^{-4}.
[b]Differentiation in 'revertant' myotubes, see text.
ND, not determined.

morphological and biochemical differentiation (Falcone et al 1985). Moreover, no detectable accumulation of muscle-specific transcripts was observed in myoblasts transformed by v-*myc* (G. Falcone et al, unpublished work). Two characteristics distinguish v-*myc*-bearing myoblasts from cells transformed by cytoplasmic oncogenes: 1) polyclonal and clonal populations of v-*myc*-transformed myoblasts maintain desmin expression (M. Grossi et al 1990); 2) the incidence of spontaneous differentiation is at least three orders of magnitude lower than that in, for instance, v-*src*-transformed myoblasts (see Table 1).

In principle, transformation may influence differentiation either directly, by affecting the expression of tissue-specific genes without being dependent on the abrogation of growth control, or indirectly, by compelling cells to cycle and

TABLE 2 Phenotypes of mouse C2 myoblasts transformed by viral cytoplasmic oncogenes

Oncogene	Anchorage independence	Desmin in cycling cells	Terminal differentiation	MyoD1/myogenin expression
none	−	+	+ + + [a]	+
v-*src*	+	+	− [b]	−
v-*ras*	+	−	− / + [c]	−
polyoma middle T	+	+	− / + + [c]	− / + [d]

[a]Phenotypically normal myotubes.
[b]Incidence of differentiated cells less than 10^{-3}.
[c]Differentiation of transformed cells in low concentrations of mitogens.
[d]Low expression only in polyclonal populations.

thus preventing the progression into the last compartment of the lineage. Myogenic differentiation appears to be equally susceptible to both mechanisms. In contrast to their normal counterparts, myoblasts expressing high constitutive levels of v-*myc* do not withdraw from the cell cycle and fail to differentiate terminally even when exposed to low levels of mitogens (a situation that can be considered inductive of differentiation). We postulate that v-*myc*-induced abnormal growth control is responsible for blocking differentiation in myogenic cells by preventing the required withdrawal from the cell cycle. This indirect mechanism of blocking differentiation should be ineffective in cells that express their differentiation programme while still replicating (see Alemà & Tatò 1987). Indeed, proliferating cells such as avian definitive chondroblasts (Alemà et al 1985a) and neuroepithelial cells (Casalbore et al 1987) maintain expression of tissue-specific genes after transformation by v-*myc*.

One of the testable predictions of the indirect mechanism hypothesis is that the normal phenotype of v-*myc*-bearing myoblasts should be retrieved by restriction of their proliferative capacities. A set of observations indicates that surrounding normal cells may limit the proliferation of transformed cells (La Rocca et al 1989 and references therein). We have used mixed cultures of normal fibroblastic cells and transformed quail myoblasts to investigate whether normal cells could selectively suppress the phenotype of transformed quail cells. In these co-cultivation assays only v-*myc*-transformed cells were growth arrested, whereas v-*src*-transformed myoblasts were essentially unaffected. Growth arrest appeared to reflect a *bona fide* phenotypic reversion from the transformed state, since it was accompanied by re-expression of the myogenic differentiation programme, as shown by fusion and synthesis of muscle-specific proteins (La Rocca et al 1989). The v-*myc*-transformed myoblasts were induced to differentiate also by normal quail myoblasts, as shown by the formation of hybrid myotubes containing nuclei from both cell types.

When considering the possible mechanisms underlying this phenomenon the following points should be noted. Firstly, since terminal differentiation and fusion of myogenic cells requires previous withdrawal from the cell cycle, we suggest that, in mixed cultures, normal cells first re-programme the abnormal growth of v-*myc* transformed myoblasts and this in turn allows withdrawal from the cell cycle and terminal differentiation to occur. Secondly, the phenotypic reversion is not dependent on a stable down-regulation of v-*myc* expression, as the oncoprotein was retained at ostensibly normal levels in the nuclei of differentiated cells. From the data (La Rocca et al 1989), it was not possible to determine whether v-*myc* was subjected to some form of down-regulation in the early stages of reversion. We are left with the speculation that normal cells provide 'functions' capable of overriding the effect of v-*myc*. Thirdly, the different behaviours of transformed cells closely paralleled the efficiency with which they established metabolic cooperation with adjacent normal cells (La Rocca et al 1989).

How do the conclusions drawn from the behaviour of quail myoblasts expressing exogenous v-*myc* compare with what is known of the role of c-*myc* as a nuclear mediator of extracellular stimuli or with the temporal expression of c-*myc* during myogenesis? Elevated c-*myc* expression is associated with many naturally occurring neoplasms and in several cell systems c-*myc* transcription was shown to be repressed concomitantly with induction of terminal differentiation. The question whether irreversible suppression of c-*myc* is required for terminal differentiation to occur has been addressed in a study with a rat muscle cell line (Endo & Nadal-Ginard 1986). Although a reduction in steady-state levels of *myc* was observed after terminal differentiation, the oncogene remained inducible in myotubes and transient expression of c-*myc* was insufficient to interfere with the differentiated phenotype. However, the possibility that a transient reduction of c-*myc* expression levels, at a critical stage in the commitment process, is required for irreversible exit from the cell cycle was not explored.

v-*src* represses expression of differentiation independently of cell proliferation

Primary cultures of quail embryo muscle cells can be transformed by v-*src*-bearing retroviruses and transformation prevents the formation of multinucleated myotubes. By the use of temperature-sensitive (ts) conditional mutants it was shown that at the permissive temperature the majority of replicating transformed myoblasts failed to withdraw from the cell cycle, initiate the synthesis of muscle-specific products, assemble myofibrils and fuse into multinucleated myotubes. However, when shifted to the restrictive temperature, the majority of these cells withdrew from the cell cycle and formed spontaneously contracting myotubes that expressed muscle-specific proteins (Falcone et al 1985). Thus, expression of pp60[v-src] kinase does not irreversibly erase the terminal differentiation programme, but only arrests it by blocking the transition from a replicating precursor cell compartment to the terminally differentiated one.

In terminally differentiated myotubes, reactivation of pp60[v-src] causes disruption of myofibrils and degeneration of myotubes that can be visualized by immunostaining with antibodies against muscle-specific proteins. Clonal populations of quail myoblasts infected with ts mutants of Rous sarcoma virus were used to study the effect of the v-*src* gene reactivation on muscle-specific mRNA accumulation. The levels of muscle-specific myosin light chain, myosin heavy chain and cardiac α-actin mRNAs, high at the restrictive temperature, dropped strikingly after shifting the myotubes to the permissive temperature for 1–2 days. On the other hand, no appreciable decrease of β-actin and vimentin mRNAs was observed after reducing the temperature. The accumulation of β-actin transcripts, on the contrary, was reduced in myotubes formed at the restrictive temperature as a consequence of terminal differentiation (G. Falcone et al, unpublished work). These experiments, and previous ones by West & Boettiger (1982) at the protein level, clearly indicate that v-*src* interferes with muscle differentiation by directly affecting muscle gene expression, whether or not the cells are proliferating. Consistent with this scheme is the finding that in ts-*src*-transformed chick myoblasts treated with mitomycin C, an inhibitor of DNA synthesis, arrest of proliferation at the permissive temperature was not sufficient to induce the expression of the differentiated phenotype, while inactivation of the v-*src* gene function at the restrictive temperature was essential (Falcone et al 1984). This implies that cell division is not required to express the terminal differentiation programme at the restrictive temperature and rules out an effect of pp60$^{v\text{-}src}$ on cell-cycle related commitment. All these observations suggest that v-*src* can reversibly affect both the initiation of the expression of the myogenic programme in replicating myoblasts and the maintenance of the differentiated state in post-mitotic myotubes. Again, the behaviour of other cell types such as chondroblasts, neuroretina cells and fibroblasts, whose specific differentiation programmes are fully suppressed by v-*src*, further substantiates the proposal of a direct mechanism (Alemà & Tatò 1987). Activation (Alemà et al 1985b) or inactivation of complex differentiation programmes in different specialized cell types suggests that v-*src* acts at a higher regulatory level than individual genes, presumably by altering pathways common to the transcription of unlinked lineage-specific genes which are usually coordinately expressed.

Muscle-regulatory genes as potential targets for oncogene action

The superficial similarity in the phenotypes attained by mammalian myoblasts treated with FGF and TGF-β or transformed by *src*, *ras* or middle T (see Table 2) suggests that the latter, known to be involved in signal transduction, may be sending persistent growth-related signals to the nucleus (Gosset et al 1988). While waiting for the elucidation of the putative biochemical pathways involved, an attractive hypothesis is that oncogenes might block differentiation by acting on the expression of master genes devoted to the specification of the skeletal muscle lineage or simply responsible for the transcription of muscle-specific genes.

Genes whose products possess many of the properties expected of master genes have recently been described. Transfection of cDNA encoding the *MyoD1* gene (Davis et al 1987) is sufficient to convert naive 10T½ cells and a variety of mesodermal cell lines into stable proliferating myoblasts, that can in turn differentiate into post-mitotic muscle cells when starved of mitogens. *MyoD1* is constitutively expressed only in skeletal muscle cells and some myogenic cell lines (Davis et al 1987, Tapscott et al 1988). Two additional muscle-specific genes named *myogenin* (Wright et al 1989, Edmonson & Olson 1989) and *Myf5* (Braun et al 1989) possess the same property of converting fibroblasts into muscle cells.

MyoD1 driven by retroviral promoters induces transcription of endogenous *MyoD1* and *myogenin* and *vice versa*, indicating that both genes are subject to a positive autoregulatory loop (Thayer et al 1989). The proteins encoded by *MyoD1* and *myogenin* show a nuclear localization, turn over rapidly and bind specifically to the muscle creatine kinase enhancer as well as to 5′ sequences of *MyoD1* itself (Thayer et al 1989). Thus, one may conclude that *MyoD1* and *myogenin* are tissue-specific transcriptional regulators (Murre et al 1989).

In an attempt to clarify the intracellular pathways whereby oncogenes negatively regulate myogenesis, we chose to examine the effect of some of them on the expression of myogenic regulatory genes at the mRNA level. C2 cells transformed by either v-*ras* or v-*src* were dramatically restricted in their ability to express *MyoD1* and *myogenin* under conditions permissive for proliferation or differentiation. Polyclonal populations of middle T-transformed C2 myoblasts expressed reduced levels of both these genes, in accordance with the phenotype of these cells (see Table 2) (M. C. Gauzzi et al, unpublished work). Similar results have been reported for the 23A2 mouse myogenic cell line transformed by *ras* or treated with FGF or TGF-β (Vaidya et al 1989).

It therefore seems likely that, beside the positive autoregulatory loop mentioned above, *MyoD1* and *myogenin* are subject to negative regulation by oncogene products. However, a number of observations are consistent with the existence of other regulatory networks of myogenic differentiation and these may be differentially affected by growth factors or cytoplasmic oncogenes. 1) Stable transfectants of fibroblasts with *MyoD1* and *myogenin* require removal of serum before they can terminally differentiate, thus indicating that their action is antagonized downstream by mitogenic signals (Davis et al 1987). 2) Middle T oncogene down-regulates transcription of *MyoD1* and *myogenin*, yet a sizeable proportion of cells in clonal strains of C2 cells transformed by middle T are capable of differentiating terminally (M. C. Gauzzi et al, unpublished work). 3) It has been shown that over-expression of *MyoD1* driven by an exogenous promoter in 23A2 cells can override the effect of *ras* but not that of FGF and TGF-β (Vaidya et al 1989).

MyoD1 and *myogenin* gene products share a region of about 50 amino acid residues that is structurally related to that encoded by the *myc* genes; by virtue of this homology they belong to the *myc* superfamily that also includes *Drosophila* genes with cell determination functions, two human Ig-ϰ enhancer binding proteins and the newly discovered *lyl-1* oncogene (Murre et al 1989). The *myc* homology domain appears to be sufficient for both DNA binding and conversion of 10T½ cells to myoblasts (Tapscott et al 1988); interestingly it lies within a region of *myc* known to be important for two-step transformation (Stone et al 1987). The potential role of c-*myc* as a regulator of myogenesis is fostered by this homology. Since *MyoD1* and *myogenin* are sequence-specific DNA-binding proteins, it is possible that they compete with *myc* for common targets and that *myc* function involves specific binding to DNA (Murre et al 1989).

Acknowledgements

Work by the authors described in this article was supported by grants from CNR-Progetti Finalizzati 'Oncologia' and 'Biotecnologie', the Fondazione Cenci-Bolognetti and AIRC.

References

Alemà S, Tatò F 1987 Interaction of retroviral oncogenes with the differentiation programme of myogenic cells. Adv Cancer Res 49:1–28

Alemà S, Tatò F, Boettiger D 1985a *myc* and *src* oncogenes have complementary effects on cell proliferation and the expression of specific extracellular matrix components in definitive chondroblasts. Mol Cell Biol 5:538–544

Alemà S, Casalbore P, Agostini E, Tatò F 1985b Differentiation of PC12 phaechromocytoma cells induced by v-*src* oncogene. Nature (Lond) 316:557–559

Braun T, Buschhausen-Denker G, Bober E, Tannich E, Arnold HH 1989 A novel human muscle factor related to but distant from *MyoD1* induces myogenic conversion in 10T½ fibroblasts. EMBO (Eur Mol Biol Organ) J 8:701–709

Casalbore P, Agostini E, Alemà S, Falcone G, Tatò F 1987 The v-*myc* oncogene is sufficient to induce growth transformation of chick neuroretina cells. Nature (Lond) 326:188–190

Davis RL, Weintraub H, Lassar AB 1987 Expression of a single transfected cDNA converts fibroblasts to myoblasts. Cell 51:987–1000

Edmonson DG, Olson EN 1989 A gene with homology to the *myc* similarity region of *MyoD1* is expressed during myogenesis and is sufficient to activate the muscle differentiation program. Genes Dev 3:628–640

Endo T, Nadal-Ginard B 1986 Transcriptional and post-transcriptional control of c-*myc* during myogenesis: its mRNA remains inducible in differentiated cells and does not suppress the differentiated phenotype. Mol Cell Biol 6:1412–1421

Falcone G, Boettiger D, Alemà S, Tatò F 1984 Role of cell division in differentiation of myoblasts infected with a temperature-sensitive mutant of Rous sarcoma virus. EMBO (Eur Mol Biol Organ) J 3:1327–1331

Falcone G, Tatò F, Alemà S 1985 Distinctive effects of the viral oncogenes *myc, erb, fps* and *src* on the differentiation of quail myogenic cells. Proc Natl Acad Sci USA 82:426–430

Gosset LA, Zhang W, Olson EN 1988 Dexamethasone dependent inhibition of differentiation of C2 myoblasts bearing steroid inducible N-*ras* oncogenes. J Cell Biol 106:2127–2138

Grossi M, Verna C, Calconi A et al 1990 Retroviral oncogenes can block terminal differentiation of quail muscle cells through distinguishable mechanisms. In: Frati L (ed) Molecular pathology of gene expression. Raven Press—Serono Symposia, in press

Kaufman SJ, Foster RF 1988 Replicating myoblasts express a muscle-specific phenotype. Proc Natl Acad Sci USA 85:9606–9610

La Rocca SA, Grossi M, Falcone G, Alemà S, Tatò F 1989 Interaction with normal cells suppresses the transformed phenotype of v-*myc*-transformed quail muscle cells. Cell 58:123–131

Massaguè J, Cheifetz S, Endo T, Nadal-Ginard B 1986 Type β transforming growth factor is an inhibitor of myogenic differentiation. Proc Natl Acad Sci USA 83:8206–8210

Murre C, Schonleber McCaw P, Baltimore D 1989 A new DNA-binding and dimerization motif in immunoglobulin enhancer binding, *daughterless*, *MyoD* and *myc* proteins. Cell 56:777–783

Okazaki K, Holtzer H 1966 Myogenesis: fusion, myosin synthesis and the mitotic cycle. Proc Natl Acad Sci USA 56:1484–1488

Pearson ML, Epstein HF 1982 Muscle development: molecular and cellular control. Cold Spring Harbor Laboratory, Cold Spring Harbor, New York

Spizz G, Hu JS, Olson EN 1987 Inhibition of myogenic differentiation by fibroblast growth factor or type β transforming growth factor does not require persistent c-*myc* expression. Dev Biol 123:500–507

Stone J, de Lange T, Ramsay G, Jakobovits E, Bishop JM, Varmus H, Lee W 1987 Definition of regions in human c-*myc* that are involved in transformation and nuclear localization. Mol Cell Biol 7:1697–1709

Tapscott SJ, Davis RL, Thayer MJ, Cheng PF, Weintraub H, Lassar AB 1988 MyoD1: a nuclear phosphoprotein requiring a *myc*-homology region to convert fibroblasts to myoblasts. Science (Wash DC) 242:405–411

Thayer MJ, Tapscott SJ, Davis RL, Wright WE, Lassar AB, Weintraub H 1989 Positive autoregulation of the myogenic determination gene *MyoD1*. Cell 58:241–248

Vaidya TB, Rhodes SJ, Taparowsky EJ, Konieczny SF 1989 Fibroblast growth factor and transforming growth factor β repress transcription of the myogenic regulatory gene MyoD1. Mol Cell Biol 9:3576–3579

West CM, Boettiger D 1982 Selective effect of Rous sarcoma virus *src* gene expression on contractile protein synthesis in chick embryo myotubes. Cancer Res 43:2042–2048

Wright WE, Sassoon D, Lin VK 1989 *Myogenin*, a factor regulating myogenesis has a domain homologous to *MyoD1*. Cell 56:607–617

DISCUSSION

Nusse: Has the *myoD* gene under the control of an LTR been put into *ras*-transformed myoblasts?

Alemà: This experiment has been done in both *ras*-transformed myoblasts and myoblasts whose differentiation had been inhibited by FGF and TGF-β (Lassar et al 1989, Konieczny et al 1989). The myogenic potential of *ras*-transformed myoblasts could be rescued, but that of myoblasts treated with the growth factors could not.

Those experiments were on stable cell lines. When Lassar et al transiently transfected 10T½ fibroblasts with an activated *ras* gene, an LTR-driven *MyoD1* gene and a plasmid containing the desmin promoter coupled to a reporter gene, they found that in those cells *ras* inhibited transcription from the desmin promoter. Thus in cells expressing high levels of *ras* transiently, *MyoD1* could not compete functionally with *ras* to determine transcription of the muscle-specific gene. This suggests that there is a mutual relationship between the levels of *ras* and of *MyoD1*, which dictates the phenotype.

Norbury: In the experiment where you co-cultivate *myc*-transformed and wild-type myoblasts, is there any *trans* effect in the myotubes that are formed? Do the *myc*-positive nuclei continue to incorporate [³H]thymidine or is DNA synthesis in them repressed?

Alemà: We have not tested this.

Hunter: What happens to the expression of c-*myc* in your system?

Alemà: I don't know. I would like to make a point about the possible role of *myc* in myogenesis. As mentioned before, a DNA-binding and dimerization motif, the helix-loop-helix motif, has recently been identified in *myc* and *MyoD1* proteins and other gene products. Murre et al (1989) have examined whether this motif could mediate heterodimer formation between members of this class, and have established a tentative taxonomy. They have shown that cell-type specific proteins, such as *MyoD1* and the *Drosophila* gene *aschaete-scute*, cannot form heterodimers with each other but only with the *Drosophila* gene *daughterless* and E12, which are ubiquitously expressed. *myc* proteins do not form heterodimers with other members of this family. Since these are all DNA-binding proteins specific for particular sequences, one could speculate that they compete with *myc* for binding to DNA.

Vande Woude: What is the effect of v-*ski* in these cells?

Alemà: v-*ski* was claimed to induce myogenesis in quail embryo cells (A. Barkas et al, personal communication; Colmenares & Stavnezer 1989). However, we could not get any transformed phenotype by infecting quail myoblasts with v-*ski* under experimental conditions comparable to those used for other oncogenes.

Vande Woude: Steve Hughes' lab has made transgenic mice with *ski*. These mice have very large muscles, 2–4 times those in normal littermates. They are supermice!

Noble: So how do you keep these mice in their cages?

Vande Woude: Expression of *ski* is very specific. It is not clear why the highest levels of expression are in the muscle, because he is using an LTR that has been shown to be rather promiscuous in its tissue specificity. The frequency of generation of transgenics is low, so it's possible that there is some selection for this specific expression.

Alemà: One might expect, for instance, v-*ski* to enlarge the myoblast compartment but at the same time these cells to be very sensitive to extracellular cues.

Vande Woude: The *ski* product is nuclear.

Land: A long time ago, I showed very potent immortalization of rat embryo cells with v-*ski*.

Vande Woude: Is it equivalent to *myc*?

Land: It's different but equipotent.

Hunter: What about the pattern of expression of c-*ski*, is that known?

Vande Woude: I believe it is difficult to detect.

Heath: I would like to draw an interesting parallel. The induction of *Xenopus* mesoderm is characterized by, amongst other things, production of muscle and expression of *MyoD1*. Recently, Doug Melton has shown that polyoma middle T acts as an inducer of differentiation rather than as a suppressor. However, it is formally a different situation because it's specification rather than execution of the terminal differentiation programme, which is the situation that you are dealing with here.

Hunter: Is middle T expressed in the inducing cell rather than the recipient cell?

Heath: I don't think that is known yet, in terms of anatomical position within the embryo.

Maness: In myoblasts containing the ts-*src* gene, what happens if you shift to the non-permissive temperature, allow the cells to differentiate, and then shift down to the permissive temperature to reactivate *src*?

Alemà: Myotubes shifted to the permissive temperature for two days tend to flatten, they do not assemble sarcomeres and detach from the substrate. It is possible to rescue them by inactivating the kinase, but expression of v-*src* in a terminally differentiated, post-mitotic cell like a myotube is eventually lethal.

Maness: Did you get a similar result with *src*-infected PC12 cells?

Alemà: No.

Hunter: Do they dedifferentiate when you switch off *src*? They do when you withdraw nerve growth factor.

Brugge: We've done that in collaboration with Simon Halegoua with the ts-*src*. It is really impressive; the neurites regress and the cells behave as though they have never seen v-*src*.

Noble: But those cells never become truly post-mitotic.

Brugge: They become post-mitotic but it's reversible.

Vande Woude: Is there a time element involved?

Brugge: In PC12 cells? It depends how far you allow the differentiation to go. At the non-permissive temperature the temperature-sensitive mutants that we are using are somewhat leaky and you see small neurites, but the cells can divide, even though they are partially differentiated.

Vande Woude: If you add v-*src* or *ras* to PC12 cells, do the cells continue to grow?

Alemà: We do not know whether fully differentiated cells do divide, but by autoradiography we showed that a significant proportion of PC12 cells expressing v-*src* with long neurites continued to incorporate thymidine.

Brugge: It depends on the level of expression of *src*—at high levels of expression there is extensive differentiation and those cells won't divide, but at lower levels of expression they will. The cells that we maintain at the non-permissive temperature have about 1/40th the level of the v-*src* protein compared to the level of c-*src* that we can express and not get differentiation. The cells that we are able to maintain have very low levels of v-*src* protein. Most of the original colonies infected with the v-*src*-carrying virus differentiated and were unable to be passed.

References

Colmenares C, Stavnezer E 1989 The *ski* oncogene induces muscle differentiation in quail embryo cells. Cell 59:293–303

Konieczny SF, Drobes BL, Menke SL, Taparowsky EL 1989 Inhibition of myogenic differentiation by the H-*ras* oncogene is associated with the down regulation of the *Myod*1 gene. Oncogene 4:473–481

Lassar AB, Thayer MJ, Overell RW, Weintraub H 1989 Transformation by activated *ras* or fos prevents myogenesis by inhibiting expression of *MyoD*1. Cell 58:659–667

Murre C, McCaw PS, Vaessin H et al 1989 Interactions between helix- loop-helix proteins generate complexes that bind specifically to a common DNA sequence. Cell 58:537–544

Cellular proteins that are targets for transformation by DNA tumour viruses

Karen Buchkovich, Nicholas Dyson, Peter Whyte* and Ed Harlow

Cold Spring Harbor Laboratory, P. O. Box 100, Cold Spring Harbor, NY 11724, USA

Abstract. Small DNA tumour viruses produce proteins that redirect cellular gene expression and growth control. The E1A polypeptides of adenovirus perform the functions of transcriptional activation and cellular transformation. These two functions are carried out by different domains within the E1A protein. The E1A protein associates with several cellular proteins, including the product of the retinoblastoma gene, p*Rb-1*. Mutational analysis correlates transformation with the sites required for binding p*Rb* and two other cellular proteins, p107 and a 300 kDa polypeptide. This correlation suggests that these proteins are targets for E1A-mediated transformation. Transforming proteins from other small DNA tumour viruses interact with p*Rb*, raising the possibility that a common event in viral transformation is the inactivation of proteins that inhibit cellular proliferation. The role of the E1A-associated 60 kDa protein, p60, in transformation is being investigated. In the absence of E1A, p60 binds to the human homologue of the *Schizosaccharomyces pombe cdc2* gene product, p34, to form a complex that has kinase activity that oscillates during the cell cycle. Ongoing studies of the effect of adenovirus infection, and specifically E1A expression, on this cellular kinase may provide clues to how E1A overcomes cell cycle controls and transforms cells.

1990 Proto-oncogenes in cell development. Wiley, Chichester (Ciba Foundation Symposium 150) p 262–278

When adenovirus infects cells, the first virus-specific proteins to be synthesized are the E1A polypeptides. As the first viral protein that appears after infection, E1A plays a key role in initiating the viral life cycle. It begins the virus-mediated modifications of cellular processes, the culmination of which converts an infected cell into an efficient factory for virus production. E1A's functions in initiating these changes can be divided into two classes, those of transcriptional regulation and those of cell proliferation control. The presence of E1A in infected cells activates transcription from the viral early promoters, and E1A has become

---

*Current address, Fred Hutchinson Cancer Research Center, 1124 Columbia Street, Seattle, Washington 98104, USA

one of the best studied transcriptional transactivators (reviewed by Berk 1986). The other extensively studied activity of E1A is its ability to act as a potent oncogene (reviewed by Green 1989). In cooperation with certain other oncogenes, E1A converts cells to a transformed state. These two activities of E1A, transactivation and transformation, have made this viral gene an excellent model system for investigating cellular regulation of transcription and proliferation.

Conserved regions within E1A

The E1A proteins are multifunctional, having separate protein domains that accomplish different functions in infected cells (reviewed by Moran & Mathews 1987). Early work from Stabel et al (1985) and Kimelman et al (1985) pointed out that when the E1A protein sequences from various serotypes of adenovirus are compared, conserved regions within the E1A proteins could be identified. Although these regions do not precisely delineate the domains for E1A functions, they do serve as useful landmarks in the discussion of important amino acid sequences. Three conserved regions have been identified (Fig. 1). Conserved region 1 is found roughly between amino acids 40 and 80, conserved region 2 between amino acids 120 and 139, and conserved region 3 between 140 and 189.

Transcriptional transactivation by E1A

The amino acids in conserved region 3 have been shown to be both necessary and sufficient for E1A-mediated transactivation (Fig. 1; Lillie et al 1986, 1987, Moran et al 1986, Schneider et al 1987). The E1A proteins will transactivate not only the other viral early promoters but also several cellular promoters. Except under unusual circumstances, mutations in E1A that destroy trans-activation also inhibit viral infection. Perhaps surprisingly, mutations that completely destroy transactivation seem to have no effect on E1A's ability to function as an oncogene. As described below, the converse is also true: mutations that destroy E1A's transforming ability do not affect transactivation, leading to the suggestion that these biochemical events are carried out by independent domains within the E1A protein (Lillie et al 1986, Moran et al 1986, Schneider et al 1987).

Transformation by E1A

Adenovirus infects cells that are not rapidly dividing. Since adenoviruses do not carry enough genetic material to perform all of the tasks needed for viral replication, one of the roles of the viral early proteins is to alter cell metabolism to ensure efficient viral replication. Many of these events are carried out by the E1A proteins, and one of the most fundamental is the modification of cell cycle control mechanisms. Consequently, E1A is an effective mitogen, stimulating mammalian cells that are arrested in G0 to progress to DNA synthesis (Kaczmarek et al 1986, Nakajima et al 1987, Spindler et al 1985, Stabel et al 1985, Bellet et al 1985, Braithwaite et al 1983). When E1A is introduced into primary rodent cells without the remainder of the viral genome, the cells overcome senescence and will grow indefinitely in culture (Houweling et al 1980). Furthermore, co-transfection of an E1A gene together with an activated *ras* oncogene produces fully transformed cells that can grow in soft agar and cause tumours in animals (Ruley 1983).

Transformation by E1A appears to require multiple functions, since sequences from both conserved regions 1 and 2 are needed for full transformation (Fig. 1; Haley et al 1984, Lillie et al 1986, Moran et al 1986, Moran & Zerler 1988, Schneider et al 1987, Smith & Ziff 1988, Whyte et al 1988b, Zerler et al 1986), and these regions are needed for its interactions with cellular proteins (see below). Although the exact functions encoded by these regions are not known, at least a portion of E1A's role in transformation is thought to be provided by its ability to interact with these cellular proteins (Whyte et al 1989).

Protein complexes between E1A and cellular polypeptides

E1A interacts with several cellular proteins

When antibodies specific for E1A are used in immunoprecipitations from adenovirus-infected or transformed cells, in addition to the E1A antigens, several

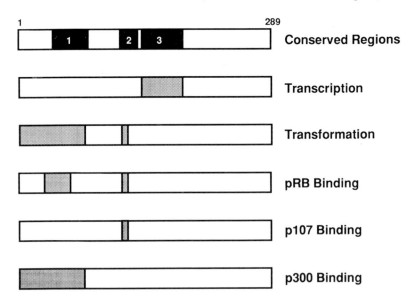

FIG. 1.   Diagram of the largest E1A polypeptide showing the conserved regions and domains responsible for particular functions.

cellular proteins are co-precipitated (Yee & Branton 1985, Harlow et al 1986). These cellular proteins have been shown to form stable protein–protein complexes with E1A, as demonstrated by several lines of experimental evidence. First, six different monoclonal antibodies that recognize six distinct epitopes on E1A all immunoprecipitate the same set of cellular proteins. These antibodies can recognize E1A directly on immunoblots but do not bind directly to the cellular proteins in either their native or denatured conformations. Second, in at least one case a monoclonal antibody raised against one of the cellular proteins will co-precipitate E1A from adenovirus-infected or transformed cells, but this antibody cannot recognize E1A directly. Third, when the cellular proteins are purified by gel filtration or on sucrose gradients, they always co-purify with a subset of the E1A proteins. Fourth, when extracts of radiolabelled cells that contain these host proteins but not E1A are mixed with a source of unlabelled E1A, complexes between the cellular proteins and the E1A polypeptides can be detected using monoclonal antibodies specific for E1A.

At the present level of detection, at least ten cellular proteins have been shown to interact with E1A. These cellular proteins are known by their relative molecular weights of 300, 130, 107, 105, 90, 80, 60, 50, 40 and 28 kDa. Of these, the easiest to detect are the 300, 107, 105 and 60 kDa species. Recently, two of the cellular proteins that bind to E1A have been identified. The 105 kDa protein has been shown to be the product of the retinoblastoma tumour suppressor gene, *Rb-1* (Whyte et al 1988a), and the 60 kDa protein has been shown to form complexes with the product of the mammalian homologue of the yeast *cdc2* gene (Giordano et al 1989).

To determine in which functions of E1A these cellular proteins might be involved, a series of point and deletion mutations were prepared to map the regions on E1A required for association. Regions of E1A within amino acids 1 to 76 and 120 to 127 were needed for binding to the most abundant of these cellular proteins, the 300, 107 and 105 kDa proteins (Fig. 1; Whyte et al 1989). When these same mutants were used in E1A functional assays, there was a strong correlation between the binding sites and the regions required for E1A-mediated transformation (Whyte et al 1988b, Whyte et al 1989). The mapping of the transformation domains to amino acids 1 to 76 and 120 to 127 extended earlier observations from other laboratories that showed that two regions of E1A were necessary for transformation (see above). The correlation between the binding sites and the regions needed for transformation suggests that these proteins are the cellular targets for E1A-mediated transformation.

The E1A-associated 105 kDa protein is the product of
the retinoblastoma tumour suppressor gene

While characterizing the E1A-associated proteins, we compared them with other proteins that were known to play a role in some type of transformation.

Following a report from Lee et al (1987) that identified the product of the retinoblastoma tumour suppressor gene as a 110 kDa nuclear phosphoprotein (now known as pRb), we compared the retinoblastoma protein with the E1A-associated 105 and 107 kDa proteins. These studies showed that the 105 kDa protein was identical to the product of the *Rb-1* gene (Whyte et al 1988a).

The retinoblastoma gene is the best studied of the tumour suppressor genes, whose protein products are thought to act as negative regulators of cell proliferation (review by Klein 1987, Knudson 1987). This suggestion derives from the observation that the loss or inactivation of both alleles of the *Rb-1* tumour suppressor gene is necessary for the development of certain tumours. The demonstration of a complex between E1A and pRb has several interesting implications. It is the first indication that oncoproteins and tumour suppressor proteins can act within the same biochemical pathway rather than functioning in a completely independent manner. It also suggests a simple model to explain the function of the E1A and pRb association. Since the introduction of E1A into cells stimulates them to divide, but it is the loss of the *Rb* protein that leads to cellular proliferation, it has been suggested that E1A binds to and inactivates the *Rb* protein.

Transforming proteins from other small DNA tumour viruses also interact with pRb. These associations are seen with the early proteins of polyoma viruses, papilloma viruses and adenoviruses (Table 1; DeCaprio et al 1988, Dyson et al 1989b, 1990). Associations between the early proteins of DNA tumour viruses and the products of tumour suppressor genes are not unique to pRb interactions, as similar types of interactions have been found with the other well characterized tumour suppressor gene, p53 (Lane & Crawford 1979, Linzer & Levine 1979, Sarnow et al 1982). The prevalence of these interactions between viral transforming proteins and tumour suppressor gene products suggests that one of the key events in the early phases of the viral life cycle is the inactivation of proteins that inhibit cell proliferation, such as pRb.

The 300 and 107 kDa E1A-associated proteins

Although genetic evidence suggests that binding of E1A to pRb is important for transformation, this interaction is not sufficient for all of the E1A-mediated changes. Mutations that inhibit interaction of E1A with other cellular proteins also block E1A-mediated transformation. Recently, several of the other E1A-associated cellular proteins have been partially characterized. Both the 300 and 107 kDa proteins bind to regions of E1A that are needed for transformation, suggesting that they are also important targets for transformation (Whyte et al 1989). Little is known about the possible functions of these proteins, although the 107 kDa protein has several characteristics that are similar to pRb (Dyson et al 1989a, Ewen et al 1989). The 107 kDa protein binds several of the same viral oncogenes as pRb (Table 1).

TABLE 1   Binding of the retinoblastoma gene product, p*Rb* and the E1A-associated cellular protein, p107, to viral proteins

Virus protein	pRb	p107	Reference
Adenovirus type 2 E1A	+	+	unpublished
Adenovirus type 3 E1A	+	+	unpublished
Adenovirus type 5 E1A	+	+	Whyte et al (1988a), Harlow et al (1986)
Adenovirus type 7 E1A	+	+	unpublished
SV40 T Ag	+	+	DeCaprio et al (1988), Dyson et al (1989a), Ewen et al (1989)
Mouse polyoma T Ag	+	nd	Dyson et al (1990)
Monkey lymphotropic polyomavirus T Ag	+	nd	Dyson et al (1990)
SA12 T Ag	+	nd	Dyson et al (1990)
Human JC virus T Ag	+	+	Dyson et al (1989a, 1990)
Human BK virus T Ag			
Dunlop strain	+	nd	Dyson et al (1990)
MM strain	+	nd	Dyson et al (1990)
Human papilloma virus 6b E7	+	nd	Munger et al (1989)
Human papilloma virus 11 E7	+	nd	Munger et al (1989)
Human papilloma virus 16 E7	+	nd	Dyson et al (1989b)
Human papilloma virus 18 E7	+	nd	Munger et al (1989)

nd, not determined.

The E1A-associated 60 kDa protein binds to the human homologue of the cdc2 gene product

During the course of our studies on the interaction of E1A with cellular proteins, we have prepared a series of monoclonal antibodies specific for these cellular proteins. These antibodies were raised against E1A–cellular protein complexes that were purified on immunoaffinity columns using an anti-E1A monoclonal antibody bound to a solid phase matrix. The C160 antibody was prepared using this strategy, and it specifically recognizes the E1A-associated p60 protein. Its specificity has been confirmed both by immunoprecipitation and by immunoblotting.

One of the best characterized proteins known to control progression through the cell cycle is the p34 product of the *cdc2* gene. p34 is a catalytic subunit which associates with other cellular proteins to form multimeric protein kinases that are activated at particular points of the cell cycle. During mitosis, p34 forms a complex with the cyclin B protein, p62, and the *suc1* gene product, p13. This complex forms an active kinase demonstrated to be a component of the M phase promoting factor (MPF) in frog eggs and the M phase-specific histone H1 kinase in starfish. When the p34 protein is precipitated using an anti-peptide antibody and the immune complexes are separated on high-resolution two-dimensional gels, the known proteins that bind to p34, p62, cyclin B and p13^{suc1} can be easily detected. However, in addition to these proteins, another polypeptide of approximately 60 kDa is also seen. A number of experiments have shown that this p60 protein is the same polypeptide that interacts with the adenovirus E1A proteins (Giordano et al 1989). This has been confirmed by comparing p60 precipitated by reagents specific for either p34 or E1A. The p60 proteins prepared in these ways migrate to identical positions on two-dimensional gels, and both proteins are recognized specifically by the C160 monoclonal antibody. As predicted from these results, the C160 anti-p60 monoclonal antibody co-precipitates the p34 proteins through its interaction with p60. Interestingly, only the underphosphorylated forms of p34 bind to p60. No evidence is available which suggests that p60/p34/E1A can form multimeric complexes.

Because other p34-containing complexes were known to have cell cycle-regulated kinase activity, the p60–p34 complex was checked for similar activity. Since *in vivo* substrates were not known, histone H1 was used as a test substrate in the kinase reactions. When the p60–p34 complex was immunoprecipitated using the C160 monoclonal antibody, the immune complexes were shown to exhibit kinase activity. When similar experiments were performed using extracts from cells in various stages of the cell cycle, it became clear that the p60–p34 kinase was regulated in a cell cycle-dependent manner, with the peak of activity occurring during interphase. Other p34-containing complexes such as the p62–p34 kinase, are maximally active in mitosis. Although the exact timing of activation of the p60–p34 complex has not been determined, it is clear that its activation precedes the p62–p34 peak of kinase activity.

Several experimental results argue that the p60–p34 complex will be important in cell cycle regulation. First, the p34 protein is known to be important in the regulation of the cell cycle. Experiments using conditional mutations in yeast, microinjection of MPF in *Xenopus* oocytes, and microinjection of anti-p34 antibodies in mammalian cells, all suggest that p34 is an essential element in progression into and through mitosis. Genetic experiments have shown that the p34 protein is also necessary for progression through the G1/S boundary in yeast. Second, the p60–p34 complex has a cell cycle-regulated kinase activity. A temporally regulated kinase implies that some targets must be phosphorylated in some stages of the cell cycle and not in others. Third, one of the members of this complex, the p60 protein, binds to the adenovirus E1A proteins, and the E1A proteins are known to disrupt cell cycle control.

Clearly, without genetic evidence that the p60–p34 complex has an important role in cell cycle regulation, any connection between these observations is circumstantial. However, the circumstances are provocative. p34 can play a major role in cell cycle control at stages other than mitosis, and the p60–p34 kinase is activated at other points in the cell cycle. Thus, the p60–p34 complex becomes an excellent candidate for more detailed study of potential p34-regulated pathways.

Summary

Small DNA tumour viruses face similar problems during their life cycles. They bring a limited amount of genetic material into cells, an amount insufficient for synthesizing the number of viral proteins needed to perform all of the steps of viral replication. Therefore, these viruses must redirect certain cellular processes to perform essential steps in viral replication. One of these key events is viral DNA synthesis. While small DNA viruses often have some of the enzymes needed for DNA synthesis, they do not have all of these. A common strategy used by these viruses is to produce effective mitogens that drive infected cells into S phase, and thus induce the production of the enzymes and substrates needed for cellular and viral DNA synthesis.

Since small DNA tumour viruses such as the adenoviruses, papilloma viruses and polyoma viruses generally infect cells that are not rapidly dividing, one of the key functions of the viral early proteins must be to remove proliferation constraints and to drive cells into the cell cycle. These properties of viral early proteins probably provide the explanation for their oncogenic potential. Proteins from all three groups of small DNA tumour viruses appear to use similar strategies for some of their oncogenic activities. Large T antigens from the polyoma viruses, E1A proteins from the adenoviruses, and E7 proteins from the papilloma viruses share structural motifs that allow interaction with pRb. Mutations within this motif that destroy interaction of these viral proteins with pRb also destroy their transforming ability. This correlation suggests that these viruses employ at least one common mechanism in transformation.

Acknowledgements

We thank N. Williamson, C. McCall, L. Duffy for technical assistance and Regina Whitaker, Jeanne Wiggins and Bob McGuirk for reliable tissue culture supplies and cells. N. D. is supported by Amersham International. K. B. was a recipient of a traineeship from the U. S. Public Health Service. This work was supported by Public Health Service grant CA 13106.

References

Bellett AJD, Li P, David E, Mackey E, Braithwaite A, Cutt A 1985 Control functions of adenovirus transformation region E1A gene products in rat and human cells. Mol Cell Biol 5:1933–1939

Berk AJ 1986 Adenovirus promoters and E1A transactivation. Annu Rev Genet 20:45–79

Braithwaite AW, Cheetham BF, Li P, Parish CR, Waldron-Stevens LK, Bellett AJD 1983 Adenovirus-induced alteration of the cell growth cycle: a requirement for expression of E1A but not of E1B. J Virol 45: 192–199

DeCaprio JA, Ludlow JW, Figge J et al 1988 SV40 large T antigen forms a specific complex with the product of the retinoblastoma susceptibility gene. Cell 54:275–283

Dyson N, Bernards R, Friend S et al 1990 The large T antigens of many polyoma viruses are able to form complexes with the retinoblastoma protein. J Virol 64:1353–1356

Dyson N, Buchkovich K, Whyte P, Harlow E 1989a The cellular 107K protein that binds to adenovirus E1A also associates with the large T antigens of SV40 and JC virus. Cell 58:249–255

Dyson N, Howley PM, Munger K, Harlow E 1989b The human papillomavirus-16 E7 oncoprotein is able to bind to the retinoblastoma gene product. Science (Wash DC) 243:934–937

Ewen M, Ludlow J, Marsilio E et al 1989 An N-terminal transformation-governing sequence of SV40 large T antigen contributes to the binding of both p110^{Rb} and a second cellular protein, p120. Cell 58:257–267

Giordano A, Whyte P, Harlow E, Franza R, Beach D, Draetta G 1989 A 60 kd cdc2-associated polypeptide complexes with the E1A proteins in adenovirus-infected cells. Cell 58:981–990

Green M 1989 When the products of oncogenes and anti-oncogenes meet. Cell 56: 1–3

Haley KP, Overhauser J, Babiss LE, Ginsberg HS, Jones NC 1984 Transforming properties of type 5 adenovirus mutants that differentially express the E1A gene products. Proc Natl Acad Sci USA 81:5734–5788

Harlow E, Whyte P, Franza BR Jr, Schley C 1986 Association of adenovirus early-region 1A proteins with cellular polypeptides. Mol Cell Biol 6:1579–1589

Houweling A, van den Elsen PJ, van der Eb AJ 1980 Partial transformation of primary rat cells by the leftmost 4.5% fragment of adenovirus 5 DNA. Virology 105:537–550

Kaczmarek L, Ferguson B, Rosenberg M, Baserga R 1986 Induction of cellular DNA synthesis by purified adenovirus E1A proteins. Virology 152:1–10

Kimelman D, Miller J, Porter D, Roberts B 1985 E1a regions of the human adenoviruses and of the highly oncogenic simian adenovirus 7 are closely related. J Virol 53:399–409

Klein G 1987 The approaching era of the tumor suppressor genes. Science (Wash DC) 238:1539–1542

Knudson AG 1987 A two-mutation model for human cancer. Adv Viral Oncol 7:1–17

Lane D, Crawford L 1979 T antigen is bound to a host protein in SV40-transformed cells. Nature (Lond) 278:261–263

Lee W-H, Shew J-Y, Hong FD et al 1987 The retinoblastoma susceptibility gene encodes a nuclear phosphoprotein associated with DNA binding activity. Nature (Lond) 329:642–645

Lillie JW, Green M, Green MR 1986 An adenovirus E1A protein region required for transformation and transcriptional repression. Cell 46:1043–1051

Lillie JW, Loewenstein PM, Green MR, Green M 1987 Functional domains of adenovirus type 5 E1A proteins. Cell 50:1091–1100

Linzer D, Levine A 1979 Characterization of a 54K dalton cellular SV40 tumor antigen present in SV40-transformed cells and uninfected embryonal carcinoma cells. Cell 17:43–52

Moran E, Mathews MB 1987 Multiple functional domains in the adenovirus E1A gene. Cell 48:177–178

Moran E, Zerler B 1988 Interactions between cell growth-regulating domains in the products of the adenovirus E1A oncogene. Mol Cell Biol 8:1756–1764

Moran E, Zerler B, Harrison TM, Mathews MB 1986 Identification of separate domains in the adenovirus E1A gene for immortalization activity and the activation of virus early genes. Mol Cell Biol 6:3470–3480

Munger K, Werness BA, Dyson N, Phelps WC, Harlow E, Howley PM 1989 Complex formation of human papillomavirus E7 proteins with the retinoblastoma tumor suppressor gene product. EMBO (Eur Mol Biol Organ) J 8:4099–4106

Nakajima T, Masuda-Murata M, Hara E, Oda K 1987 Induction of cell cycle progression by adenovirus E1A gene 13S- and 12 S-mRNA products in quiescent rat cells. Mol Cell Biol 7:3846–3852

Ruley HE 1983 Adenovirus early region 1A enables viral and cellular transforming genes to transform primary cells in culture. Nature (Lond) 304:602–606

Sarnow P, Ho Y, Williams J, Levine A 1982 Adenovirus E1b-58kd tumor antigen and SV40 large tumor antigen are physically associated with the same 54 kd cellular protein in transformed cells. Cell 28:387–394

Schneider JF, Fisher F, Goding CR, Jones NC 1987 Mutational analysis of the adenovirus E1A gene: the role of transcriptional regulation in transformation. EMBO (Eur Mol Biol Organ) J 6:2053–2060

Smith DH, Ziff EB 1988 The amino-terminal region of the adenovirus serotype 5 E1A protein performs two separate functions when expressed in primary baby rat kidney cells. Mol Cell Biol 8:3882–3890

Spindler KR, Eng CY, Berk AJ 1985 An adenovirus early region 1A protein is required for maximal viral DNA replication in growth-arrested human cells. J Virol 53:742–750

Stabel S, Argos P, Philipson L 1985 The release of growth arrest by microinjection of adenovirus E1A DNA. EMBO (Eur Mol Biol Organ) J 4:2329–2336

Whyte P, Buchkovich KJ, Horowitz JM et al 1988a Association between an oncogene and an anti-oncogene: the adenovirus E1A proteins bind to the retinoblastoma gene product. Nature (Lond) 334:124–129

Whyte P, Ruley HE, Harlow E 1988b Two regions of the adenovirus early region 1A proteins are required for transformation. J Virol 62:257–265

Whyte P, Williamson NM, Harlow E 1989 Cellular targets for transformation by E1A. Cell 56:67–75

Yee S, Branton PE 1985 Detection of cellular proteins associated with human adenovirus type 5 early region 1A polypeptides. Virology 147:142–153

Zerler B, Moran B, Maruyama K, Moomaw J, Grodzicker T, Ruley HE 1986 Adenovirus E1A coding sequences that enable *ras* and *pmt* oncogenes to transform cultured primary cells. Mol Cell Biol 6:887–899

DISCUSSION

Land: Can E1A still repress enhancers in *Rb*⁻ cells?

Harlow: I don't think that experiment has been done. Repression is a very complicated phenotype in E1A. E1A can repress several different transcription units, and there are at least two and probably three different types of repression. These probably require different sorts of events, some of which are inhibited by mutations that knock out binding to p*Rb*, others which don't seem to matter at all.

Vande Woude: Is anything known about the complexes with, for example, p107 and p105? Are they ternary complexes?

Harlow: No, E1A binds to either p105 or p107. Their associations are mutually exclusive.

Vande Woude: So you find two different classes of molecules?

Harlow: Yes. That can be tested physically by separating them on columns and looking at the E1A, and by using the anti-p*Rb* antibody.

Vande Woude: What about displacing one or the other?

Harlow: I don't know. We need a source of the protein to do the experiment convincingly.

Hunter: Do you have peptides that compete?

Harlow: Peptides bind better to p*Rb* than to p107.

Hunter: If you take an E1A–p*Rb* complex, can you compete off E1A?

Harlow: Yes, that actually works better with the SV40 large T antigen. David Livingston's group at Dana Faber, Boston, has been using an immunoaffinity purified complex. They add a peptide, let it incubate for a while and ask whether the peptide will displace p*Rb* from the complex. That does happen.

Hunter: That raises the issue of how phosphorylation of p*Rb* apparently displaces large T from the complex. If that binding was very tight, you might expect the complex to remain intact throughout the cell cycle.

Harlow: David Livingston has been working on that. His general model is that there is a site near the binding area that's phosphorylated, which changes the shape of *Rb* protein and large T comes off.

Hunter: Have the sites for phosphorylation been mapped?

Harlow: Jean Wang's lab is probably as close as any to identifying the actual sites. She thinks there may be 11–12 sites, but none of those are known in any detail yet.

Hunter: Since p*Rb* is a substrate for the *cdc2* kinase, and has several predicted consensus phosphorylation sites, do any of those lie in the region involved in binding T antigen?

Harlow: *In vitro* p*Rb* is a good substrate for the p34^{cdc2} kinase. If you compare by peptide mapping p*Rb* phosphorylated *in vivo* versus *in vitro*, many of the spots are the same. That's the level of distinction at the moment. No one knows exactly what sites are being hit and which *cdc2* complex is doing it.

There are consensus *cdc2* phosphorylation sites in p*Rb*, one, seven or nine, depending on how strict you want the consensus to be. One of the seven is near this region that is necessary for binding to large T antigen and E1A.

A number of other kinases have been tried to see whether they would phosphorylate p*Rb in vitro* and, as far as I know, nothing else has worked.

Waterfield: Has the possible role for TGF-β been ruled out?

Harlow: The experiments were done in a collaboration between Joan Massague's lab and Bob Weinberg's lab. They looked at retinoblastoma cells and found 100% correlation between human retinoblastomas and lack of TGF-β receptors. However, Wen Hwa Lee's group (University of California at San Diego) has reported that if you introduce the *Rb* polypeptide into those retinoblastoma or osteosarcoma cell lines, there is no re-expression of the TGF-β receptor. There are also retinoblastoma-like cell lines, human retinoblasts that have been transformed with adenovirus, and those still express TGF-β receptors. I think at the moment the 100% correlation suggests that something in the natural development of that tumour requires that it loses TGF-β receptors, but it doesn't seem to be related to the *Rb* protein, at least not directly.

Waterfield: Was it a particular type of TGF-β receptor?

Harlow: There are three types there and they all go away.

Hunter: It is partly a question of what's on a normal retinoblast.

Harlow: One would argue that if retinoblasts that have been transformed by adenovirus have the receptor, then retinoblasts can have them, so they are either on normal retinoblasts or induced by E1A.

Vande Woude: How were those experiments done? What was the time between introducing p*Rb* and assaying for TGF-β receptor?

Harlow: Well it's been several months and they don't seem to be expressing the receptors.

Vande Woude: Are these the cells that quiesce when you add *Rb*?

Harlow: No, they are stable cell lines that have *Rb*.

Sherr: I had the impression that there was a controversy about re-introducing the *Rb* gene into cells, in that Weinberg's clone doesn't seem to work. The implication was that replacement with *Rb* might be toxic to cells. There was even the suggestion that Weinberg's clone and Lee's clone might be different.

Harlow: I think the current story is that the clones are identical. Everybody who tries to express *Rb* from commonly used promoters after transfection finds that the *Rb* is toxic to cells, whatever mammalian cell is used. In the experiments done in Lee's laboratory, they used a retroviral vector that seems to have overcome the transfection problem. The consensus seems to be that Lee's experiments are correct (Huang et al 1988). He sent his vectors to George Klein's lab and they get the same results. Wen Hwa Lee looked at about 19 different constructs that didn't work and the 20th did, so there is a magic combination of promoter, vector and method. I believe that they have actually selected for a low level of expression.

In Lee's original experiment, the cells infected with the *Rb*-carrying retrovirus eventually deleted *Rb* and started growing rapidly again and then became tumorigenic again. So there is selective pressure against expression of *Rb*. One therefore assumes that the mechanisms that control expression of *Rb* are very fine controls. If you overexpress *Rb*, it overwhelms those controls.

Noble: When you say toxic, do the cells just stop dividing or do they die? In all of these experiments, you are treating terminal differentiation as a lethal event, whereas terminal differentiation is the interesting event to someone like me!

Harlow: Experimentally, it's difficult to make that distinction.

Noble: Why not carry out microinjection studies? With mRNA for *Rb*, for example?

Harlow: I don't know if anybody has done that. Again I don't know exactly what one would look for. Presumably, if you wait long enough, the message is degraded and the cells grow. You need a transient single cell assay to be able to score for something that happens.

Noble: Or you could work on a cell which moves into a differentiated phenotype that you can identify.

Heath: There could be sequestration of transcription factors by *Rb* protein or something trivial.

Harlow: There could be all sorts of reasons. If people have cells in which *Rb* is overexpressed and those cells die, it doesn't tell you very much. A lot of people are interested in where in the cell cycle those cells are arrested. It's important biologically, but experimentally it is not approachable at the moment.

Norbury: *Rb* doesn't seem to be toxic in yeast. I did the experiment with Jon Horowitz from Bob Weinberg's laboratory. Whatever *Rb* protein interacts with, it doesn't seem to be one of the more highly conserved cell cycle regulators.

Vande Woude: Is that in *S. cerevisiae* or *S. pombe*?

Norbury: *S. pombe.*

Hunter: Was there a reasonable level of expression of the *Rb* protein?

Norbury: Yes; it was readily detectable by immunoblotting.

Land: Does the *Rb*-induced toxicity correlate with the known functional domains of the *Rb* protein?

Harlow: We don't know yet.

Waterfield: Ed, what were the results in the different tumours or cell lines on mutation of the *Rb* protein?

Harlow: It's been done in several ways, depending on the cell type and how easy it is to isolate tumour material. People have looked at cell lines that come from tumours, which raises the problem of selection for a loss of *Rb* in tissue culture. But in most of the cases I know about there is at least one study using fresh tumour tissue. Those percentage changes were all gross chromosomal rearrangements.

Waterfield: What fraction of primary tumours show gross structural changes, compared to the percentage of cell lines showing such changes?

Harlow: Most of these percentages have been confirmed in studies on fresh tissues as well as cell lines.

Alemà: Is there an endogenous ligand for the *Rb* protein, doing whatever E1A does?

Harlow: I don't know. I would predict that there is a cellular protein that has a site that resembles the binding site on E1A or large T antigen. That site may be linked to a completely different type of function. The best suggestion at the moment is that the virus has stolen a small region of the cellular protein and mutated it to a high affinity form which blocks this site. That doesn't mean that the cellular protein with the original sequence would necessarily have the same function.

If you look in cells for something that binds *Rb* with high affinity, you don't find anything that binds as well as E1A or large T.

Hunter: You said there were no mutations of *Rb* that are dominant in *trans*, but we are talking about potential ligands for *Rb*, which might suggest that one would find transdominant mutations in *Rb*.

Harlow: That's one of the reasons I think a potential cellular equivalent of E1A or large T is not a ligand but a target site for an enzyme or something along those lines.

Hunter: So the only class of mutations you have, by inference, prevent binding because they have lost large T antigen binding sites. Maybe it's difficult to make a mutation that is dominant in *trans* because most mutations simply inactivate *Rb*.

Harlow: It's certainly possible, but not the most likely solution I think.

Hunter: No one has taken *Rb* and made random mutations?

Harlow: I don't know how you would select for that.

Maness: You said that *Rb* was expressed in almost every cell that's been looked at. Does that include mature neurons?

Harlow: I don't know. Everywhere people have looked they have found it.

Maness: Is it possible that the antibodies recognize related *Rb* family members in other tissues, and authentic *Rb* is expressed only in the retina?

Harlow: The family member would have to be the same size. It is possible, but if a related protein was picked up that easily, one would expect the hybridization to work and it hasn't yet.

Maness: Has the *Rb* clone been put into PC12 cells?

Harlow: I bet it has but I haven't heard anything—that means it's either very interesting or very boring!

Sherr: Why do you think that deletion of *Rb* is preferentially associated with induction of retinoblastoma and not other tumours? I suppose that p*Rb* is detected in other tumours—a percentage of osteosarcomas, for example—but why this preference for retinoblastoma?

Harlow: There are no suitable explanations. It is clear that retinal cells are privileged but that's meaningless, it just tells you that they are different.

Sherr: Immunologically privileged?

Harlow: No. For some reason they are more dependent on *Rb* than other cells are. When one inherits an *Rb* mutation in one allele, presumably mutations occur in several tissues; there is no evidence that the retina is a hot spot for mutation. In patients with hereditary retinoblastoma, there is an average of three tumours per patient, so the second event needed to cause tumorigenesis is happening three times in the developing retina—it must happen in other tissues, but you don't see tumours in these tissues. For some reason, the *Rb* protein is very important in the retina.

This is a common feature of tumour suppressor genes. Patients with Wilm's tumour get kidney disease and renal tumours. In familial adenopolyposis you get tumours of the colon. For some reason the products of these genes must be specific to that particular disease in that particular tissue. Everybody suggests that there is something involved in differentiation of those tissues. I think that makes the most sense right now, however, it would make more sense to me if *Rb* expression weren't so widespread, if there wasn't an *Rb* protein in every cell that one looks at.

Sherr: Doesn't that argument suggest that the *Rb* protein is not playing a general role in cell cycling, but, rather, a more specific role?

Harlow: I think it says that retinal cells rely heavily on control by *Rb*. Your argument is right, but when you are talking about repression I think you have to consider it the other way around—that the other controls aren't as important in retinal cells. Whatever those controls are, p53 or unknown genes, they aren't involved in regulating retinal cell division. Consequently, when you lose *Rb* the cell is left with no regulatory system.

Hunter: There are beginning to be examples of nuclear regulators, particularly transcription factors, which have positive and negative functions. Perhaps *Rb* acts negatively in the retinal cell lineage and positively in other lineages.

Harlow: That's certainly possible. There are reports in the literature that say p53 must have a positive role in driving cell division, as well as a negative role. One example is from microinjection of antibodies done by Ed Mercer and Renato Baserga. They arrest Swiss 3T3 cells by serum deprivation, then stimulate with serum, and they find that there is a window in which you can inject antibodies against p53 and block DNA synthesis. The window is 2–4 hours post-serum stimulation.

Hunter: The criticism of those experiments is that it's not clear whether those Swiss 3T3 cells have a normal or mutant p53.

Harlow: I would be surprised if they didn't have a mutant p53.

The other experiments are from Varda Rotter's lab. They use antisense oligos to knock out p53, and find that the cells don't progress through the cell cycle. She has also used an Abelson virus-transformed cell line, L-12, which is tumorigenic but doesn't metastasize. These cells do not express p53. She re-introduces p53 into those cells and they metastasize. However, it probably was a mutant p53 in the cell used for the antisense experiment and a mutant that was re-introduced into the L-12 cells, so it's pretty hard to understand the biochemistry.

Hunter: Have you injected monoclonal antibodies against p*Rb* into serum-stimulated cells?

Harlow: We did that when we really didn't understand what to look for. We are now taking peptides that correspond to the binding sites on E1A for p*Rb* and injecting those into cells. We have coupled these peptides to other proteins and are injecting those as well. We have made monoclonal antibodies against the peptides, which we hope will block the action of E1A.

The antibodies against the binding sites cross-react with cellular proteins. We are now faced with the problem of deciding what's an interesting cross-reaction and what's a boring cross-reaction.

Vande Woude: Can you do peptide competition experiments and select the one that competes most strongly?

Harlow: I don't know if the one with the highest affinity is necessarily the one we want. Some of these monoclonal antibodies cross-react with the large

T antigen and E7, so they see structural similarities between the different proteins, which is something you would predict.

Méchali: Do you know how much *Rb* protein there is per cell, in normal cells and in those infected with adenovirus or SV40?

Harlow: When we purify pRb from HeLa cells, we get about 0.05 µg/10^8 cells. The amount is not as low as that of many transcription factors.

McMahon: Presumably it's quite easy to get large quantities of retina from a cow, for example. Has anybody looked at the biochemistry of the protein interactions?

Harlow: Not that I know of. We have screened a large bank of human tumours to look for changes in *Rb* from other tumours. We have a few cell lines that might overexpress the *Rb* protein which we are using for purification.

Maness: Are there any other *Rb*-like genes that are picked up by low stringency screening?

Harlow: The Weinberg lab has worked quite hard on that, and they haven't found anything. We are trying to use the domain that is important for interacting with E1A as a probe. p107 I think is going to be a distant relative.

The mouse *Rb* gene has been cloned and it's very similar in sequence to the human one, particularly in the regions that are necessary for binding to E1A. There are at least two other polypeptides in the cell that interact with the same region on large T and E1A as do p107 and p105, so I think the family is larger than two members.

Vande Woude: Where p105 binds in the large domain, is that the same consensus or is it a different one?

Harlow: In the upstream region, the consensus between E1A, large T and E7 is not as good as in the downstream region. We have made peptides that correspond to the different regions. A peptide containing the two regions linked together blocks the binding of E1A and pRb well at about 1µM. A peptide corresponding to the second, downstream region blocks 10-fold less well; one corresponding to the upstream region blocks, but 100-fold less well. So there is at least a 2 log spread in the blocking concentrations, but all peptides block and that makes us feel that both the upstream and downstream consensus regions are real contact sites. We think the downstream region forms the core binding region. It is also where p107 binds.

Vande Woude: Are there then two binding sites on pRb?

Harlow: We don't know. We tried to get one of the peptides to interact with various regions independently but we couldn't detect binding.

Hunter: Is that using some of the natural *Rb* deletion mutants or those you have created?

Harlow: The *in vitro* mutants. Our feeling is that these regions of pRb are folding into some sort of pocket; E1A and large T antigen have the right structures to slide into it. You can covalently bind the peptides to Sepharose beads and use those as an adsorbant to pick p105 and p107 out of cell extracts.

Hunter: Have you changed the spacing between the two domains in p*Rb*?

Harlow: You can reduce the size of the intervening segment but you can't remove it completely and retain binding.

Hunter: Is that using random linkers?

Harlow: We haven't used random sequences. We have only used nucleotides from the native intervening sequence at the moment. This gives you the idea that there is a turn or coil that produces the structure needed for correct folding.

Vande Woude: Are there any sequences that predict certain structures?

Harlow: There are some putative β-pleated sheets, but nothing one could use for serious modelling.

Hunter: Can you use PCR to amplify between those two E1A consensus regions on cellular mRNA in the hope of finding *Rb*-binding proteins?

Harlow: The second region shows the strongest consensus; we calculated that the degeneracy would be one in 130 000, at which point I gave up.

Hunter: You should do it, because Steve Gould has used a millionfold degenerate oligos and got very nice results (Gould et al 1989). Do the regions have to be in that order? Have you tried to reverse those regions in E1A?

Harlow: It probably has been done, but I don't know the answer.

References

Gould SJ, Subramani S, Schleffler IE 1989 Use of the DNA polymerase chain-reaction for homology probing. Isolation of partial cDNA or genomic clones encoding the iron sulfur protein of succinate-dehydrogenase from several species. Proc Natl Acad Sci USA 86:1934–1938

Huang H-JS, Yee J-K, Shew J-Y et al 1988 Suppression of the neoplastic phenotype by replacement of the RB gene in human cancer cells. Science (Wash DC) 242:1563–1566

Final discussion

The role of proto-oncogenes in differentiation and development

Wagner: We have used the embryonic stem (ES) cell system to overexpress proto-oncogenes, aiming to look at two things. First, whether ES cells *in vitro* have altered growth parameters; secondly, what is the consequence of that altered expression on development. In terms of *in vitro* growth parameters, several proto-oncogenes seem, to some extent, to abrogate the requirement for leukaemia inhibitory factor (LIF)—usually necessary to maintain stem cell viability.

Noble: Does it bypass the requirement for exogenous LIF, or do the cells not need to make it any more?

Wagner: That's difficult to answer. It is not known whether active LIF is made by these cells.

Heath: Embryonal carcinoma (EC) cells, in which the ability to differentiate spontaneously is blocked, don't secrete LIF. It is more difficult to assay this in ES cells, because they have to be grown in the presence of LIF, then this has to be washed off, and you look for secreted LIF while they are differentiating.

Wagner: The consequence of expression of exogenous proto-oncogenes is that ES cells can grow without LIF quite happily. In the clonal assay that several people have used, these ES cells prevent the spontaneous differentiation that normally occurs when wild-type ES cells are grown without LIF. However, when these cells are put back in an embryo they differentiate normally. So it is not a permanent effect in that sense.

In terms of nuclear oncogenes, such as *fos*, we do not observe a dominant phenotype when c-*fos* is expressed at various levels in these ES cells. This is surprising, I would not have expected to see embryos developing normally if they are making large amounts of *fos* protein in the ES cells. Dominant effects on embryonic development are seen with middle T and with v-*src*, provided that v-*src* is expressed at very high levels.

The most detailed study is our experiments with the polyoma middle T gene. Normal proliferation of endothelial cells is altered by the mid-gestation stage, leading to embryonic arrest, because normal vascularization in the yolk sac is prevented.

With v-*src* expression in ES cells, Catherine Boutles has done a lot of experiments in our lab and recently in Martin Evans' lab with Rosa Beddington. We have analysed the effects of high levels of v-*src* expression on early development, using three or four different ES cell clones which express very high levels of v-*src*. Catherine sees a dominant effect at E8 or E8.5. Preliminary data suggest that v-*src* causes a twinning phenotype with an enlargement of the visceral yolk sac. A duplication of the body axis would be one interpretation, although we are not sure about that.

Heath: We have similar very preliminary results with SV40 large T suppressing differentiation of ES cells (or rendering them independent of LIF). But it slightly bothers me that with these oncogene experiments there might be a difference between messing up differentiation and specifically inhibiting it. We don't know whether the oncogene is just overriding the endogenous mechanisms, which would be in a sense mechanistically irrelevant, or whether you are manipulating a particular pathway, which would be more interesting. I feel that some of these oncogene/ES cell experiments may be rather difficult to interpret at the end of the day.

Noble: One study that speaks against these genes screwing up normal differentiation is experiments done by Parmjit Jat. He put temperature-sensitive SV40 large T antigen into rat embryo fibroblasts. At 33 °C, these transformed cells will grow forever. He grew them at 33 °C past the time when they should have senesced, then switched to 39 °C and they senesced immediately. The cells seemed to remember how old they are. It looks as though the biological clock, in the presence of SV40 large T antigen, has continued to run normally, the cell achieves its end stage decision but the large T antigen overrides the capacity to enact that decision.

Heath: That in a way illustrates my point. SV40 large T is probably the worst possible example of oncogenes in this category, because as you describe, it seems to override some endogenous mechanism, which isn't going to tell you much about that endogenous mechanism.

Noble: If you have large T present in the cell, the cell is clearly altered, as shown in wonderful experiments done in Hucky Land's lab.

Land: Anne Ridley in my lab has shown that nuclear oncogenes, in particular SV40 large T and adenovirus E1A, can alter the cellular response to *ras*-induced signals (see Ridley et al 1988). In Schwann cells, *ras* alone induces proliferation arrest in G1 and G2 phases of the cell cycle; SV40 large T alone induces a slight increase in growth rate and, as expected, a reduction of the stringent growth factor requirements of normal cells. However, *ras* and large T together lead to a conversion to the fully transformed, very rapidly proliferating cell phenotype which is independent of exogenous growth factors.

Anne's results have shaped our current models of oncogene cooperation. From her work it became clear that the constitutive activation of *ras*-controlled signalling does not necessarily always contribute to cellular transformation. Only when the cellular response to *ras* is altered by a nuclear oncogene does *ras* act as a potent oncogene. This indicates that the effect of the *ras* oncogene is context dependent, while it is probable that the biochemical function of *ras* remains the same.

From these findings, one could postulate a model whereby cells utilize the same signalling pathways for different purposes in a context-dependent manner.

Furthermore, it would follow that nuclear oncogenes, such as SV40 large T and adenovirus E1A, are able to alter the utilization of signalling pathways. Proteins of this kind might control a cellular gearbox that normally controls switching between different signalling pathways to enable a cell to use its signalling network for multiple purposes.

Vande Woude: Is it known what the viral genes are doing? Don't they usually force entry of the cell into S phase? It is essential to start the cell's replication machinery so that it can be used for viral replication.

Land: Yes, nevertheless these genes alone will not make a cell totally unresponsive to outside signals. A cell that expresses only a nuclear oncogene will not form a tumour.

Harlow: These viral proteins are under incredible evolutionary pressure to drive cells into S phase. You are talking about them setting up a gearbox; it is clear that they are doing that, but it's not clear that they are doing it by one function, they may be doing it in 15 ways that are completely unrelated. Viruses may not care about *ras* as a signal.

Land: I'm not implying that they do. I agree, these viral nuclear oncogenes may do a hundred things, maybe they control the gearbox because they change the genetic programme in ways that alter the targets for the cellular signalling pathways. It is also possible that the viral oncogenes do this better and more efficiently than cellular genes. Maybe part of their action is to allow cells to see signals as a proliferative stimulus because the virus requires cellular replication for its own propagation. The cell might use a distinct, more complicated machinery to do similar things in a more sophisticated way.

Brugge: Hucky, from your studies, would *myc* have the same effect as a viral oncogene which has as many pleiotropic effects as large T?

Land: When we introduce *ras* and *myc* into Schwann cells, this also results in full transformation. However, there are differences between the actions of SV40 large T and *myc*. For instance, cells that express a *ras* oncogene and a temperature-sensitive SV40 large T gene cannot be rescued by *myc* at the non-permissive temperature. However, we can rescue these cells with wild-type large T or E1A genes. Why we cannot rescue these cells or we rescue them only extremely rarely with the *myc* gene, I don't know.

Brugge: So large T is doing more than *myc*?

Land: Yes, these two viral oncogenes are more potent than *myc* and their spectrum of action is slightly different.

Brugge: In fibroblasts, you can overcome the *ras* growth inhibitory effect with *myc*, is that right?

Land: No, not in the same way. For example, REF52 cells behave similarly to Schwann cells (Hirakawa & Ruley 1988). In addition, in Parmjit Jat's system cited earlier by Mark Noble, non-permissive temperature-induced senescence cannot be rescued by *myc* but it can be rescued by large T.

Harlow: Can you immortalize Schwann cells with *myc* alone?

Land: Schwann cells *per se* seem to be immortal in culture. This sounds surprising, but most of what we know about the lifespan of cells in culture comes from experience with fibroblasts. Indeed, it may be that we inhibit the proliferation of fibroblasts by the serum we put in the culture medium. There has been a report from D. Barnes' laboratory claiming that embryo cells continue to grow indefinitely in the absence of serum. However, as soon as these cells are exposed to serum, they senesce (Loo et al 1987). Unfortunately, this observation is very difficult to reproduce because the media conditions were not well-defined. Nevertheless, the Schwann cells might be a cell type which is not susceptible to the serum factor that induces senescence of fibroblastic-type cells.

Harlow: Those Barnes' studies use epithelial cells.

Land: In their publication they claim that their cultures are not different from the cultures grown in the presence of serum. However, our experience is that different cell types are selected, depending on the exact media conditions used.

Harlow: Yes, but the resulting cells that grow are epithelial, by a number of different markers.

Land: I never saw this documented anywhere. In our cultures using medium prepared according to Barnes' recipe, a number of neuronal-type cells proliferate rapidly in the first few days. This is something we never see in the presence of serum.

Hunter: What is the current state of work on knocking out specific proto-oncogenes?

Wagner: Three methods have been employed to do this. These are positive/negative selection, microinjection of genes carrying small alterations in the targeting vector and the method using the promoterless neomycin resistance gene, which so far has been the most successful. Using these methods, at least a dozen genes have been inactivated in ES cells and all of those have given rise to chimaeras. Germline chimaeras have been reported in only two cases. One is with c-*abl*, but I am not sure what alteration was made in the gene and whether it will really lead to an *abl*⁻ phenotype.

Hunter: Pamela Schwartzberg, Liz Robertson and Steve Goff (1989) made a deletion in the C-terminus, because in the v-*abl* oncogene that region is dispensible for fibroblast transformation but not for lymphoid transformation. They hoped to get a subtle phenotype rather than a lethal phenotype.

Wagner: The other gene to have been passed through the germline is *engrailed2* (Joyner et al 1989).

Nusse: β_2 microglobulin has been passed through the germline in mice in Rudi Jaenisch's lab (personal communication).

Wagner: He told me the only evidence for that is by coat colour. Pups have been born but they have not done the DNA analysis yet.

Vande Woude: If the right coat colour comes out, do they express a phenotype as well?

Wagner: One would expect them in the next generation.

Hunter: One could see a phenotype in the F1 generation if there are dosage effects.

McMahon: It's worth noting that *decapentaplegic*, a *Drosophila* TGF-β-like gene, is haploinsufficient; that is, two copies have to be present to produce the normal phenotype.

Nusse: People expect knocking out their gene to be a gold mine of information. We found a mutation in *Drosophila* where our favourite gene was knocked out. It is a very good phenotype to study and many other genes give the same phenotype, but we still do not know what *wingless* or *int-1* is doing! So knocking out the gene is only the beginning.

Hunter: I agree, but it is one way of showing that the gene is essential for some process. Whether it tells you what the gene does is another issue.

There is the possibility of making dominant-negative mutants where you know of a protein–protein interaction that is essential for gene function, such as that described by Inder Verma for *fos* and *jun*. A dominant-negative mutant gene could be introduced into a transgenic mouse, then one wouldn't have to do a gene replacement. However, there may be a problem of getting the gene expressed in every cell type you would like it expressed in.

Heath: Another point is that so far you can only see the first event that is affected by knocking out a particular gene. There is a lot of information in later events.

Hunter: One would like to have inducible expression of an exogenous copy of the gene that could be turned on after the early events have occurred if one wants to look at late development.

Wagner: People are also talking about making subtle changes in the regulatory region or in binding domains.

Heath: Yes, one possible way is to play around with the promoter so you can specifically repress transcription by putting in regulatory elements.

Vande Woude: Luis Parada at Frederick has been studying *trk* and *trkB* expression in early mouse development. He has found that *trk* is expressed specifically in the trigeminal ganglion and dorsal root ganglion during midgestation. He plans to do knock-out experiments to study the effect of lack of expression of this gene on ganglion development.

Hunter: And the patterns of N-*myc* and c-*myc* expression are apparently mutually exclusive during embryogenesis.

Vande Woude: Yes, that was surprising. Expression of N-*myc* is much higher than that of c-*myc* in early embryos. The patterns of expression of N-*myc* and c-*myc* appear to be mutually exclusive.

References

Hirakawa T, Ruley HE 1988 Rescue of cells from ras oncogene-induced growth arrest by a second, complementing, oncogene. Proc Natl Acad Sci USA 85:1519–1523

Joyner AL, Skarnes WC, Rossant J 1989 Production of a mutation in mouse En-2 gene by homologous recombination in embryonic stem cells. Nature (Lond) 338:153–156

Loo DT, Fuquay JI, Rawson CL, Barnes DW 1987 Extended culture of mouse embryo cells without senescence: inhibition by serum. Science (Wash DC) 236:200–202

Ridley AJ, Paterson HF, Noble M, Land H 1988 Ras-mediated cell cycle arrest is altered by nuclear oncogenes to induce Schwann cell transformation. EMBO (Eur Mol Biol Organ) J 7:1635–1645

Schwartzberg PL, Goff SP, Robertson EJ 1989 Germ line transmission of a c-*abl* mutation produced by targeted gene disruption. Science (Wash DC) 246:799–803

Index of contributors

Non-participating co-authors are indicated by asterisks. Entries in bold type indicate papers; other entries refer to discussion contributions.

Indexes compiled by Liza Weinkove

285

Subject index